Reducing Urban Pove
the Global South

Urban areas in the Global South now house most of the world's urban population and are projected to house almost all its increase between now and 2030. There is a growing recognition that the scale of urban poverty has been overlooked – and that it is increasing both in numbers and in the proportion of the world's poor population that live and work in urban areas.

This is the first book to review the effectiveness of different approaches to reducing urban poverty in the Global South. It describes and discusses the different ways in which national and local governments, international agencies and civil society organisations are seeking to reduce urban poverty. Different approaches are explored, for instance, market approaches, welfare, rights-based approaches and technical/professional support. The book also considers the roles of clientelism and of social movements. Case studies illustrate different approaches and explore their effectiveness. *Reducing Urban Poverty in the Global South* also analyses the poverty reduction strategies developed by organised low-income groups especially those living in informal settlements. It explains how they and the federations or networks they have formed have demonstrated new approaches that have challenged adverse political relations and negotiated more effective support. Local and national governments and international agencies can become far more effective at addressing urban poverty at scale by, as is proposed in this book, working with and supporting the urban poor and their organisations.

This book will be an invaluable resource for researchers and postgraduate students in urban development, poverty reduction, urban geography, and for practitioners and organisations working in urban development programmes in the Global South.

David Satterthwaite is a Senior Fellow at IIED and a Visiting Professor at the Development Planning Unit, University College, London, UK.

Diana Mitlin is an economist and social development specialist working at the International Institute for Environment and Development (IIED), and a Professor at the University of Manchester, UK, working at the Global Urban Research Centre, the Institute for Development Policy and Management and the Brooks World Poverty Institute.

'This important book is the first systematic study to compare and evaluate the wide range of paradigms underlying recent efforts to mitigate urban poverty, and makes an eloquent empirical and analytic case for the importance of poor urban communities in defining, catalyzing and implementing reducing urban poverty. It is indispensable for all policy-makers, scholars and activists concerned with urban poverty and inequality in the cities of our world.'

Arjun Appadurai, New York University, USA

'A comprehensive book on how to reduce urban poverty – invaluable for students, teachers, policy actors and agencies wanting to understand and help cities and communities. The authors are brilliantly equipped to steer us all towards better outcomes for low-income urban settlements.'

Anne Power, London School of Economics, UK

In memory of Perween Rahman and her extra-ordinary courage and contribution to the struggle for social justice both in and beyond Karachi.

Reducing Urban Poverty in the Global South

David Satterthwaite and
Diana Mitlin

LONDON AND NEW YORK

First published 2014
By Routledge
2 Park Square, Milton Park, Abingdon, Oxon OX14 4RN

Simultaneously published in the USA and Canada
by Routledge
711 Third Avenue, New York, NY 10017

Routledge is an imprint of the Taylor & Francis Group, an informa business

British Library Cataloguing in Publication Data
A catalogue record for this book is available from the British Library

Library of Congress Cataloging-in-Publication Data
Satterthwaite, David.
Reducing urban poverty in the global South / David Satterthwaite, Diana Mitlin.
pages cm
Includes bibliographical references and index.
1. Urban poor--Developing countries. 2. Poverty--Developing countries.
I. Mitlin, Diana. II. Title.
HV4173.S38 2013
362.5'6091724--dc23
2013004088

ISBN13: 978-0-415-62462-6 (hbk)
ISBN13: 978-0-415-62464-0 (pbk)
ISBN13: 978-0-203-10433-0 (ebk)

Typeset in Baskerville
by Taylor & Francis Books

Contents

Illustrations

Figures

Tables

Boxes

Acknowledgements

This book draws heavily on what we have learned from grassroots organisations and federations, and from the staff of local NGOs who work with them. This includes so many people and groups who have shared their knowledge and experiences with us. We are particularly grateful to grassroots leaders for the time and hospitality given to us and for what we learned as they showed us their homes and neighbourhoods, explained their struggles and experiences (both positive and negative) and the very real achievements they secured. We are also particularly indebted to the staff of the five organisations and their associated networks whose work is described in Chapter 4. This includes:

From Ghana: Braimah R. Farouk, Mensah Owusu, and others from Dialogue on Human Settlements and the Ghana Homeless People's Federation.

From India: Jockin Arputham, President of the National Slum Dwellers Federation (NSDF) and of SDI, Savita Sonawane, Parveen Sheikh and many other Mahila Milan members, and Sheela Patel and Sundar Burra (SPARC).

From Kenya: Joseph Muturi, Benson Osumba, Irene Karanja, Jack Makau, Jane Weru, and others from Muungano wa Wanavijiji, the Muungano Support Trust and the Akiba Mashinani Trust.

From Malawi: Sarah Jameson, Mphatso Banda, Sikhulile Nkhoma, and others from the Homeless People's Federation in Malawi and the Centre for Community Organization and Development.

From Namibia: Edith Mbanga, Anna Muller, Heinrich Amushila, and others at the Shack Dwellers Federation of Namibia and the Namibian Housing Action Group.

From the Philippines: Sonia Fadrigo, Ruby Papeleras Haddad, Rolly and Josie from the disaster response team, Father Norberto Carcellar, Jason Rayos, May Domingo Price, and others in the Philippines Homeless People's Federation and the Philippine Action for Community-led Shelter Initiatives (PACSII).

From South Africa: Patrick Magabula, Rose Molokoane, Joel Bolnick, Ben Bradlow, Water Ffiew, Bunita Kohler, Greg van Rensberg, and others at the Federation of the Urban and Rural Poor, the Informal Settlements Network, the Community Organization Resource Centre (CORC) and uTshani Fund.

From Tanzania: Husna Shechonge, Tim Ndezi, Mwanakombo Mkanga, and others at the Homeless People's Federation of Tanzania and the Centre for Community Initiatives.

From Uganda: Lubega Edris, Katana Goretti, Hassan Kiberu, Sarah Kiyimba Nambozo, Joseph Sserunjogi, Skye Dobson, and others from the National Slum Dwellers Federation of Uganda and ACTogether.

From Zambia: Joyce Lungu, Margaret Lungu, Nelson Ncube, and others at Zambian Federation and the People's Process on Housing and Poverty.

From Zimbabwe: Catherine Sekai Chirembe, William Hwata, Davious Muvindi, Jessy Nziramasanga, Beth Chitekwe-Biti, Patience Mudimu, Sheila Magara, and others from the Zimbabwe Homeless People's Federation and Dialogue on Shelter.

From Pakistan: Arif Hasan, Perween Rahman and Anwar Rashid, and others at the Orangi Pilot Project and Muhammad Younus (Urban Resource Centre, Karachi).

And from the Asian Coalition for Housing Rights family: Somsook Boonyabancha, Tom Kerr, Maurice Leonhardt, Fr. Jorge Anzorena (Sophia University, Japan), Le Dieu Anh (ACCA, Vietnam), Lajana Manandhar (Lumanti, Nepal), Enkhbayer Tsedendorj (Urban Development Resource Center, Mongolia), Somsak Phonphakdee (UPDF Cambodia) and Rupa Manel (Women's Co-Op, Sri Lanka).

Thanks are also due to staff at IIED-América Latina, especially those who work in the support programme to informal settlements and their residents, including Ana Hardoy, Jorgelina Hardoy and Florencia Almansi,

Thanks are also due to our colleagues in the Human Settlements Group at IIED, especially for their interest in our work, willingness to engage us in discussion and patience while we wrote this and the previous volume. We would also like to thank our many academic colleagues and students at the Development Planning Unit (University College London) and the Institute for Development Policy and Management at the University of Manchester who have helped us with substantive ideas, recommendations for literature, discussions and helping to connect our thinking to wider academic debates. Beyond these organisations, we are particularly grateful to the authors of papers in the journal we edit – *Environment and Urbanization* – who keep us up to date and grounded on the work that is ongoing in towns and cities across the Global South.

We would also like to thank the staff from development assistance agencies who have shared their work with us, including the challenges they face in getting more attention to urban poverty and in designing and supporting effective interventions.

Acronyms

ACCA	Asian Coalition for Community Action
ACHR	Asian Coalition for Housing Rights
ADB	Asian Development Bank
ANC	African National Congress
BSUP	Basic Services for the Urban Poor (in India)
BUILD	Bombay Urban Industrial League for Development
CARE	An international charity founded in the USA in 1945 after World War II to send food aid and basic supplies to Europe; originally known as the 'Cooperative for American Remittances to Europe' (hence the acronym)
CARITAS	A charity made up of a confederation of Catholic relief, development and social service organisations
CCT	Conditional Cash Transfer
CDC	Community Development Council
CDN	Community Development Network
CDN	Credit du Niger
CENVI	Centro de la Vivienda y Estudios Urbanos (Mexican NGO)
CEUR	Centro de Estudios Urbanos y Regionales (Research institute in Argentina)
CFAF	CFA franc, currency used in some African countries (*Communauté Financière Africaine* franc)
CLIFF	Community Led Infrastructure Financing Facility
CODI	Community Organizations Development Institute (National government agency in Thailand)
COHRE	The Centre for Housing Rights and Evictions
CDF	Community Development Fund
COPEVI	Centro Operacional de Vivienda y Poblamiento (Mexican NGO)
DAC	Development Assistance Committee
DESCO	Centro de Estudios y Promoción del Desarrollo (Peruvian NGO)
DFID UK	Government Department for International Development

FEGIP	Federação de Inquilinos e Posseiros do Estada de Goiás (the Goiás or Goiânia Federation for Tenants and *Posseiros*)
FONHAPO	Fondo Nacional de Habitaciones Populares (Mexican National Popular Housing Fund)
HIV/AIDS	Human immunodeficiency virus/acquired immunodeficiency syndrome
IDB	Inter-American Development Bank
IDS	Institute of Development Studies, University of Sussex
IFRC	International Federation of Red Cross and Red Crescent Societies
IIED	International Institute for Environment and Development
IIED-América Latina	Instituto Internacional de Medio Ambiente y Desarrollo – América Latina (Latin American NGO based in Buenos Aires)
ILO	International Labour Organization
JNNURAM	Jawaharlal Nehru National Urban Renewal Mission (India)
KHASDA	Karachi Health and Social Development Association
KWSB	Karachi Water and Sewage Board
KWWMP	Korangi Waste Water Management Project
MDC	Movement for Democratic Change (political party in Zimbabwe)
MDGs	Millennium Development Goals
NGO	Non-government Organization (or as Jockin Arputham suggests, 'Not Grassroots Organization')
NSDF	National Slum Dwellers Federation (India)
NULICO	National Union of Low-income Community Organisations (Thailand)
OECD	The Organisation for Economic Co-operation and Development
OECF	Overseas Economic Cooperation Fund (a major bilateral aid donor that was part of Japan's bilateral aid programme and was absorbed into the Japan Bank for International Cooperation)
OPP	Orangi Pilot Project (Pakistan)
OPP-KHASDA	Orangi Pilot Project Karachi Health and Development Association
OPP-RTI	Orangi Pilot Project – Research and Training Institute
PB	Participatory budgeting
PHP	People's Housing Process
PPP	Purchasing power parity
PRI	*Partido Revolucionario Institucional* (Political party in Mexico)
PT	*Partido dos Trabalhadores* (Workers Party) in Brazil
RAY	*Rajiv Awas Yojana*, a programme set up by the Government of India Ministry of Housing & Urban Poverty Alleviation
RedR	Register of Engineers for Disaster Relief, an international disaster relief charity

ROSCA	Rotating Savings and Credit Association
SACDA	Southern African Catholic Development Association
SAP	Structural Adjustment Programme
SCP	Sustainable Cities Programme
SDI	Shack/Slum Dwellers International
SELAVIP	Servicio Latinoamericano, Africano y Asiatico e Vivienca Popular (the Latin American, African and Asian Social Housing Service), a private foundation that supports housing projects for very low-income families
SEWA	The Self-Employed Women's Association
SIDA	Swedish International Development Cooperation Agency
SPARC	Society for the Promotion of Area Resource Centres (Indian NGO)
TAP	Training and Advisory Program run by the Asian Coalition for Housing Rights
UCDO	Urban Community Development Office (Thailand)
UMP	Urban Management Programme
UN HABITAT	United Nations Human Settlements Programme
UNCHS	United Nations Centre for Human Settlements (former name of the United Nations Human Settlements Programme)
UNDP	United Nations Development Programme
UN ESCAP	United Nations Economic and Social Commission for Asia and the Pacific
UNICEF	United Nations Children's Fund (formerly United Nations International Children's Emergency Fund)
UN ISDR	United Nations International Strategy for Disaster Reduction
UPDF	Urban Poor Development Fund
UPFI	Urban Poor Fund International (international fund managed by Slum/Shack Dwellers International)
URC	Urban Resource Centre (first developed in Karachi and then with similar centres developed in other urban centres in Pakistan)
US AID	United States Agency for International Development
WIEGO	Women in Informal Employment: Globalizing and Organizing (a global action-research-policy network that seeks to improve the status of the working poor, especially women, in the informal economy)
ZANU-PF	Zimbabwe African National Union – Patriotic Front (political party)

1 Introduction

Why this book and this theme?

This book explores what has been attempted in the effort to reduce urban poverty in the Global South over the past 30–40 years, what have been the experiences of the projects and programmes that have been introduced, what new innovations have emerged from such experiences and what current practices suggest for our understanding of effective interventions. In writing this book we are drawing together our work on this topic over the past 20 years – the ideas to which we have been exposed, the literature we have reviewed and contributed to, the groups we have visited and the experiences and histories that have been shared with us. We have written this book both to share what we have learnt – and to catalyse a wider debate.

Why this theme? Urban poverty is and will be a continuing global challenge. As the world continues to urbanise, the pressure for effective solutions will become more intense. We are now some five years after the year when the world's urban population came to exceed its rural population and this trend towards increasing urbanisation is continuing. Most of the world's population increase is taking place in urban areas in the Global South (United Nations 2012). Despite this, there is remarkably little discussion in much of the development literature about the problems of urban poverty and little attention is given in many development assistance agencies and national governments about how it might be addressed. Indeed, at the international level, many international donors still have no urban policy and some that had such a policy have been moving away from an urban programming capacity in the past ten years – for instance, the bilateral agencies of Sweden, the Netherlands and Switzerland. National governments in the Global South have a range of responses – as might be expected – but, as we argue below, too few have done enough in this field. The companion volume to this (Mitlin and Satterthwaite 2013) whose main points are summarised below argued that urban poverty has been misunderstood and under-estimated, due to the use of models and measurement methods based on rural poverty. This book reports on the ways in which urban poverty reduction programmes have been conceptualised and realised. We argue that the record of success is mixed and limited – but that a small number of innovative programmes have stuck at the task of learning, fine-tuning their interventions to advance their methods and secure increasingly impressive results.

Why now? The international development community is preparing for a renewed round of poverty reduction commitments and targets. As we discussed in the previous book and summarise in Chapter 6, progress on urban components of these targets has been limited. However, we believe that such commitments continue to be significant in framing international vision and ambition, and influencing the ways in which national governments set priorities. But the attention given to preparing the post-MDG (Millennium Development Goals) agenda is small in comparison to that of managing the economic crisis that is continuing in much of the Global North. This crisis appears systemic, reflecting the pressures on both state and capital in the search for continuing economic growth, secure investment opportunities and, for individual companies, competitive advantage. The consequences for the urban poor are acute, and remind us that despite considerable progress urban poverty is not a global problem that has been solved.

In this context, what we hope, above all, is that this volume will encourage debate. We view this book as our contribution to that debate – we believe that academic institutions, government agencies, professionals (wherever they are located), specialist organisations, and organisations of the urban poor themselves need to be more vocal. We acknowledge that there is much that we do not know – in part because many activities have not been documented, or if they have been documented that it is only in respect of some aspects of their work. We argue that there is now a need to analyse experiences, articulate ideas and share development histories in a wide-ranging discussion that considers what has been achieved, why successes and failures have been realised and how we can achieve both scale and depth in efforts to reduce urban poverty.

Understanding the problem: the nature of urban poverty and inequality

This book focuses on reducing urban poverty; the companion volume published in 2013 focused on understanding and assessing its scale and nature. This chapter outlines the key points from the previous book and then explains how the rest of this book is structured.

The previous book documents how the scale and depth of urban poverty in Africa and much of Asia and Latin America is greatly under-estimated because of inappropriate definitions and measurements. How 'a problem' is defined and measured obviously influences how the 'solution' is conceived, designed and implemented – and evaluated. The use of inappropriate poverty definitions that understate and misrepresent urban poverty helps explain why so little attention has been given to urban poverty reduction by most development assistance agencies. This also helps us understand the paradox of so many statistics apparently showing relatively little urban poverty in nations (or globally) despite other evidence showing the very large numbers living in poverty. Around one in seven of the world's population lives in poor-quality and usually overcrowded housing in urban areas. Most of these areas lack provision for safe, sufficient water, sanitation and many other needs. This includes very large numbers of urban dwellers who are malnourished and suffer premature death or disease burdens that are preventable. A significant proportion of these are actually defined not to be poor by many poverty lines.

Almost all official measurements of urban poverty are also made with no dialogue with those who live in poverty and who struggle to live on inadequate incomes. It is always 'expert' judgement that identifies those who are 'poor'. In many countries, poverty is still defined and measured with no poverty reducing initiatives to respond to these statistics. At best, sections of the urban poor are 'targeted' and become 'objects' of government policy which may bring some improvement in conditions. But too rarely are they seen as citizens with rights and legitimate demands who also have resources and capabilities that can contribute much to more effective poverty reduction programmes.

The dollar-a-day poverty line (and its adjustment to US$1.25) is one example of the use of overly simplistic income-based 'nutrition plus'[1] poverty lines. This was chosen as one of two indicators for monitoring progress on the Millennium Development Goal of eradicating extreme poverty and hunger. Such simplifications lie at the core of why urban poverty is under-estimated. If we are to use a monetary measure for defining and measuring whose income or consumption is insufficient (and from this determining who is poor), this measure has to reflect the cost of food and of non-food needs. If the costs of food and non-food needs differ – for instance, by nation and by location within each nation – this monetary measure has to be adjusted to reflect this. But most income-based poverty lines do not do this. Many urban centres (especially the more successful ones) are places where the costs of non-food needs are particularly high, especially for low-income groups who live in informal settlements where costs such as rent, water (from vendors or kiosks) and access to public toilets are particularly high. In addition, dollar-a-day and most other poverty lines are set with no consideration of living conditions – for instance, of those who do not have reliable, good-quality and not too costly access to water, sanitation, health care and schools, as well as not having a voice and being served by the rule of law. Aid and other forms of development assistance are legitimated on the basis that they meet the needs of 'the poor' but decisions about the use of development assistance do not include any role for 'the poor', nor are those who make such decisions accountable to 'the poor'. Similarly, poverty lines are set without dialogue with those who struggle to live on inadequate incomes and without needed data – and so inaccurate poverty lines based on wholly inappropriate criteria are being used to greatly overstate success in urban poverty reduction.

For instance, applying the dollar-a-day poverty line[2] in 2002, less than 1 per cent of the urban populations of China, the Middle East and North Africa, and East Europe and Central Asia were poor. In Latin America and the Caribbean, less than 10 per cent of the urban population was poor. For all low- and middle-income nations, 87 per cent of their urban population was not poor (Ravallion *et al.* 2007). As the companion volume to this shows, there is a very large volume of work that shows how these figures are inaccurate and the associated methodologies inappropriate.

This companion volume includes a review of the approaches used in the World Bank poverty assessments and the national poverty programming processes associated with international development assistance and the Millennium Development Goals. It presents examples of how different criteria used for defining poverty in a nation can show almost no urban poverty – or 30 to 50 per cent of the urban population in poverty. It identifies hidden influences and assumptions within poverty definitions that often help under-count who is identified as being poor – for instance, inappropriate assumptions about the costs of meeting the needs of infants and children which assume that these are only a small proportion of the cost of adult needs. Inappropriate poverty lines also help explain why urban populations that apparently have very little poverty still have high levels of under-nutrition and very high infant, child and maternal mortality rates.

One puzzle here is the refusal of those setting poverty lines to acknowledge that the costs of food and non-food needs vary not only between nations but within nations. When international 'experts' and consultants work abroad, they get daily allowances to cover their accommodation and living costs that are adjusted by country – and by city within that country. This shows recognition that daily food and non-food costs for such experts vary by up to a factor of five, depending on location. So why is there no such recognition accorded to low-income groups?

The companion volume addresses the question of why health is so poor among low-income urban dwellers. It describes the very large health burdens associated with urban poverty, including very high infant and child mortality rates, large percentages of malnourished children, and large and easily prevented health burdens for children, adolescents and adults. It discusses the causes, including very poor-quality and overcrowded living conditions and the lack of provision for safe water, good-quality sanitation, health care, schools and emergency services. These in turn are linked to local governments who may refuse to work with those living in informal settlements, even when they house one-third or more of a city's population – or who have substantial under-provision for formal and informal residents alike. It also discusses the inadequacies in available data on illness, injury and premature death, provision for water and sanitation, and the impact of disasters on urban poverty.

The inadequacies in official data on development are another reason why urban poverty is understated. Much of the data used in policy-making rely on national sample surveys with sample sizes that are too small to reveal the inequalities within national urban populations or within individual cities. Urban averages for health-related statistics are improved by the concentration of middle- and upper-income groups in urban areas. Hence figures hide how low-income urban dwellers living in informal settlements can be facing comparable health problems to those faced by low-income rural dwellers – or in some instances with worse health problems. The concentration of people and housing in cities provides many potential agglomeration economies for health as the costs per person or household served with piped safe water, good-quality sanitation and drainage, health care, schools and the rule of law are lowered. But in the absence of a government capable of addressing these issues, this same concentration brings profound health disadvantages. In addition, the data collected in most nations on provision for water and sanitation do not show who has provision to a standard adequate for good health.

To survive, the urban poor have to find work that provides a cash income. In the urban context, basic services are commodities that have to be purchased. Finding income-earning opportunities that are more stable, less dangerous and provide an adequate return is central to reducing their poverty or moving out of poverty. Yet we actually know very little about the difficulties facing low-income urban dwellers in securing sufficient income and what would help them to do so. This is all the more remarkable when poverty is defined by income-based poverty lines.

In part, this lack of knowledge is because such a high proportion of low-income groups work in what is termed the 'informal' economy on which little or no data are available. In part, it is because the official data collected on employment have never been able to capture the variety, complexity and diversity of income-earning sources, working conditions and hours, and their implications for health and income levels. But there are case studies that show the struggle of households to earn sufficient income (often also involving children and having to withdraw them from school), the often devastating impact of illness, injury or premature death on household income, and the societal limits faced by women in labour markets (especially formal jobs other than low-paying domestic workers). Of course, this is also part of a larger global picture in which enterprises reduce their costs by employing temporary or casual workers (or day labourers) and drawing on suppliers and services from the informal economy. There are a few detailed studies that provide us with insights into the difficulties faced by those working in the informal economy – for instance, the importance of social networks for getting employment and the more powerful local people who prey on street traders and other own-account workers or demand payment from them. What is clear is that the relative scale of the informal economy has increased.

For many households, the home has great importance as the location for income-earning work too – especially for women – and in this context the lack of adequate basic services exacerbates the difficulties in earning a living. We know remarkably little about the ways in which income circulates in low-income settlements and how this is influenced by relations with the wider city and the drivers of economic growth. We also know remarkably little about what best supports low-income groups in getting higher incomes – although case studies highlight important factors such as the availability of credit and being able to have a bank account, the extension of a reliable supply of piped water and electricity to the home (so useful for many income-earning opportunities), good social contacts, literacy and the completion of secondary school.

Most measures of poverty applied in low- and middle-income nations are for absolute poverty. They do not concern themselves with inequality. But it is not possible to understand poverty without an engagement with inequality and what underlies it. Studies of inequality, like studies of poverty, also focus on income; they show the increasing stratification of many urban labour markets, particularly those in countries that have secured economic growth. What is evident is that those at the lower end of the labour market, in unskilled and/or informal employment, do not share equally in the benefits. However, many of the most dramatic (and unjust) inequalities are in relation to the other deprivations on which our first book concentrates – housing and living conditions, access to services, the rule of law. These are also reflected in the very large inequalities in health status and in premature mortality. It is clear that inequalities in access to infrastructure and services with cities also reflect inequalities in political power, voice and capacity to hold government agencies to account. In some nations, those living in settlements with no legal address cannot register as voters, while in most informal settlements, residents face difficulties obtaining the official documents needed to get on the voter's register, access entitlements and hold government or private service providers to account. In all nations, the inequalities faced by those living in informal settlements are reinforced by the stigma associated with living there.

Understanding these issues requires attention to the spatial aspects – i.e. inequalities in these between neighbourhoods and districts within cities; so often, the data collected on incomes, living conditions or service provision are from too small a sample to show these spatial inequalities. An understanding of the many different factors that create or exacerbate inequality also means more routes by which inequality can be reduced. There are also other aspects of inequality that throw light on deprivation – for instance, inequalities in household assets or capital, or inequalities caused or exacerbated by social or political status and relations including discrimination. A study of inequality also needs to consider the implications for low-income groups of a larger and wealthier elite within the city – for instance, as their demands and influences restructure city planning to serve their priorities. They can separate themselves from 'the poor' through gated communities and highways that link their homes, places of work and places for leisure.

There are also some nations where governments have reduced some of the most profound inequalities among the urban population – for instance, through extending provision for water, sanitation, schools and health care to a larger proportion of the low-income population; or through transfer payments that reach large sections of the low-income population with supplements to their income – for instance, including pensions, conditional cash transfers and child allowances. Where these reach low-income groups, they certainly reduce absolute poverty, although they may not reduce income inequality as incomes may rise more among high-income groups. These cash transfers also do nothing to address the inequalities in provision for infrastructure and services.

Thus, in summary, the companion volume to this showed how the scale and depth of urban poverty can be greatly understated if inappropriate poverty lines are used. In addition, that all nations need poverty lines that take into account the actual costs faced by low-income groups with regard to food and non-food needs and how these vary by location. And that all nations need a consideration of other aspects of poverty and what underlies them.

For poverty specialists who have long focused on income- or consumption-based poverty lines, this broader view of poverty may be hard to understand.

Many of its aspects are not easily measured. For instance, as this second book makes clear, two critical underpinnings of urban poverty reduction are the extent to which the urban poor are organised and the nature of their relationship with local government; but indicators that track organisational capacity and relational quality are not easy to find. Many aspects of urban poverty may be considered to be state failures – mostly the incapacity or unwillingness of local governments to meet their responsibilities – although the experience in the Global North shows that much more than 'active' government is required to tackle deprivation and inequality. And in some cases it is also the result of the lack of attention given to such investments by aid agencies and development banks. But a broader understanding of poverty also means more entry points and more scope for intervention. City and municipal governments may have limited capacity to increase incomes for the poorest groups but they have more scope and capacity to address many other deprivations. This is also true for NGOs and for grassroots organisations.

If poverty is defined only by income or consumption then little attention is given to the multiple roles that housing and their immediate surrounds (or neighbourhoods) can have in reducing deprivation. A focus only on income poverty can mean that a low-income household with a secure home with good-quality provision for water, sanitation and drainage and with their children at school and access to health care is considered just as poor as a low-income household with none of these provisions.

Reducing urban poverty – and the structure of this book

Reducing urban poverty demands a functioning state in each urban centre or district that addresses its responsibilities. This is more likely if and when government is accountable to its low-income population. It needs such relations if it is to act in the interests of the public good. This second book explores what it takes to make the state act in ways that support at least some of the multiple routes to poverty reduction. It considers how international agencies can learn how to support this – and how much this also means a need to work with and support representative organisations of the urban poor, and setting up funding streams that are accessible and accountable to these organisations and their members. The discussion considers what has been tried in terms of approaches to urban poverty reduction, what has succeeded, what has failed and some of the reasons that help to explain these outcomes. It describes how new initiatives have sought to create a new pro-poor form of urban development that goes beyond the outcomes dominated by clientelist politics and/or elite interests. This volume explains how a more equitable inclusive urban future can be secured, the kinds of activities that are needed and the ways in which new strategies that respond to this have been defined and realised. The discussion offers a measured optimism. The challenges, summarised above, are considerable and the scale of need very large. However, as described below, we know what works, we understand why it works, and we have some idea about what might be done to scale it up.

The chapters that follow describe and analyse the different ways in which national and local governments, international agencies and civil society organisations are seeking to reduce urban poverty. Chapter 2 discusses the different ways in which poverty reduction has been understood, Chapter 3 examines the ways in which different agencies translate such approaches into programmes of action, and Chapter 4 considers in more detail five interventions chosen because of their impact, in terms of the large numbers of households reached, the breadth of their activities across sectors and the ways in which they made a difference to families. Chapter 5 analyses these interventions in terms of similarities in their underlying strategies and explores reasons for their effectiveness. The concluding chapter (6) highlights the implications for governments and international agencies in moving forward and reducing urban poverty.

Chapter 2 identifies eight approaches which underlie most agency interventions and programmes for urban poverty reduction. The eight approaches are defined by distinct arenas of action and the implied causal relations between activities and poverty reduction. For example, some seek to address the consequences of poverty through an emphasis on redistribution, while others challenge the causes of poverty and inequality and provide for social mobility, improved access to basic services and/or economic growth. These eight approaches are divided into three sections according to their major emphasis. The first four are predominantly the domain of the state – welfare-based interventions, urban management, rights-based approaches and participatory governance. Although emphasis is placed on the state, we recognise that particular initiatives may involve non-state agencies seeking to catalyse improvements in government policy and programming changes. The next two approaches are led by non-state sectors; the first is a market-based approach to poverty reduction, and the second a political transformation through the efforts of social movements. In the first of these, the state may be either leading or facilitating activities. The final two approaches – clientelism and self-help – give particular emphasis to ongoing practices of urban development and propose approaches that augment these practices (or, in the case of clientelism, asks that its benefits be recognised). Together they make the case that the most effective strategies are likely to be incremental improvements to existing structures, rather than more ambitious efforts to challenge other economic, political and social forces. Of the eight approaches, the first five are managed by professionals and all but two (social movements and self-help) encourage vertical relationships between the urban poor and local and/or national elites.

In most nations, more than one of these approaches is or has been used. Experiences to date demonstrate some of the limitations of each approach in terms of producing the institutional framework for a long-term reduction in urban poverty. Most struggle with the institutional complexity of urban areas (especially larger cities) that usually include government roles and responsibilities scattered across different sectoral agencies and many different levels of government. All struggle to generate the resources needed for providing the trunk infrastructure (for water, waste water, roads, paths, electricity) to underpin universal provision. And all struggle to address the diverse scale and depth of urban poverty.

Chapter 3 examines a number of key programmes and directions in programmes for urban poverty reduction that agencies have taken up over the past 20 years – from national governments, local and metropolitan authorities, official development assistance and NGOs, and other civil society agencies. The chapter discusses agency policies, programmes and practices to understand how these approaches are translated into attempts at poverty reduction. This review is necessarily selective; it is not possible to cover all such activities in the available space. Nevertheless, the chapter illustrates the work of agencies, and shows how the approaches introduced in Chapter 2 are embedded within specific programmatic interventions, and what such interventions have offered to the reduction of urban poverty. The discussion begins with a focus on two government programmes: finance for housing improvement and social protection through cash transfers. While much housing finance has been orientated to higher income households, many governments have made particular (although often limited) efforts to provide finance that reaches low-income groups. In a very different and more individualised approach to poverty reduction, recent social protection initiatives have concentrated on cash transfers to targeted households.

The chapter then examines the contribution of local government whose proximity and traditional responsibilities mean that they have or should have a significant role for the well-being of the urban poor. Most of the initiatives or programme interventions that could reduce poverty fall within their jurisdiction – but the range of roles and responsibilities allocated to them is rarely matched by the institutional and financial base needed to do so. The importance of local government in supporting improvements in people's settlements and livelihoods is recognised. The chapter also considers the nature of the relations between state and civil society, and how the partnerships they form can help reduce the disadvantage faced by low-income households and communities. The analysis of local government focuses on three themes: the upgrading of informal settlements, access to basic services and participatory budgeting. This is followed by a section considering which official development assistance agencies have sought to contribute to poverty reduction and what they seek to do. Finally, the chapter considers initiatives of civil society in development including NGOs within three thematic directions: campaigning for greater attention to be given to rights; technical support by professional development agencies; and the development of a model based on alliances between NGOs and social movements.

Thus, Chapters 2 and 3 identify and explore reasons for the limited success of particular approaches and particular agencies and, in so doing so, help readers to understand the reasons for the emergence of a new generation of approaches to poverty reduction. Chapter 4 discusses five particular programme interventions that are notable because they have reduced poverty and achieved a degree of pro-poor political change. Each has done so through an alliance between grassroots organisations and local NGOs that support their work (although, in one, it is a national agency that supports this). The interventions are discussed in chronological order and we make the case that they are best understood in relation to one another as their development has drawn on an engagement with a common set of earlier experiences (in part those discussed in Chapters 2 and 3) as well as learning between them. These case studies illustrate evolving strategies and modalities of action as well as new emerging roles for the different agencies involved in them. In Chapter 4, each programme is introduced and their historical trajectory is summarised as well as their strategy explained and their impacts identified. The five programme interventions are: the Orangi Pilot Project (Pakistan); the National Slum Dwellers Federation, *Mahila Milan* and SPARC (India); the Asian Coalition for Housing Rights (Asia-wide); the Community Organization Development Institute (Thailand); and the slum/shack/homeless people's federations that form Shack/Slum Dwellers International (primarily sub-Saharan Africa and south Asia).

Chapter 5 explores these programme interventions in the context of wider perspectives on pro-poor social change. The discussion examines the underlying rationales and shows why these interventions have been successful in challenging existing power relations to shift the balance in favour of the 'urban poor', creating new options where previously none existed. This success rests on their ability to advance the interests of the urban poor in six dimensions. First, these processes advance the search for solutions that are inclusive and universal; when all can benefit, a political coalition for change is possible and scarcity associated with clientelism can be replaced by more accountable and transparent government. Second, they are strategic interventions of the urban poor in political processes that avoid contestation with local government where possible and instead seek to build an effective collaboration that legitimates the continuing contribution and presence of the organised urban poor in governance. Third, they have developed approaches to urban development based on co-production, the joint planning, financing and implementation of shelter improvements in ways that involve the residents of informal settlements (and their organisations) and local government – what Castells refers to as the creative powers of labour for the common good used in a different way. Fourth, a central concern of these interventions is to nurture the practice and potential of gendered empowerment and women's leadership. Fifth, they are aware of the need to work both at the city and the national level, seeking a strong city-based process but also recognising the need for central government support and redistribution. Finally, they invest in the development of collective political capabilities among the urban poor both through learning processes for particular communities, and through nurturing an institutional practice of federating and networking. This enables lessons to be identified and embedded in improved practices by other groups.

This book ends with discussions of how local governments, national governments and international agencies can become far more effective in addressing urban poverty at scale – and how this depends on them being able to listen to, learn from and work with low-income populations and also be far more accountable to them. It considers the different ways in which citizen-led and community-led poverty reduction can be supported by national and local governments and international agencies. This includes funding channelled through national and city funds set up and managed by the urban poor federations. It also discusses the role of two international funds that support grassroots initiatives in urban areas that also seek to get the engagement and support of local governments – the Urban Poor Fund International and the Asian Coalition for Community Action. It also contains a short section on how urban poverty reduction needs to take account of the increasing risks and new risks that climate change brings or will bring to the urban poor and their health, homes, assets and livelihoods. As yet, little is being done on this – but the combination of well-organised and representative federations of the urban poor and local governments that want to work with them are among the essential underpinnings of effective climate change adaptation. There is a real danger that the international funding for adaptation will not be structured to support such local organisations and partnerships. The book ends with a discussion of where and how to insert what we know about effective urban poverty reduction into the international discussions about aid effectiveness, about 'the future we want' and about the post-2015 development agenda – so that these measures help produce a future that low-income urban dwellers also want.

Notes

1 Nutrition plus because they are based on the cost of food with an upward adjustment in acknowledgement that there are also non-food needs that have to be paid for. However, usually there is little or no attempt to gauge how much these non-food needs cost. As the companion volume to this described in detail, in many poverty lines the upward adjustment (for instance increasing by 20–50 per cent the amount assigned to food costs) is far too little to allow many urban individuals or households to meet their non-food needs.
2 Here, the poverty line was set at $1.08/day (in 1993 PPP); see Ravallion *et al.* (2007).

2 Approaches to poverty reduction in towns and cities of the Global South

How is urban poverty reduction understood by those seeking to act on it? This chapter reviews eight approaches to urban poverty reduction from a range of sectors and related disciplines. These are summarised in Box 2.1.

Box 2.1 Different approaches to poverty reduction

STATE DIRECTED

Welfare assistance to those with inadequate incomes; usually takes the form of income supplements and/or free or lower-cost access to certain goods and services. Universal access to health care and schools is also a characteristic of a 'welfare-state'.

Urban management to improve 'local government' with a focus on efficiency, technical competence, a stronger fiscal base and implementing local regulations to get effective planning and land-use management, and to address the inadequacies in basic infrastructure and services.

Participatory governance includes greater accountability, transparency, and scope for citizen and community participation – i.e. improved processes of democratic government to ensure that urban governments are more responsive to the needs and interests of low-income and disadvantaged citizens. Its scale and effectiveness may increase through co-production where state support for community action is added.

Rights-based approaches that extend rights and entitlements to those who lack these – and that usually focus on low-income groups and those living in informal settlements.

MARKET BASED

Market-based approaches that seek to support higher incomes and livelihoods through access to financial markets and to support infrastructure and service provision or improvement that recovers costs.

SOCIAL MOVEMENT BASED

Social and urban movements supported because of their representation of urban poor groups and their capacities to negotiate pro-poor political change.

WORKING WITHIN THE STATUS QUO

Aided self-help with support to households and community groups to address their own needs – for instance, through bulk supplies, equipment loan, technical assistance and loans.

Clientelism: in many urban centres, despite the negative connotations of clientelism, this does provide an avenue for low-income disadvantaged citizens to access state services, albeit within vertical relationships that are often exploitative and that provide only limited support for some.

For each approach, the discussion elaborates on the need that the approach seeks to address and the rationale that lies behind the way in which the approach intervenes in the lives of the urban poor or, alternatively put, the 'theory of change'.[1] The first six approaches pick up on major themes within development efforts over recent decades. The first four (welfare assistance, professional urban management, the rights-based approach and participatory governance including co-production) each have the government or state as the primary agency. The next two (the market and social movements) focus on the contribution of alternative institutions. The final two build on the realities of struggles for livelihoods and housing faced by the urban poor: aided self-help and community development; and clientelism. Chapter 3 elaborates on experiences with these approaches as they are used to design and implement particular programmes of action.

In addition to describing each approach, this chapter explores some of the emerging problems each has in terms of its effectiveness in addressing absolute poverty and inequality. This analysis draws on the nine dimensions of urban poverty elaborated in the companion volume to this and summarised in Chapter 6, and on inequality with regard to income, and to its social, spatial and political dimensions. The discussion reminds us that the approaches are ideological and historically determined. An analysis of the approaches shows us that in many cases they reinforce patterns of exclusion and adverse incorporation into state and market processes. Even well-intentioned interventions may risk exacerbating difficult situations because they have not understood the levels of commodification, the highly stratified nature of urban livelihoods and the high levels of contestation over resources.

Introduction

The importance for governments and international agencies of acting on urban poverty has long been realised, although there is much less agreement on the most strategic ways in which to achieve an effective impact. This chapter considers how agencies and individuals tasked with poverty reduction have approached this challenge in towns and cities of the Global South. Our discussion is wide-ranging, including some approaches that have been contested in terms of the significance of their contribution, but which nevertheless remain in the mix of approaches. The objective of this chapter is to enable readers to understand the ways in which poverty reduction interventions have been understood, and the assumptions about causal relations that are a part of the explanation for the effectiveness of different approaches. Many of these approaches seek to address only one or two of the facets of poverty summarised in Chapter 1 and elaborated at length in our previous volume. While recognising the importance of integrated approaches, we have included here some significant approaches that seek to address a single aspect of urban poverty.

As noted above, the first six approaches pick up on major themes within development efforts over recent decades. The first four each have the government or state as the primary agency and include welfare assistance, professional urban management, the rights-based approach and participatory governance including co-production. The next two focus on the contribution of alternative institutions, the market and social movements. The final two approaches are embedded in the status quo: the ways in which communities provide themselves with self-help to address a multitude of individual and collective needs, and the extent of the contribution of clientelist relations to addressing the needs of those living in informal settlements.

Many of these approaches are not specifically urban. Arguably only urban management and the final two approaches (aided self-help and clientelism) are 'urban'. These two final approaches are embedded in the status quo and distinctively adapted to the urban context – and thus to the commodification of all non-family aspects of everyday life, the high level of institutional engagement by the state, the influences of high population densities on housing and basic services, and with livelihood dependence on formal and informal labour markets. However, the remaining five approaches are also used widely in urban areas and have taken on a particular modality due to the realities of urban lives and the nature of urban institutions. It is through the form that they commonly take in urban areas and the realisation of the approaches within an urban context that they are considered here.

In terms of this adaptation to the urban context, several features stand out. The market is significant as cities have formed, for the most part, because private capital chose to invest there as cities offered them economics of scale (and other agglomeration economies) in manufacturing and services. There are also agglomeration economies for much of the infrastructure and public services that meet the needs of enterprises and residents, although these require government action.

Welfare provision has moved beyond self-help and mutual relations of care and responsibility within neighbouring families and communities. The three more political approaches (rights-based approaches, social movements and participatory governance) are attuned to the dynamics of urban government. This means that they are attuned, to varying degrees in different places, to city and national elections, councillors and their local constituencies, and developers and land speculators and associated conflicts. Also relevant are the ways in which urban regulations are used which is often by a range of self-interested elites at various levels to penalise the less powerful and deem illegal their homes and livelihoods.

Over time, the balance of preferences between approaches has shifted as new explanations have emerged to account for the failure of development and 'modernisation' to address problems of poverty and inequality. For example, emphasising the state was the favoured strategy of anti-colonial political forces as liberation movements in Africa and much of Asia sought to challenge their peripheral role in the major European economies as they came to form the new independent governments. In addition, during the 1980s, structural adjustment programmes sought to reinvigorate the market with concerns that economic and public policy were damaging market growth. As we illustrate below and in Chapter 3, all eight approaches to poverty reduction are found among the agencies currently supporting urban development, reflecting a range of understandings about the reasons for urban poverty and hence the most effective intervention strategies.

Our argument is that each approach represents a deeper ideological position, although how this emerges in practice is influenced by circumstance. A market-based approach, for example, takes a different form in an advanced formalised large city economy than in a small town with much informal production. At the same time, we believe each is limited in its capacity to address urban poverty reduction; in part, because they are, almost universally, designed and realised without the very populations they are intended to benefit. As a result of this, ideas, plans and actions may not be attuned to the realities of the urban poor. More significantly, because of the lack of engagement and connection between conceptualisation/realisation of the intervention and the organised urban poor, there will be no pressure to change political relations and enable social transformation. The failings of the existing electoral system are well recognised in North and South. As Charles Tilly (2004, p. ix) commented after a lifetime of studying social movement activities and organisations in the United States: 'I regard my own American regime as a deeply flawed democracy that recurrently de-democratizes by excluding significant segments of its population from public politics, by inscribing social inequalities in public life, by baffling popular will, and by failing to offer equal protection to its citizens.' The exclusion of the urban poor from decisions about their lives reflects the reality that democracy is concerned with electoral choices rather than encouraging a sustained engagement by citizens in decisions that affect the quality of lives. This is particularly true of the urban poor who are, as shown in Mitlin and Satterthwaite (2013), systematically excluded from the definition and realisation of development options – and even from the definitions of what constitutes poverty. The reality today – and for the past decades – is that almost all development interventions are designed by professionals and do not involve the local communities that are struggling to achieve urban development (in its multiple forms). Without this local knowledge and accountability, many are relatively superficial in their analysis of the problems that the urban poor face. In some cases, at the local level, local NGOs have sought to engage local communities in the planning and implementation of activities. However, for many reasons they have found this difficult to do. One reason is that most NGOs secure their funds primarily from official development assistance agencies and the controls and limits placed on the use of these funds make it difficult for a locally responsive and flexible process to emerge. Chapter 3 discusses problems with the system of aid that are evident in the urban context. Given that electoral democracy does not result in the effective political inclusion of the disadvantaged, alternative political processes need to ensure that the urban poor engage in influencing political decision-making and associated processes: without that pressure, interventions are likely to remain partial and insignificant in addressing the scale of need.

The first five approaches described in this chapter are primarily designed and realised by professionals. Such professionals may be located in the official international development assistance agencies, national and subnational governments (that include urban governments), and international, national or local NGOs. The remaining three are more clearly hybrids between formal and informal urban livelihoods, and ways in which professional interests have sought to engage with and represent more grounded experiences. While this experience may be incomplete and unsuccessful in terms of a sustained and substantive impact on urban poverty, it represents the attempts of local populations to secure development (Myers 2011). In this case, the approaches are realised both by professionals and by a wide range of agencies and other individuals as they attempt to create new development options. Arguably this is also true of the market approach which, while deliberately promoted by some, also includes the main modalities by which individuals and households accumulate income and wealth. Chapter 3 discusses the ways in which these ideas are taken forward into planning and programming. As is evident in the discussion in that chapter, there are few actual interventions that are modelled on a single approach. In practice, the ideas described below are taken up and blended into agreed sets of activities that reflect the preferences of those planning the intervention. In a positive sense, such a blending recognises the multidimensional nature of urban poverty and hence the need for multifaceted programming. However, it may also reflect the lack of clarity about the underlying rationale in urban development programming and/or the need to make compromises to advance particular policies and programmes.

Following the analysis of the approaches in this chapter and their realisation through agency programmes in Chapter 3, Chapter 4 turns to several programme interventions that have sought to implement a more integrated programme of change. These are programmes of activities that have been particularly influential in an emerging generation of urban programming. These interventions emerged from experiences that faced the limitations of the approaches discussed in this chapter and involve a series of experiential activities that have refined and continue to refine practice.

Approach 1: Welfare-related assistance

There are a range of approaches to urban poverty reduction that are, broadly speaking, led by the state. Organisations using these approaches seek to address urban poverty by influencing the ways through which government acts, the priorities it has and the modalities used. They share a belief that the state is an important influence on and contributor to development, and hence are aware that the state is a political agency, influenced by electoral realities, group interests and the distribution of power. Notwithstanding simple political pressure, states may also be influenced by a professional and/or compassionate discourse. The first four approaches we consider share the belief that the state is critical and that neither market nor citizens alone can address some fundamental issues of urban poverty. It follows that the nature of government is important and, while democracy is not essential to pro-poor government, it certainly helps. As Heller and Evans (2010, p. 437) argue, '[D]emocracy ... allows subalterns to contest control of the state apparatus. ... Transformation of the role of the state is at the heart of democratization and the citizenship that goes with it.' However, as elaborated later, democracy is not always supportive of inclusive and pro-poor urban development.

The first approach considered here is that of welfare. The need for both individuals and households to have some form of state assistance available if required is widely recognised. This is mostly framed in the context of absolute poverty or absolute needs. As summarised in Chapter 1, the failure to address absolute needs is related to multiple factors but primarily a failure to consider the complexity of urban poverty. One cause of poverty is inadequate income (and hence inadequate consumption of key goods) and a lack of capital assets – and welfare measures have considered both (although with more attention given to income). As countries grow richer, relative poverty (and other aspects of inequality) also becomes important with the acceptance of concepts such as relative deprivation in determining levels of redistribution, and the introduction of poverty lines linked to median income. At the same time, as countries grow richer the provision of services and infrastructure also generally helps to address other causes of poverty.

The needs addressed by welfare are usually around the income required to acquire essential goods (for instance, food and housing) or around access to needed services (for instance, health care and schools). There is rarely a single deprivation and this partly explains why there is a wide range of welfare-based measures for urban poverty reduction. Welfare interventions may respond to any one of a number of perceived needs. These needs may in part arise from an unexpected event (e.g. a local flood or a global economic recession) in the context of inadequate long-term infrastructure to protect against such events. Needs may result from the structure of the economy (e.g. wages that are barely sufficient or insufficient for survival) and no alternative employment options for some people given existing labour market skills and capabilities. Needs may also be related to the individual life cycle, and it is evident that children and old people are disadvantaged in the labour market. One of the difficulties facing people as they age and their strength and physical capacity diminishes is the lack of pension provision in the Global South. While, in many cases, households can support individual members who have limited earning capacity, a proportion of individuals have no such support and have to manage alone.

Within the welfare approach there are a multiplicity of different strategies for intervention. In terms of understanding the perception of needs that underpin welfare measures, it is helpful to recognise that this may be either charitable and empathetic or rights/entitlement based. As described by Britto (2005, p. 17) in a discussion of conditional cash transfers in Brazil, there has been an implicit and unresolved tension between two distinct notions of the programme: as a basic right or as a response to a need.

We include access to basic services as support is commonly made for welfare reasons, although in this case other rationalities are also important in influencing provision, including the significance of health, education, water and power to economic development, and the rights of citizens to a package of basic entitlements (see below). State investment in health and education may be legitimated through the need for a healthy workforce or alternatively such provision may be seen as an essential component of citizenship. The reasons for the subsidy of services are multiple, and state funding for these services may be justified by market failure (i.e. as public goods they are under-provided for by the market) and in terms of the externalities arising from infrastructure provision such as those related to public health. However, an important part of the discourse that secures investment in infrastructure and services is the recognition that these are 'basic needs' essential for human well-being.[2] We discuss the significance of the rights-based approach below and its emphasis on the establishment and claiming of rights that are available to all according to citizenship. Such entitlements make explicit who is and who is not recognised as eligible to receive support. While in general welfare measures reach out to particular individuals, households or groups, the nature of the discourse related to the transfer differs depending on the rationale used: a charitable response is within a vertical relationship from grantor to grantee while a rights- or an entitlement-based response recognises the underlying equity between individuals and the duty of the state to provide. Hence welfare approaches may be underpinned by the conceptual understanding associated with other approaches discussed in this chapter, such as rights and the importance of the market (Barrientos and Hulme 2008). However, we believe that welfare approaches are significant enough for poverty reduction to be discussed separately here. Moreover, drawing on Chatterjee's (2004, p. 34) interpretation of Foucault and the investment of the state in the population and its needs, we believe that welfare approaches have important broader consequences for state programme and citizen responses, and this also emphasises the need to consider them directly.

As noted above, income transfers are the most common form of welfare, although welfare approaches are sometimes orientated to asset acquisition.

Income transfers are often cash while asset transfers are generally specific to a particular need such as housing. Income transfers are what is most commonly recognised to be 'state welfare assistance', although this may be for livelihood protection or livelihood promotion, and even this distinction (although made) is hard to maintain in practice (Devereux 2002). Such support is targeted (i.e. focused on those most in need) so as to be effective. Recently, greater emphasis has been placed on more transparent systems that are seen to be fair, and hence to ensure such redistribution is viewed as legitimate by the electorate.

In understanding the specificities of particular welfare programmes, it is important to identify the role of broader social norms and values that define who is considered to be entitled or worthy to be included and who is excluded. Income-based support commonly includes social insurance measures that provide funds designed to protect workers and their households against life cycle- and work-related contingencies, such as maternity, old age, unemployment, sickness and accidents (Niño-Zarazúa *et al.* 2010, p. 5). Individuals directly contribute funding to these contingencies (often through deductions from their salaries) and then may call on the scheme as and when required. Governments may provide additional funding for these due to the inability of individuals to contribute sufficiently and/or to encourage them to contribute something. Social insurance measures recognise that there are events that can occur to everyone for which assistance is needed. The basic idea is that through making provision with some form of insurance, such events can be better managed for individuals and for the broader community. One of the most common forms of social insurance is a pension, paid to older people to replace incomes that were previously earned from work. One major distinction is how it is financed – and the role of those who become pensioners in contributing to pension funds during their working life. It is common for those in formal employment to contribute to pension funds – but this is less common for those in informal employment (and there are rarely pension schemes available to them). As a consequence, large numbers of old people do not have access to a pension. For example, in Brazil 32 per cent of those over 60 years of age receive a pension, while in Bangladesh the figure is 37 per cent of eligible older people (DFID 2005, p. 10). In South Africa, the government pays out to everyone regardless of their previous employment and 80 per cent of those over age 60 benefit (nearly 100 per cent of the elderly black population). In South Africa, pensions are widely recognised to have a significant impact on reducing poverty both for the pensioners themselves and for children whom the older women may care for (Niño-Zarazúa *et al.* 2010, p. 9). In addition to pension provision the International Labour Organization adds the legal frameworks to ensure minimum standards for employment, including working conditions and the safeguarding of workers' rights (Niño-Zarazúa *et al.* 2010).

A second type of income-based welfare support is social assistance that includes tax-financed policy instruments designed to address poverty and vulnerability and available to those in acute need (Niño-Zarazúa *et al.* 2010, p. 5). Available assistance does not require contributions from those who benefit. Social protection support may be divided into 'nets' (ex-post) and 'ladders' (ex-ante). Safety nets provide immediate assistance to those in acute need. 'Ladders' help those in need move out of poverty in the medium term, and reduce the need for emergency support in the future. In recent years, social assistance has increasingly taken the form of cash transfers for households in poverty (see Chapter 3). Previously greater emphasis was placed on the direct transfer of goods, particularly food. Since the current round of cash transfer programming began in the 1990s, over 30 countries have introduced large-scale programmes incorporating millions of beneficiaries and numerous smaller pilot schemes (Barrientos *et al.* 2010). Arguably cash transfers to households with adults of working age represent a new modality of poverty reduction as increased efforts are made to address long-term needs in addition to providing immediate support (Niño-Zarazúa 2010, p. 3). They are also seen as more effective than food aid; cash transfers recognise the need to provide a foundation that enables households to develop capabilities and improve their livelihood strategies through regular and reliable assistance (Niño-Zarazúa 2010). Considerable efforts have been made to target those households most in need and to ensure that such systems are both robust and transparent – and to avoid the partiality of previous systems used to repay political favours and reinforce clientelist politics. The emergence of these programmes is attributed to regionally specific factors: in Latin America to address the persistence of poverty after adjustment and liberalisation and to extend state benefits beyond the formally employed who were previously targeted; in Asia to reach those who benefit little if at all from rapid economic development or to help those in greatest need during times of financial crises; and in Africa to move beyond emergency or humanitarian aid to more predictable assistance that enables poverty to be reduced (Barrientos and Hulme 2008). While the programmes have scale in some Latin American and Asian countries, they remain small in sub-Saharan Africa (excluding South Africa). Moreover, in much of sub-Saharan Africa they are dependent on international development assistance – only in southern Africa are they financed primary by domestic resources (Niño-Zarazúa *et al.* 2010).

As discussed in Chapter 3, a new generation of social protection measures has emerged from the mid- to late 1990s (ibid.).[3] In many cases these have taken an emphasis on capability development a step further through associated conditionalities for particular households and women and children within these households: these are referred to as Conditional Cash Transfers (CCTs) to distinguish them from programmes that simply distribute small cash transfers with no associated requirements. As elaborated by Levy (2006) in the case of Mexico, they were introduced so as to be more effective and efficient than food aid and to address the need to enhance human capital to reduce poverty in the longer term. There may also be a variety of income-for-work schemes that undertake some public works and provide (usually very poorly paid) incomes to those who participate or those who these schemes target.

Asset accumulation policies provide opportunities for households 'to accumulate and consolidate their assets in a sustainable way', enabling them to reduce their future welfare needs (Moser 2009, p. 253). Moser (2009, pp. 254–256) argues that one key difference between asset accumulation and social protection such as safety nets lies in the orientation of the former towards risk reduction. Hence while the latter is concerned to protect households from risk, the former seeks to reduce the risks they face by, for example, improving shelter (leading to improved health), investing in education and supporting savings accumulation. Conditional cash transfers that incentivise education blend assets and incomes interventions through providing income on condition that education takes place. Food for work programmes may also been seen as contributing to assets by providing the necessary infrastructure (Devereux 2002, p. 664). Other approaches discussed in this chapter also support the consolidation of assets and create and/or strengthen development options that are considerably more than welfare interventions: for example, financial accumulation within the market approach and political capabilities within the social movement approach.

Asset transfers (rather than cash) in support of poverty reduction are most notable in support for income generation and housing and asset accumulation (Moser 2007). In a small number of cases, households may be assisted with grants (either in cash or in kind) to start up businesses. However, this is unusual following the greater success of micro-credit and there is now a strong argument made that subsidising business investment does not lead to success. Support for accessing housing is much more commonplace in part due to the recognition that securing adequate housing if often unaffordable without some form of support. In many northern countries this does not add to assets as it is simply access to subsidised rental provision but in other countries there may be support for home ownership both through capital subsidies and interest rate subsidies. In low-income countries, support for low-income groups to obtain or afford housing is very limited or non-existent, although there may be support for temporary accommodation following an emergency. Subsidised housing or subsided loans to help low-income households secure housing are more common in middle-income nations, as described in Chapter 3, although the nature of such programmes varies considerably, as does their scale and effectiveness.

State support may be offered to ensure or improve access to basic services. Access to a number of infrastructural services, including piped water, sewers, paved roads and drainage, is important for household well-being and health – and more broadly for the health of others. In wealthier and well-governed cities, there is often universal or close to universal provision of these services. These are not seen as 'welfare' but as what all dwellings (and other buildings) should be provided with; here, welfare assistance focuses support on helping those with limited incomes to pay the costs (or the costs of connection). In most urban centres in the Global South, the deficits in provision for piped water, sewers and drains may mean that programmes to address these deficits are seen as welfare. Financing for water provision may be redistributive through an incremental block tariff structure that provides water at lower unit costs for low-consumption users – or in the case of South Africa, 'free-basic water' up to a set amount (see Muller 2008).

Two fundamental services that are seen as needs and that are usually provided by the state free or below cost are health care and schools (and often provision for pre-school children too). For these, the issues are around whether these services actually serve or are available to low-income groups, the quality of provision, and the costs that have to be paid (if any). Our previous book looked in some detail at the deficits in provision (for instance, the large numbers living in informal settlements who could not get places for their children in public schools or access to health care facilities). In addition, the many instances of poor-quality services (our previous book noted the comment of one resident in Delhi on the government-provided health care that 'only our servants go there' – Lama-Rewal 2011, p. 576) and the hidden costs (such as those facing low-income groups keeping children at school that include transport, books and uniforms).

Provision for emergency services (for instance, fire services, ambulances and access to hospitals, and early warning systems for approaching extreme weather) would not be considered part of welfare in high-income nations but these are usually part of state provision for basic needs. We have found little discussion of the extent and quality of provision for these other than particular instances of where they were not provided.

In terms of the understanding of how welfare is provided, while this discussion focuses on the activities of the state (at all levels), much welfare support has been provided (at least in part) by civil society and particularly by faith-based organisations. Their role continues to be significant in many countries and contexts, and may be supported by state policies that offer tax concessions and/or directly subsidise these activities in addition to direct state provision. Chapter 3 considers the work of different agencies in poverty reduction, and discusses the broader contribution of civil society agencies.

Contribution to poverty reduction

The diversity of modalities for welfare provision between nations (and also within nations over time) means that it is not possible to draw general conclusions about the significance of the contribution of such measures to poverty reduction. However, there are some consistent challenges that emerge across this family of programmes. One challenge, discussed in detail in Chapter 3, is the mismatch between government responsibilities and the capacity to fulfil these responsibilities that is often most evident in city or municipal governments. Many of these issues and challenges are returned to in the discussion about specific programmes and activities of local and national governments and international agencies in Chapter 3. Here we discuss the scale of inclusion, the specificities related to who is included, the politics behind such redistribution and the likelihood that gains will be maintained or not, and the reinforcement of the government's role as an (independent) arbitrator between competing needs.

In terms of the effectiveness of programmes, their scale and the extent of inclusion is critical. In Chapter 5 we discuss the importance of universality for transformative political relations and inclusion. In summary, programmes that are too narrowly framed are likely not to reach many of those in need. They may also lack political support and hence, even if useful, are unlikely to be maintained. In some cases, rules may exclude some of those most in need. As noted above, there have been efforts in some programmes to make the rules that govern entitlement fairer and most transparent but this remains an important issue. Even when the rules that govern access to entitlements may appear fair, some of those most in need may be denied access. For example, the right of people living in vinyl (informal housing) in Korea to receive state benefits was only recognised in 2010 after a long struggle (ACHR 2011a). The exclusion of those with 'informal' addresses is a recurring programme in urban areas. Legal addresses are often requirements for accessing state provision (for instance, education and health care). In India, ration cards allow low-income groups to access subsidised food and fuel but some of those in need find it difficult or impossible to obtain these cards. Mumbai pavement dwellers were not able to obtain ration cards because they could not provide a legal address.[4] Britto (2005) explains that the targeting in cash transfers in Mexico is only as good as the survey data used to identify beneficiaries – and there are many reasons why such surveys may not have included all the households entitled to benefit.

In other cases, people are entitled to benefits but are reluctant to take these up. Even if they fit within the rules, some of the lowest-income citizens may struggle to establish their entitlements due to the associated social process and anti-poor attitudes among officials. Sabry (2008) illustrates these problems when she discusses the intrusive questioning and observation that widows face in Egypt when they apply for social assistance. It is for such reasons that, when Gomez-Lobo and Contreras (2004) analyse the Chilean (means-tested) subsidy system for water, they find that even under the most optimistic of assumptions only half of those entitled to receive the subsidy in the poorest groups actually receive it. Auyero (2010) explains how social welfare systems exhaust and so control the urban poor in Argentina and in so doing develops a more substantive critique of the ways in which welfare approaches may be designed to work with more abusive state practices to reduce political opposition in a context of growing poverty and inequality.

Achieving redistribution at scale through effective modalities that reach the lowest-income and most vulnerable groups is critical, but the politics of such a process is complex. As Chapter 3 will describe in more detail, responsibilities for some aspects of social welfare are with local governments. Although this may help increase the scope for citizens to influence these responsibilities, local government capacities to respond depend on whether higher levels of government permit them to develop a strong local revenue base or provide them with financial support. Obviously, among the local governments with the weakest economies, the most limited revenues and high concentrations of low-income groups, the possibilities of fully meeting their responsibilities depend on external support.

There are other difficulties, including the risks that programmes will be used to further clientelist relationships (i.e. they will become trades for votes with subsequent narrowing of scale) or political interference that influences who benefits (i.e. the entitlements only reach favoured clients). Or the programme is seen as linked to a particular party or government – and hence is unlikely to survive any change in government. Whatever the nature of the programme, it is critical to think about its contribution to poverty reduction in the longer term. This includes its ability to create a process that manages these propensities through instigating political dynamics that support rather than undermine the commitment of the state to effective poverty reduction. Nelson (2005) warns about the dangers of very selective policies being discarded due to their lack of political popularity.

Underlying alternative practices related to welfare provision are different perspectives on what should be the relations between citizens and the state. Both income and asset transfers are recognised as influencing such relations. There are concerns that they are increasing dependency and reducing initiative among citizens, and Britto (2005) argues that one reason for the popularity of conditional cash transfers in Mexico and Brazil is that they are considered to reduce this danger by requiring behavioural changes. A second concern is that they are undermining radical collective politics. Posner (2012) illustrates this latter concern in the context of housing support in Chile when he discusses how groups compete against each other to access the subsidies rather than work together to improve the system of housing support. This resonates with a Foucauldian understanding of the ways in which power functions; the argument is that as the state increases the scale of its intervention in areas of human well-being, people look to the state to regulate and control these activities. At the same time, these services and those of cash transfers strengthen individual relationships between citizens (households) and the state – at the expense of collective action and a more substantive political challenge. Collective activities are weakened, as is people's consciousness of their own powerlessness and subjugation. Whether or not the state intends to weaken political protest, citizen action becomes less likely and, when evident, it is more likely to be frustrated and personalised protest rather than a more substantive critique of such programmes and their redesign. Gledhill and Hita (2009) explore these concerns and argue that, at least in Brazil, there are still enough politicised associational activities to suggest that these concerns are misplaced. Research in Salvador leads them to conclude that conditional income transfers encourage rather than reduce aspirations for state support with community organisations combining with each other to increase pressure on the state. However, as explained by Britto (2005), the *Bolsa Familia* (the most significant programme for the families in Salvador, Brazil) considers itself, at least sometimes, to be providing universal rights to those who are citizens. Such a representation may reduce the risk of a more passive citizenry. We return to this issue in Chapter 3 when cash transfer and other poverty reduction programmes are considered, and again in Chapter 5 when we examine the challenges involved in securing pro-poor politics.

Approach 2: Urban management: technical professional approaches and state-led urban development

The previous approach, namely welfare, has been characterised as being driven (at least in part) by empathetic compassionate feelings that may be realised either by voluntary sector charitable activities or by the state (although with public support). The urban management approach falls unambiguously within the ambit of government, although here it is city or municipal government that has the central role. This approach, modelled on observations of what it takes to be a modern city, places central emphasis on the state as the implementing agent and relies on state intervention to improve and upgrade physical space and, in some cases, economic activity. This urban management approach recognises the critical importance of government in ensuring provision of urban infrastructure and services and, more fundamentally, in establishing, regulating and influencing the development of urban space. It generally focuses on essential trunk infrastructure such as electricity, piped water, drains, roads and transport networks, conscious of the additional expenses that enterprises face if these are lacking and/or inadequate in scale and quality. Usually, it also includes a more ambitious urban vision. It has received more support in the past 10 to 15 years as the key role of cities and urban systems in economic growth came to be more widely appreciated. See, for example, the recommendations by McKinsey on infrastructure needs in Mumbai and elsewhere (see McFarlane 2008, p. 429; also McKinsey 2011). The McKinsey Report on 'Urban World: Mapping the Economic Power of Cities' (McKinsey 2011) highlights just how much the focus is on economic growth, as this report does not mention health issues or environment issues, or climate change or disasters, or even local governance – and seems to assume that the finance needed for city infrastructure will be there.

As summarised in the introduction and elaborated at greater length in Mitlin and Satterthwaite (2013), there are massive deficiencies in urban management (and in urban infrastructure) in much of the Global South, particularly in the context of informal economies and informal settlements. From the early 1980s, public investments also came to be cut. Increasing public debt and economic mismanagement resulted in financial and exchange rate crises in the early 1980s. As a condition of crisis-management and recovery-related lending and financial support, the global financial institutions (the International Monetary Fund and the World Bank) required governments to adopt a package of measures that became known as Structural Adjustment Programmes (SAPs). These reduced state expenditures, resulting in lower employment, particularly in the public sector and economic recession (at least in the short term). These often did not produce the intended boost to economic expansion which was also meant to help fund poverty reduction. While social investment funds were introduced to alleviate increasing social problems in many cases, these funds had little impact in terms of improving basic infrastructure. They were not sufficient to make good the deficiencies in state infrastructure and services in informal settlements, and other areas with concentrations of low-income groups. Meanwhile, urban populations continued to grow, although the rate of growth may have been reduced in part due to the lack of economic opportunities. The withdrawal of the state became associated with increasing private sector involvement in the provision of basic services and, for a decade or more, this was held out as what governments should support. More recently this trend has stalled. For private sector enterprises, this is perhaps because anticipated profits have not been realised (Von Weisächer *et al.* 2005); for governments and citizens, it relates to many anticipated benefits (greater efficiency, lower costs, extension of services) being unrealised.

The urban management approach to development emphasises the importance of adequate infrastructure for economic growth (Keivani and Mattingly 2007). There have been the high costs for private enterprises that need to invest in private infrastructure and services – for instance, private investment in water supplies (including water treatment plants), liquid and solid waste removal, and treatment and electricity generators. Individual enterprises are less able to make up for deficiencies in roads, highways and ports, although here, locations may be chosen that address this in part. Private infrastructure and service provision also became increasingly common for industrial and commercial centres (for instance, shopping malls) and residential areas (as middle- and upper-income groups come to pay for and be provided with infrastructure and services within gated communities).

Of course the same opportunities do not exist for low-income residents. The approach extends beyond individual infrastructure investments to emphasise the importance of adequate investment in infrastructure networks and urban planning. There is recognition that households cannot make sensible decisions about local development if the broader planning for the city is not taking place; emphasis is placed on the inefficient choices that urban households take due to a lack of such investments. For example, households may invest in pit latrines which offer only limited benefits if the land is low-lying and frequently floods. The approach is encapsulated within the Urban Management Programme itself, funded by bilateral assistance and managed by the United Nations Human Settlements Programme, UNDP and the World Bank (see Box 2.2).

Box 2.2 Urban Management Programme (UMP)

The UMP was a global technical assistance programme which began in 1986 and which was designed to strengthen the contribution that cities and towns in the Global South make towards human development, including economic growth, social development and the reduction of poverty. The UMP was a partnership of the United Nations Development Programme (UNDP), the United Nations Centre for Human Settlements (UNCHS-Habitat, later to be renamed the UN Human Settlements Programme) and the World Bank. Supported by bilateral external support agencies, it claimed to be the largest global multi-agency technical assistance programme in urban development but this was only so because so few development assistance agencies were giving any attention to urban development.

The first phase (1986–1991) developed urban management frameworks and tools on the issues of land management, municipal finance and administration, infrastructure and urban environment. The second phase (1991–1996) used the frameworks and lessons learned to build capacity at the regional level, using mechanisms such as regional panels of experts and workshops and consultations to introduce new policies and tools. The third phase (1997–2001) built on and re-focused the work of the first two phases to the local level, emphasising city consultations and institutional anchoring. Phase 3 had three themes: urban poverty alleviation, urban environmental sustainability and participatory urban governance, with gender as a cross-cutting issue. Phase 4 brought a stronger focus on pro-poor governance and knowledge management activities that have direct impacts on the living conditions of the urban poor: it maintains the three themes of Phase 3 and adds HIV/AIDS. The UN Habitat website reports that from January 2004, the UMP successfully managed the transition from Regional Offices to Regional Networks, led by the UMP's regional networks of Anchor Institutions. The current status of this Programme is unclear; its last reported activity was in 2008.

Another United Nations programme, the Sustainable Cities Programme (SCP), sought to develop an operational framework for urban environmental management, primarily through developing pilot experiences of consultative and participatory processes in environmental planning and management in a number of cities. But neither this nor the Urban Management Programme had the resources to be able to support the investment and action needed on the issues they identified.

Source: Wegelin (1994) and http://www.unhabitat.org/categories.asp?catid=374 (accessed 12 August 2012)

Contribution to the reduction in poverty and inequality

The importance of competent, accountable local government and the successful delivery of basic services and infrastructure (either by local authorities or some other state agency) is obvious. If there is the funding and capacity to ensure the needed (often very large) expansion of infrastructure investments that includes informal settlements and other areas in which low-income households are concentrated, then there will be benefits secured by low-income and disadvantaged households. However, in terms of the engagement of the urban management approach with the reduction of poverty and inequality, there are three primary lines of concern. A first concern is that the orientation to economic growth and enterprise development places insufficient emphasis on issues of equity and the needs of the urban poor. Considerable emphasis may be placed on ensuring that there is a water supply, for example, but the price of water may be unaffordable for a significant proportion of low-income residents. A second concern is that this approach places great emphasis on the contribution of professionals, assuming neutrality in their actions (or even a commitment to the public good) but this is usually not warranted. A third concern is that the approach is naïve about what it takes to catalyse economic growth and fails to engage with the structural complexities of economic transformation which necessarily require interventions beyond urban management. Each of these concerns is discussed in more detail below.

While some of the tools and methods used by the urban management approach date from the times when government intervention in urban planning for the public good was unquestioned, the approach was more orientated to a market economy, and preparing cities for a successful engagement with such a market economy (either at a national or global level). While Curitiba has been acclaimed for its approach to urban planning and management, Klink and Denaldi (2012) argue that it is now not possible for this and other metropolitan councils to plan and deliver services because of the pressures of global capital. One problem has been the reduced role for state metropolitan planning (ibid., p. 549). The result (among other things) has been infrastructure investment favouring private sector developments and adverse housing opportunities for low-income households. The orientation to urban management tended to focus more on the infrastructure that served formal sector enterprises, not residents, but by presenting such exclusion in a 'neutral' light, it created additional difficulties for those who are disadvantaged. For example, while the importance of basic infrastructure and services is acknowledged by all, as the emphasis has shifted towards corporatisation and in some cases the involvement of the private sector, so prices have risen. It is not always the case that access is more difficult than before; in many cases low-income households have never been served both because of affordability and because of their location in informal settlements. But while in theory the reform of urban basic services includes the extension of provision to low-income settlement, in practice the emphasis on cost recovery has jeopardised the equity that was intended (Myers 2011, p. 104). As elaborated in Chapter 3, utilities, whether public or private, have had to balance objectives related to economic prosperity, equity and ensuring a basic standard of health and well-being. The key point is that these processes are not neutral and while at one level it is easy to represent infrastructure improvements as being in the interests of all, in practice what investment takes place and where it takes place is highly contested. For example, the reconstruction of city centres to benefit enterprises almost universally seems to involve the displacement of large numbers of low-income groups and/or informal vendors (see Bhan (2009) for Delhi; Fernandes (2004) for Mumbai; Crossa (2009) for Mexico City; and Simone and Rao (2012) for Jakarta). Increasingly it is the discourse of prosperity for the many that provides the legitimation for the eviction and displacement of some of the lowest-income urban residents. Compensation is generally inadequate or non-existent.

Moreover, it is not clear that the city planning is in the collective interest. Rolnik (2011), in an account of the evolving urban policy and planning context in Brazil, argues that businesses are particularly powerful in influencing urban issues due to the significance of such developments and associated state investments for their interests. McFarlane's (2008) account of the campaign for improved sanitation in Mumbai highlights the complex relations between infrastructure investments, city vision and the relations between the residents of informal settlements and others in the city. This was a context, historically and today, in which the residents of informal settlements were denied access to sanitation with consequences for morbidity and mortality. The investments that have been made have, until recently, been focused on the interests of the elite (see ibid., pp. 425–426 for the historical analysis). Hence, on the one hand, such investments are critical to the well-being of the urban poor; but, on the other, in practice, such investments have been partial and exclusionary.

It is not only control over space that is a source of conflict, as shown by the work of the Orangi Pilot Project Research and Training Institute in Karachi. The successful support to the residents of 'lanes' or streets in informal settlements to design, implement and fund covered sewers and drains shows how the style of infrastructure development is also not neutral (Hasan 2008). With this system, communities in informal settlements could afford the costs but they could not do so if this was implemented by government. At a larger scale, this NGO also argued against drawing in international funding (and 'expertise') for installing or improving the trunk sewers and drains as it demonstrated designs and methods through which this could be done far more cheaply. After arguing with consultants and contractors who stood to benefit from the high-cost externally funded plans, the government accepted the alternative approach and the sanitation programme for Orangi cost P Rs 38 million (US$0.63 million) rather than the initial estimate of P Rs 1,300 million (US$21.67 million). The proposed Korangi Waste Management Project costing US$100 million was shelved and the loan from the Asian Development Bank cancelled when the OPP-RTI demonstrated that it could be done for US$25 million. This did not cost the government more because they were expected to contribute a similar amount to the financing of this project and it saved them having to repay a US$60 million loan. Benjamin (2004) makes a somewhat different but related critique, arguing that urban management approaches have favoured formal elite development processes but penalised the informal enterprise sector that is more likely to support pro-poor growth, and he argues in favour of approaches that support more municipal control over urban neighbourhoods as, at least in India, they are more responsive to the needs and interests of low-income informal households.

In regard to the second concern, the approach gives a central role to the contribution of 'urban' professions such as architecture, planning and engineering. Notwithstanding the importance of a technically competent urban management, it has long been recognised that professionals may unwittingly disadvantage those urban citizens whose lives are embedded in informality. As articulated by Escobar (1992), the development of professionals and professionalism is related to the nature of the modern state, associated problems and the expansion of the state into new areas of social provision. The growth of laws, rules and regulations in areas such as settlement planning, household construction and medicine has resulted in a 'need' for planners, architects and a range of health officials to address problems that people previously solved for themselves (Illich *et al.* 1977). In a context in which the state dominates the choices for those living in informal settlements, and the state exercises its practice through professionals, their interventions and associated norms, standards and regulations are often a means to enforce social stratification and segregation (Burgess 1978; Myers 2003; Yahya *et al.* 2001). Once communities become more centrally involved they can negotiate outcomes that work for them, and in so doing build their own capability and reputation (see Chapter 5). In addition, as Chapters 4 and 5 illustrate, they also bring a range of cost-saving techniques and methods which greatly reduce the costs of what is needed and so increase what can be implemented.

In regard to the third concern raised above, implicit in the approach is the belief that urban management can address deficiencies in development. For so many UN supported urban initiatives, there is the massive gap between what is actually needed and what is provided. This includes but also goes beyond the initiatives supported by the Urban Management Programme and the Sustainable Cities Programme that were discussed in Box 2.2. They also include Child-Friendly Cities initiatives that for a while received support from UNICEF and Healthy Cities initiatives in low- and middle-income countries that received support from the World Health Organisation (WHO). But these are not initiatives that can respond to numerous city or municipal governments wanting to act on these issues and seeking advice and support.

Often it seems as if these organisations expect a document, seminar or report to fundamentally change the competence, capacity and accountability of city government. None of these programmes brought funding to address deficiencies in infrastructure and service provision. City governments will often sign up to these initiatives in the expectation that it will produce funding – but this is rarely the case. There is also little recognition of macro-economic constraints despite the history of structural adjustment and the recessions catalysed in many countries because of the lack of demand in the economy. In addition, tools and methods for urban management drawn from experience in the Global North were applied in totally different contexts – for instance, in cities where there is little or no transparency and accountability within local government, and where there is little capacity or willingness within local government to manage land-use changes.

Although improved infrastructure reduces costs of production and enhances the viability of enterprise activity by increasing profitability, it does not necessarily bring the expectations of increased demand for goods and services that underpin investment. Watson's (2007, pp. 225–226) analysis of the Urban Management Programme recognises both that municipal finances have been too weak and the global economy too strong for the success of this specific programme to be achieved.

More generally, while the local impact of improved infrastructure may be impressive,[5] it is less clear that this approach supports greater economic growth for the city as a whole. Equally, while there is a lack of recognition of the importance of such broader issues of economic structure, there is also a lack of appreciation of the complex social linkages that lie behind many successful enterprise activities. Simone and Rao (2012) illustrate this in their exploration of livelihoods in Jakarta (Indonesia). They explain the complexity of family livelihoods in the Karanganyar and Kartini districts and discuss the ongoing threat to relocate the largest 'traditional' market, Tanah Abang, despite its economic significance. They elaborate on the physical space that has enabled low-income households to resist the gentrification of their neighbourhoods with small roads not navigable by automobiles (ibid., p. 323); they also challenge a discourse which labels citizens as 'stakeholders' but plans evictions of low-income households. Their account is helpful because it illustrates the contested nature of development in southern towns and cities, and the naivety of city authorities which propose the relocation of the market and 'believe that these local economies can simply be transplanted to other areas. … There is little recognition that such productivity may depend upon the layers of sedimentation, spatial memory and diversity within the built environment that is not easily replicable in intentional designs in new locations' (ibid., p. 331). There is a similar story in Accra in the informal settlement of Old Fadama which has a large and lively economy and whose residents provide much of the labour for a very large nearby market. Here the residents have had to resist successive city governments that want to relocate them (Farouk and Mensah 2012).

Recognition of these weaknesses has contributed to the shift towards more participatory forms of governance. This was also evident in the work of the UMP office for Latin America and the Caribbean that gave strong support to city consultations and participatory governance, before being closed in 2004.[6]

Approach 3: Participatory governance

Many low-income and/or disadvantaged groups have sought more democratic political regimes in anticipation that a political solution will address injustice and exclusion (Heller and Evans 2010, p. 437). For many decades, there have been continuing attempts to define and secure democratic political alternatives against dictatorships, and this could be judged to have had some success in terms of the reduction in the number of countries ruled by non-representative governments. Forty years ago almost all nations in Latin America were under dictatorships; so too was much of Asia and Africa – and there were also the apartheid states in Namibia, what was to become Zimbabwe and South Africa, and the colonial governments in Angola and Mozambique. Whatever the subsequent political realities, resistance and liberation movements typically make reference to the need to challenge disadvantage and provide development opportunities for all, especially those most in need.

More recently these national struggles have been echoed in the attempts of city governments to secure their own right to make decisions for themselves and for their citizens. In recent decades, as democratisation has increased, and arguably because of the era of decentralisation, there has been less emphasis on the central state and more on city governments. In addition, organised urban poor groups had more success in influencing their local governments – and there are certainly more examples of democracy delivering some benefits for low-income groups within particular cities or municipalities. There are also the contestations between national and city governments – for instance, when city mayors from a different political party to that controlling national government are elected, national government limits what the newly elected mayor can do (see, for instance, when Cuauhtémoc Cardenas was elected Mayor of Mexico City or the lack of support for city development in Rosario in Argentina as the last two mayors were from a different political party to the ruling national party). This contestation may also be seen in the election of several mayors in Nairobi who were from opposition parties over the past two decades and whose possibilities for taking action are severely limited by national government – and also in abolishing the position of mayor by the national government which meant no elected mayor between 1983 and 1992. Similar restrictions on local democracy in the capital city have also taken place in other cities, including Harare (Zimbabwe) and Kampala (Uganda). Despite this repeated striving for representative democracy – mostly at national level – events in the past three decades have shown that this politics does not result in a pro-poor inclusive and equitable urban development. As noted by Moore with Leavy and White (2005, pp. 186–187),

> [T]here is no evidence that democracy leads to pro-poor policies. ... Votes are not enough. In personalistic or patrimonial political systems, the potential power of the votes of the poor is neutralized by their fragmentation among numerous, competing, particularistic networks and interests.

However, in many Latin American nations, pressures from organised urban poor communities and a new generation of mayors who were elected and who were committed to stronger local democracy did produce a shift to more participatory forms of political engagement. Participatory governance emphasises the need to introduce mechanisms to encourage the involvement of those who do not find it easy to participate in state structures and processes because these are generally far removed from their own cultures and practices. Participatory governance implies the engagement of government with a group with interests beyond those of 'its individual citizens'; i.e. it goes considerably beyond systems for more accountable government. Participatory governance seeks to provide an inclusive political space at a local level, recognising that other fora can enable the views and opinions of citizens to enter the political process and usefully augment representative democracy. It challenges the notion that widely spaced elections for representatives are a sufficient engagement in collective decision-making.

One of the most notable urban examples has been participatory budgeting, introduced when the federation of residents' associations in Porto Alegre (Brazil) found that their mayoral candidate did not deliver on his promises (see Chapter 3). Participatory budgeting enables local residents in low-income neighbourhoods to be involved collectively in setting priorities for government expenditures in their areas. The innovation in Porto Alegre evolved in a context in which the movements sought an alternative to both the authoritarian and clientelist state, and their strategy sought to ensure that their electoral influence was not rendered ineffectual by a lack of government accountability (Abers 1998). By 2004, participatory budgeting had come to be applied in over 250 urban centres around the world. Most are in Brazil, but participatory budgeting initiatives are also flourishing in urban centres in many other Latin American nations and in some

European nations (Cabannes 2004; Menegat 2002; Souza 2001). By 2012, participatory budgeting initiatives had been implemented by over 1,000 local authorities around the world (Cabannes 2013).

But the introduction and expansion of participatory budgeting in Brazil needs to be understood within the many political changes during the late 1970s and 1980s with the return to democracy, decentralisation and the strengthening of local democracy (democratisation was not limited to national institutions) and the new constitution. It was also part of an agenda for rebuilding democratic institutions to fight corruption, improve access to government and strengthen government accountability (Souza 2001). The new Constitution in 1988 gave more powers to the legislative (reducing the dominance of the executive) and mandated more revenue to municipalities and more responsibilities (including social assistance). Municipalities' capacity to intervene in land use in favour of the urban poor was also strengthened through an array of new urban planning instruments introduced by the new Constitution (Melo *et al.* 2001). Some municipalities (notably Porto Alegre and Belo Horizonte) were able to do more because they improved tax collection and this increased their budgets. Participatory budgeting was also served by the growing citizen support for the *Partido dos Trabalhadores* (PT – Workers Party) with the increasing number of mayors from this party (although participatory budgeting was also supported by some non-PT mayors). In some cities, participatory budgeting helped sustain the party in power – as in Porto Alegre and Belo Horizonte – and this also meant a greater impact. For instance, as shown in Porto Alegre, it takes time for civic organisations with a history of confrontation or dominated by clientelist practices to change (Souza 2001). The PT's electoral success in particular cities also paved the way for its electoral success at national level and, despite its limitations and difficulties (see Fernandes 2007), the national government under the presidency of Lula may be considered an example of a national government that did address poverty.

However, this is far from being the only example, and Fung and Olin Wright (2003, p. 15) suggest that there are a number of similar models which they group together under the term 'empowered participatory governance'. The impetus behind participatory governance is not just in the Global South and its use in the Global North is illustrated by Fung and Olin Wright (2003, p. 16) when they recognise that 'bottom-up neighbourhood councils [in Chicago] invented effective solutions that police officials acting autonomously would never have developed'. The importance of joint solutions being developed by state and citizens is exemplified again in Peru where women's organisations to address nutritional needs emerged following the structural adjustment programmes and the crisis in urban poverty (see Box 2.6). Such examples extend participation in political decision-making to a more intense involvement in aspects of planning and implementation, a process known as co-production. The women's kitchens in Peru illustrate the practice of co-production. Co-production has been defined as 'the joint and direct involvement of both public agents and private citizens in the provision of services' involving 'regular, long-term term relationships between state agencies and organized groups of citizens, where both make substantial resource contributions' (Joshi and Moore 2004, pp. 33 and 40). A more engaged involvement in the production of infrastructure and basic services necessarily entails collective action and organisation.

The concept was 'discovered' in the United States with the recognition that despite the scale, spread and complexity of the modern state, essential services such as security and health care could only be secured with the active involvement of local citizens (Mitlin 2008a). 'Co-production' outcomes suggest that citizen involvement has the potential to improve service delivery and associated outcomes (Parks *et al.* 1981; Whitaker 1980). For example, in the case of crime, it is widely recognised that police manage the streets through a set of negotiated interactions rather than through the authoritative imposition of order. To achieve street security, formal services are dependent on the participation of local residents.[7] Equally, state organisations are more effective when they respond to the perspectives and experiences of citizens in respect of both needs and state activities. This argument was later elaborated in Ostrom (1996) who looks at its contribution to condominial sewerage systems to address sanitation needs in the northeast of Brazil, and to education in Nigeria. She describes how low-income settlements are connected to city sewerage systems by reducing conventional engineering standards and involving local residents in local planning decisions, some financing and voluntary labour; she notes the success of these designs in improving access to sanitation for some of the lowest-income neighbourhoods.

The concept as developed in the US side-steps debates about the reduction in resources, arguing that whatever the level of funding, state agencies could provide a better quality service by involving local residents. Moreover, this is not simply a question of cultural differences between professional and subaltern groups; this body of US research also demonstrated that co-production involves higher-income residents who may also be keen to be actively engaged in local service provision. Ostrom (1996) and Evans (1996) both articulate the approach in a special issue of *World Development* and, in so doing, recognise the depth of innovation that has occurred as local government has struggled in contexts of acute resource scarcity to fulfil their mandate.

While many of the discussions of co-production have been orientated to recommendations that state-led programmes may usefully reach out to individual and organised groups of citizens (Joshi and Moore 2004; Ostrom 1996), an alternative conceptualisation is that organised citizens lobby for state support for their own development solutions. In this context, co-production involves state financial support for development strategies defined and undertaken by the poor themselves. The potential for co-production is particularly strong in the context of land and neighbourhood development activities. Groups may begin with self-help activities and come to realise that effective collaboration with the state is essential, but they require it on their own terms (see the discussion in Chapter 3). Box 2.3 describes the ways in which citizens can make pro-poor state action possible both because they can overcome the lack of state organisational capacity in key areas of city management, and because their recognised capacity enables them to influence the direction of the programme.

Box 2.3 Co-production in resettlement (India)

The upgrading of Mumbai's suburban railways required the removal of informal settlements that in many places had grown to be within touching distance of the trains. The railways had been ordered to slow their speeds in such places to try to reduce the very large numbers of fatalities each year. Those living close to the railway tracks were rehoused without force and with their support. This was only possible when the National Slum Dwellers' Federation of India provided an organisational resource (to support the Railway Slum Dwellers' Federation) that could match the state contribution of land and finance. Needs were acute. Mumbai relies on its extensive suburban railway system to get its workforce in and out of the central city; on average, over seven million passenger trips are made each day on five major railway corridors. In 1999, nearly 32,000 households lived in shacks next to the tracks at high risk and without water and sanitation. Discussions within the Railway Slum Dwellers' Federation (to which most households along the railway tracks belonged) showed that most families wanted to move if they could get a home with secure tenure in an appropriate location.

A relocation programme was developed as part of a scheme to improve the rail network. First, the negotiations reduced the number of people who had to move by cutting down the size of the space cleared (the railway authorities initially wanted close to 10 metres each side; the Federation negotiated down to 3 metres with the promise of a wall). Land sites were identified to accommodate those who had to move (and these were visited), and the Federation was given the responsibility for managing the resettlement. People to be resettled were involved in designing, planning and implementing the programme. Critically the Federation identified those with an entitlement to receive a new home. Teams of Federation leaders, community residents and NGO staff prepared maps that showed each hut, and each hut was identified with a number. Draft registers of all residents were prepared and the results returned to communities for checking. Any claims were checked against the register. Anyone could claim that they had been left out but they had to have all of their neighbours (as identified on the map) verify their claim to residency. This system prevented fraudulent claims for inclusion and speedily resolved disputes. Households were then grouped into units of 50 and each unit moved to the new site together to reduce the social costs of dislocation.

Source: Patel *et al.* (2002)

Contribution to poverty reduction

Participatory democracy is only possible at the local level. While there have been attempts to have greater accountability at the national level, in most cases this is consultative workshops that include only those able to travel to the capital city. Even if there is a consultative process at the local level to feed into national decision-making, this is a single event such as the drafting of the poverty reduction strategy. At the level of local government, even in the larger cities, there are possibilities for a more substantive process. The physical distances do not prevent people from getting together, even if not everyone can afford to travel all of the time. Moreover, it is also easier for local authority officials and political representatives to visit local communities.[8] At the same time, adjacent neighbourhoods can organise together to press their case to the local authority. At the city level, clientelist political arrangements frequently blend into more organised protests made towards the local authority and this shared political activity helps build relationships between community leaders. If there are possibilities for more systematic engagement through some kind of sustainable participatory process, then these relations can increase, strengthening the capabilities of local citizens. Equally important is that organised communities have a chance to observe policies being made, and to assess the results of these policies. This helps to increase local accountability and deepen the nature of the relationship between organised citizens and the state who have shared experiences on which to draw.

Making the democratic state work for low-income and disadvantaged citizens is difficult, and it is acknowledged that participatory processes go only some way towards addressing these problems. There is widespread acknowledgement that the lowest-income and most disadvantaged groups do not participate in these processes. The emphasis on citizen involvement and public debate resonates with Habermasian ideas about progressive development. Habermas argues that public debate in the public sphere needs to enable a rational debate such that participants can develop an understanding about the collective action that would best address the interests of all. But it is acknowledged that not everyone participates equally in such forums (Cleaver (2005) offers an example from a village in Tanzania). There are systemic exclusions including those who are not allowed to participate (e.g. on the grounds of age, gender or ethnicity), those who choose not to participate and those who attend but are silent during such processes. Roberts and Crossley (2004) also suggest that there are other public spaces, in addition to those of formal governance, occupied by less privileged citizens that are less visible but nevertheless present (although they may be ignored) (p. 12). If participatory governance is to be effective, there needs to be a way to integrate the different dialogues.

Todes and colleagues (2010, p. 419) illustrate the difficulties when they describe new attempts at spatial planning in Ekurhuleni (South Africa) and explain that there is a wide range of groups being consulted – 'formally elected councillors, local ward committees, sector representatives, business, civil society organisations, officials and lay people'. However, at the same time, groups that are not formally organised (which are likely to be those with the lowest incomes and most vulnerable) have little voice in this process. As discussed in Chapter 5, even if less powerful groups are included in such governance processes, they may be dominated by those who have better political connections.

Research studies suggest that similar problems are also evident within participatory budgeting processes, and the discussion of participatory budgeting in Chapter 3 explores these concerns in more detail. A further problem is that the scope of participatory budgeting (and participatory governance in general) is often very limited, with most decision-making and resource allocation being left within representative democratic processes (Cabannes 2004).

Despite the above discussion and its focus on professionalism, state authority and state–citizen relations, many of those writing about co-production, and particularly those writing in the context of development, view it as a secondary strategy for service delivery. It is somewhat more significant than self-help (but below full state provision) and may be used in the context of a weak (underdeveloped) state prior to it gaining in political will for redistribution and bureaucratic capacity (Leftwich 2005, p. 598). Joshi and Moore (2004) analyse a citizen's initiative to improve policing in Karachi (Pakistan) and the services provided to tanker drives by a business association of commercial vehicles in Ghana. They conclude that such organisations provide essential services where the state lacks capability, and that they offer lessons for 'other contexts where conventional public provision is under stress' (ibid., p. 38). However, they also suggest that such activities undermine the Weberian principles that public organisation has defended: public and private separation, public accountability, universality and uniformity. But not all co-production is conceptualised as a temporary strategy for service provision. The concept as originally elaborated in US inner cities was about the absolute limits of the bureaucratic state and the importance of engaging people to be pro-active in improving the services government departments were providing. In addition, as we discuss in Chapters 4 and 5, social movements have used co-production to provide a more appropriate solution which recognises both that the multiple functionalities from greater citizen involvement and the informality of everyday life requires something more appropriate than highly formalised strategies which modern urban development prescribes.

Research on co-production highlights the importance of local associations' capabilities, as is also the case with research on participatory budgeting (see Chapter 3). The importance for organised communities of both existing relations to the political system and the capacity to use these relations to good effect is also emphasised by Nance and Ortolano (2007), who study different cases within the programme of condominial sewers in the northeast of Brazil. They conclude that neighbourhood associations with good connections to officials and the ability to push their case with the responsible authorities are more likely to secure effective outcomes. Superficial participation only in micro-management does not lead to successful community involvement.

Approach 4: Rights-based approach to development

The rights-based approach to development emerged in the 1990s in the hope that this would reinvigorate discussions and activities to achieve poverty reduction. Here, the justification for development was no longer on the basis of unmet needs that the state (and donor agencies) should address but on the basis of the rights of those with unmet needs to have these needs addressed.

It had long been argued by a small number of agencies in the field that there needed to be a greater acknowledgement of the importance of social and economic rights in addressing poverty and inequality (see e.g. Audefroy 1994). Hence in this case the problem, namely a lack of rights, was not new, but rather there was a new awareness of an existing problem.

The rights-based approach draws on a number of different traditions: the international human rights frameworks, attempts by advocacy organisations to improve national legislation, autonomous movements of the poor and dispossessed, and the shift from clientelist relationships between the state and people to ones of citizenship (IDS 2003). Arguably the growing popularity of 'rights' by development agencies may be understood as a response to the need to address tensions between the experiences of both neoliberalism and political transformation through democratisation.

Supporting rationales include the ideas of democracy campaigners who feared that elected governments would start to backtrack on the promises of campaigns and betray the sacrifices of resistance movements. With established legal rights, the courts could protect citizens even if particular governments failed. Moreover, rights, and particularly the enactment of human rights legislation, guaranteed the leaders of political opposition a level of protection from the hypothetical risk of campaigning against future authoritarian states. For the NGOs that had become service providers, rights offered a complementary campaigning position that alleviated internal challenges and external questions about the legitimacy of their other work (Dagnino 2008, p. 59). For the middle class in the Global South, some of whom had entered the 'new poor' as a result of structural adjustment programmes, rights offered a way back to engaging with a state that had previously protected their interests in providing access to basic goods and services. For bilateral development agencies pressured by criticism of their neoliberal policies from their own populaces, rights offered a way of extending their model of a liberal democratic state while not significantly changing economic policies (Edwards 2001; Mayo 2005; Tomas 2005). For some agencies seeking to support economic growth through entrepreneurship, rights were seen as supporting stronger market economies, as these required a capacity to reinforce contracts and recognize the legal ownership of assets (De Soto 2000). For the UK Department for International Development (DFID), 'rights' resonated with the historic concern of the political party in power from 1997 to 2010 (Labour) for the limitations of charitable endeavour and its long-standing commitment to pro-poor programming, while also allowing the agency to continue with pro-market strategies.

One of the first examples of the rights-based approach applied to development was the attention given to housing rights, including rights not to be forcibly evicted (see Leckie 1989, 1992). At this time 'rights' were seen as a new way to put pressure on governments to fulfil their responsibilities, since arguing on the basis of needs that had to be met had produced such inadequate state responses.

The rights-based approach grew in popularity between 1995 and 2005. In 1995, it could be argued that 'rights' remained the preserve of lawyers, specialist NGOs and United Nations treaties. By 2005, rights had entered the language, commitments and promotional material of development agencies (Eyben and Ferguson 2004; Moser *et al.* 2001). Integrating rights into development work became associated with a 'package' of measures, although the emphasis and sometimes the content varied (Mandar 2005; Molyneux and Lazar 2003; Moser *et al.* 2001; Uvin 2004). Notable components include:

- (Pressure for) formal rights as laid down within some legal system, stipulation, rules or regulations.
- The implementation of such rights through legal campaigns and stronger links with the legal profession.
- A more complete system of interconnected rights, rather than single rights.
- Adherence to international rights and a hierarchy of rights at local, national and international scales.
- A perception of rights as a development goal to be achieved independent of other goals.
- The explicit acknowledgement that engaging with rights requires an overtly political approach.

In some cases, agencies emphasised a number of non-legal political processes that were important additions to the processes involved: these included seeking participatory citizens rather than passive recipients, promoting greater political transparency and accountability, and prioritising those with the lowest incomes and those most excluded (Gaventa 2002; Hinton and Groves 2004; Lewis 2007, pp. 79–81). Rights (for some) became a way of addressing each and every development challenge. The work of Molyneux and Lazar (2003) and Uvin (2004) reflects an ambition among rights activists to reclaim the development imperative and reorient development policies and activities towards addressing the needs of those facing exploitation, exclusion and dispossession.

In the urban context, rights-based approaches have been particularly notable in regard to land tenure (anti-eviction), access to basic services and, in the context of livelihoods, access to trading spaces. In the first and last of these, the strategy has been primarily defensive, reacting to protect homes and workplaces. Basic services have frequently not been provided and in this case the strategy has been to argue for inclusion, although the highest profile struggles have been defensive reactions to the privatisation of utilities. The rights approach also led to pressure for more inclusive practices, both within groups and neighbourhoods. An alternative and significant conceptualisation has been the 'right to the city', or the reclaiming of urban space, urban lives and urban political activism by the urban subaltern.

The long history of anti-eviction struggles is indicative of the difficulties faced by the urban poor. Most of the large-scale evictions during the 1970s were implemented by authoritarian governments, including many who were responsible for large abuses of civil and political rights. These were regimes that would hardly respond positively to a rights-based approach. It was hoped (or assumed) that political shifts to elected governments would ensure that these evictions ceased. In addition, struggles by those living in informal settlements to avoid eviction could be considered to be among the earliest examples of a rights-based approach as these often resorted to the courts to prevent eviction (see e.g. Arputham (2008), who describes the tactics used to stop the eviction of Janata Colony in the late 1960s and early 1970s).

In many cases, groups have been forced out of their homes as others have sought to use and to possess the land (COHRE 2006). Where there has been an offer of tenure security, negotiations can be difficult to realise. In the Philippines, strong protest movements protected squatters in an illegal settlement referred to as the National Government Centre during the final years of the Marcos regime. Racelis (2003, p. 9) describes the struggles they went through and the ways in which Sama-Sama (the local people's organisation) sought to secure benefits for all families, not just their own members. In most cases, street protests and resistance combine with legal measures to challenge the eviction through the courts.

To reduce the vulnerabilities of a simple defensive action, some groups have tried to use rights as the basis for a more pro-active strategy. For example, in Goiânia (Brazil) tenants undertook a series of land invasions to claim access to common land (Barbosa *et al.* 1997). In this case, the city had grown rapidly with little public housing provision. In the absence of affordable accommodation, tenants' movements came to explore the potential within *posseiro* rights, namely a free right over all land which has not been subject to subdivision. These lands are without formal owners or title deeds and *posseiro* is the name given to those claiming a right of use over these untitled lands. In 1984, an existing Federation renamed itself the union of *posseiros,* three years later to become FEGIP (the Goiânia Federation for Tenants and *Posseiros*). By 1991, 12.3 per cent of the population of Goiânia was living in these *posseiro* areas, and by 1997 the city had officially registered 193 *posse* areas, 75 of which had been established by FEGIP. A further example is provided in Thailand, where land sharing, the negotiated division of the squatter land between the original owner and the residents, was used to ensure that rights could be realised (Angel and Boonyabancha 1988). However, in other cases progress on land rights has proved more difficult. The National Coalition for Housing Rights in India was begun in the 1980s by the NGO, Unnayan (Mageli 2004). Unnayan supported the emergence of *Chhinnamul,* a movement organisation seeking to involve Calcutta's squatter population in the campaign through demonstrations and street protests. Participants included technical experts, social and political action groups, squatter movements, and others concerned with justice for society's marginalised groups including lawyers.

One of the main aims was to draft a people's law – the People's Bill of Housing Rights – and have it passed in Parliament as a constitutional amendment. However, this was not achieved, as both the public intellectuals and the people's movement failed to shift political developments in their favour.

Rights-based approaches have also been used to understand and protect the claims of traders and vendors to trading spaces, particularly those in the central city (Brown and Kristiansen 2009, pp. 31–32). Market trading is often controlled within the city with particular zones and sites defined as suitable for trading and associated licence fees charged. Informal traders face harassment from the local authorities who control streets and markets. Vendors who sell outside designated areas may be fined, forcibly evicted, have their goods confiscated and/or jailed. In some cases such payments are associated with the corrupt practices of officials or local strongmen. Vendors may collaborate to manage this situation. Etemadi (2000, 2001) describes the activities of one such movement in Cebu City, the Philippines. Threatened with expulsion from the city centre, vendors' associations formed a common platform to ensure livelihood security and reduce harassment. The city authorities were persuaded to establish a vendors' management study committee. The committee concluded that activities should be legalised in some areas of the city and the authorities shifted to an agreed policy that demolitions of vendor stalls would only be considered following complaints from other road users. In practice, street vending became more acceptable. Other experiences such as those in Mexico City highlight the importance of access to trading space and the denial of access being a catalyst for protest and claim-making (Cross 1998; Crossa 2009).

A further aspect of the rights-based approach is an emphasis on inclusion and equity (as rights apply to all). Drinkwater (2009) develops this theme in the context of his work with CARE. He illustrates this dimension to the approach through experiences from a drop-in centre for sex workers in Bangladesh and a related self-help group of sex workers, *Durjoy Nari Shanga* – meaning the 'difficult to conquer womens' association – see Box 2.4.

Box 2.4 Rights and equality

When women were asked how their association had helped them, the first to reply stated, 'we realised we are also human beings'. Recognising their innate equality with others had not reduced the risks the women faced, but it had equipped them to deal with these risks more effectively.

In its early days CARE's SHAKTI project worked with these women in creative ways, with staff themselves going through a profound change process as they sought to overcome their previous assumptions about these individuals and support the group. The project was piloted in the Tangail brothel within a community of some 800 sex workers and their children. Interactive discussions with the sex workers late into the night enabled the social analysis and self-analysis staff required to understand the role their own attitudes played in perpetuating the discrimination and stigma against the sex workers.

In one exercise, the sex workers were asked what their priorities were; at the top of their list was the ability to wear shoes outdoors. In the complex network of social relations of the Tangail neighbourhood, the *samaj* – modelled after traditional village councils, and consisting of landlords and originally two *sardanis* or madams – wielded tremendous power and control over the sex workers. *Mastans* – male gangs allied to local politicians and landlords – act as enforcers, regulating local economies and exploiting vulnerable groups through the use and threat of violence. Forbidding the sex workers the right to wear shoes was a way of publicly marking their status as lesser beings and restricting them to the locality.

As the women organised together and negotiated the right to wear shoes, they were more able to address their needs, for example, challenging the police when they were harassed. The women were working together to manage a health clinic, secure education for their children and the ability to save for their old age.

Source: Drinkwater (2009)

Contribution to poverty reduction

The rights-based approach has been promoted on the basis that it is an effective and strategic entry point to address exploitation, dispossession and discrimination. A positive contribution of rights is its emphasis on universality and inclusion (Hickey and Mitlin 2009). However, a number of limitations of the rights-based approach have been identified (Hickey and Mitlin 2009; Kabeer 2002; Moser *et al.* 2001); and we discuss some of those that have been found to be of particular significance to the urban context. First, the rights approach may be insufficiently sensitive to the local context as it brings a focus on an established package of rights that is difficult to realise fully. There is limited room for negotiation and this causes its own problems with difficulties in trading off gains in one area with slower progress in another, and lack of attention to a more incremental process of change. Second, although ideologically rights have been seen as a counterbalance to the market, actually property rights and the right to participate in the market have been a major theme. Rights in this context are not necessarily pro-poor and may be used to exclude as well as include. Moreover, rights reinforce the role and contribution of formal systems and processes to urban development. The problem, for the urban poor, is that they do not fit easily into formality – and whatever the intentions, in practice rights may be difficult for the most vulnerable. In South Africa, for example, the right to water is of assistance to home owners but offers little to informal tenants.

Third, where individuals and households have inadequate access to resources and are unable to secure a subsistence livelihood, they may have little interest in abstract rights which can only be realised through contestation and struggle. The National Campaign for Housing Rights in India (see above) is an example of this. While movement members participated for some time, in the end they found the rewards too few when set against their daily livelihood struggles. In some cases the approach has made sense – for example, the women's nutrition organisations introduced in Box 2.6 – but in this case success appears to have been relatively easy to achieve.

Middle- and upper-income groups are also in a better position to use the rights-based approach – as they know the law, their housing is legal and most work within the formal economy. The Bhagidari programme in Delhi has been reviewed in this light (Chakrabarti 2008; Joshi 2008). This sought to provide a collective forum for government agencies and citizen groups to address problems. It organised workshops that brought neighbourhood-level Resident Welfare Associations together with officials and political representatives. It also provided these Associations with more direct access to senior bureaucrats (and thus to bypass local politicians). But the Resident Welfare Associations are mostly neighbourhood management committees formed by the residents of apartment blocks and legal housing colonies (i.e. mostly middle- and upper-income groups). These Associations have not only become more active in making demands on local government and in protesting about increased prices and new policies; they have also opposed master plan guidelines that sought to regularise illegal commercial establishments and have also been active in filing public interest litigation against informal settlements.

A fourth critique is that the approach places too much confidence in the state (including its legal system), but in many cases states are too weak, and/or are controlled by the very forces that rights are needed to protect against. While this may be positive and encourage the state to play a role, there are other problems. Rights may be enacted but enforcement may be partial and reduce confidence in government (Molyneux and Lazar 2003, p. 82). More specifically, a rights-based approach requires a neutrality and competence within the courts and legal profession – and this is often not in place. In such cases those seeking rights are unlikely to be strong enough to contest these outcomes and change practices. See, for example, Perlman's description of the *favela* residents in Rio who are unable to challenge informal armed militia who charge them for services; she explains how the police are unaccountable (and ineffective), and the government indifferent (2010, pp. 181–182). While these communities may wish for a government that respects their right to security, this is clearly a long way from being realised.

Fifth, reducing poverty and inequality is complex in relational terms and rights-based strategies may be difficult to realise for this reason. Driven by external political processes, the rights-based approach can mean a technical, professionally managed media campaign. While the violations may be real, the subsequent confrontation both maintains 'the poor' as victims, and leaves them vulnerable as the global campaign moves on. As noted by Ruby Papeleras from the Philippines Homeless People's Federation:

> Poor people are also pulled by activists into their campaigns, but it's almost never the communities who dictate that kind of process. These activists are always people from outside the community: they aren't affected by those problems themselves and they don't really feel what the people feel, as insiders. They come in and try to nurture anger against injustice in the community people, stir them up and get them to fight. But after the protests and the barricades, those outsiders go back to their homes, while we are still here, still living with these problems, without any solutions.
>
> (Papeleras and Bagotlo with Boonyabancha 2012)

When rights fail, the urban poor communities are exposed and have to live with the consequences (see e.g. Benjamin 2004, p. 185; Patel and Mitlin 2009). Also complex are the alliances between grassroots organisations and NGO support agencies, as the former may struggle with the formal, highly professionalised complexity of the legal approach even if it is sensitively delivered (Chapman *et al.* 2009).

Nevertheless, the importance of rights and social justice issues permeates both poverty reduction strategies and attempts to reduce inequalities. We return to the complexities of securing rights in Chapter 5 when we analyse how alternative approaches have sought to overcome these weaknesses to this approach.

Approach 5: Improving market access

The rights-based approach was in part a reaction to the greater attention given to the 'market-based approach' that became more common during the 1980s. The market-based approach includes strong support for economic growth and support for market mechanisms that can improve supplies of goods and/or reduce their costs. It recognises that urban poor groups work within markets – labour markets, financial markets, markets for cheap accommodation and access to services, competitive markets for goods produced or services offered – and looks to measures that make these markets better serve them. For instance, where low-income groups rely on water vendors, a more competitive market and better access to water for the vendors could help reduce prices and improve quality (Hardoy *et al.* 1990). In addition, if some good or service that benefits low-income groups can be provided with full cost recovery, it greatly reduces the constraints on increasing scale.

The significance of economic growth for poverty reduction is widely recognised. There is a strong body of opinion which argues that economic growth is the key to poverty reduction, as it enables low-income households to improve their situation – and avoids having to compete to secure resources from more powerful groups. It is also argued that economic growth can produce more possibilities for increasing government revenues that may then be used to reduce poverty. Thus, a priority for development is a stronger and larger economy with all that this implies in terms of encouraging and supporting economic growth. But redistribution is difficult even if resources are relatively plentiful. If growth is not taking place, it is frequently impossible.

The market approach places greater emphasis on livelihoods and employment rather than on other aspects of urban poverty. Improving cash incomes is particularly important in an urban context as few households have a direct means of subsistence through which they can secure their basic need for food and other essentials. As described in the introduction and elaborated in Mitlin and Satterthwaite (2012), low-income households lack the income they require to meet their immediate basic needs for food, fuel and journeys to and from work, water, shelter, toilet access and essential health services. Lack of income means that their diet is insufficient for good health and that it is difficult for families to invest in the education and training which might enable them to obtain higher wages. Lack of income also means that it is difficult to save and secure assets, rendering households particularly vulnerable to crises and exploitative money lenders. However, while money is essential for urban livelihoods, there are many opportunities to earn incomes – due in part to the multiple pressures of people's lives.

The position of the market approach with respect to inequality is ambiguous (Mitlin and Satterthwaite 2013). While the consequence of extreme inequality for declining social cohesion, authoritarian government and minimal redistribution and hence a lack of public investment are recognised, inequalities are also equated with incentives for economic growth and prosperity. At the same time, the market has been conceptualised as a force against social discrimination (Friedman 1962).

There is no single path through which the market-based approach seeks to reduce urban poverty. The common bond between these forms of intervention is their interest in enhancing the participation of the urban poor in market transactions. From that starting point, several major strategies are followed which are discussed in turn below: the extension of financial markets with savings and loan facilities; support for housing markets with shelter micro-finance; and support for enterprise development. Many of these interventions are targeted at individual households and/or individuals, although in the case of micro-finance groups may be involved to provide social collateral. The immediate focus is micro in its scope (i.e. individual workers and micro-enterprises), although the scale of programmes varies considerably. While this section discusses professional interventions, note Benjamin's (2004) argument drawing on India that market processes in informal settlements have the potential to provide relatively inclusive urban growth as long as flexible regulatory and land-use policies prevail to enable mixed-use development.

Micro-credit and micro-finance: While higher income households are likely to save in formal institutions, households with low incomes face the reluctance of commercial banks to provide them with services. Solo (2008) reports on surveys in Mexico City, Bogotá and several Brazilian cities that show that between 65 and 85 per cent of households do not hold any kind of deposit account in a formal sector financial institution. It should also be noted that these cities are in upper-middle-income nations. Formal bank accounts are usually unavailable to residents of informal settlements, often because they fail to meet the legal requirements (for instance, proof of employment). Or formal accounts require large deposits or high commissions. Another problem is their location far from informal settlements. There are also more subtle forms of discrimination. Solo (ibid., p. 52) describes the response when a focus group member was asked why he hadn't gone to a bank when he needed a loan. 'Laughter filled the room and one voice spoke up: "Don't you see how we look, *compañera*? We just aren't the kind of people the banks would want."'

The costs of this lack of access are considerable. It makes it more costly and time-consuming to pay utility bills. Solo (2008) reports that the Central Bank of Mexico estimates that cash transactions can cost up to five times more than payments by cheque and up to 15 times more than electronic payments. In the absence of formal alternatives, low-income households turn to other forms of savings although they may be more costly and high risk.

Micro-finance emerged from an earlier and more narrowly focused tradition of micro-credit introduced to address the needs for loan finance. Micro-credit projects developed to provide low-income entrepreneurs with the capital they needed to expand their business (Johnson 2009). They responded to the high rates of interest charged by informal money lenders, 10–20 per cent a month being common. With such penal rates of interest, it is not possible for these micro-enterprises to accumulate the capital they need to grow their business and/or improve their livelihoods. The very low incomes and low levels of assets meant that micro-credit programmes had to develop alternative forms of collateral with agencies such as the Grameen Bank's pioneering group-based lending in which there are mutual guarantees with associated social pressure. Box 2.5 illustrates the significance of such approaches in one small town in Bangladesh in terms of the proportion of the population reached in informal settlements. It also shows the lack of finance available to support those in informal settlements addressing collective needs such as water, sanitation and drainage – an issue to which we will return in Chapter 3.

Box 2.5 Micro-finance activities in one Bangladeshi town

Micro-finance activities in Faridpur, a district town in Bangladesh, 130 kilometres south west of Dhaka, illustrate their scope, diversity and significance. Approximately 9 per cent of the town's population of 109,000 is living in 22 informal settlements with 52 per cent of working adults having insecure and irregular employment – mainly as casual day labourers.

Twenty-four NGOs are active in the town and 18 of them provide micro-credit. In 2006, these 18 supported approximately 3,600 households, with the average loan size across all 18 organisations varying between US $80–160 (ibid., p. 217). Terms and conditions vary. None of the NGOs were providing training to enhance livelihood skills – some training is provided around micro-credit but it is recognised that the quality is variable; collectively they recognise this weakness in what they have been doing (ibid., p. 224). This lack of provision may be because of the tendency within micro-finance in recent years to focus on financial services rather than on other activities. Some do provide assistance in animal rearing despite the urban location. Areas with micro-credit interventions have lower interest rates in the informal sector than those found in other neighbourhoods. The international NGO Practical Action recently established a city forum to assist in the coordination of NGO activities and help organisations to be more effective in their work (ibid., p. 225).

This study also shows the partiality of NGO interventions. Due to a lack of finance, there is no NGO work on improving infrastructure – but only 20 per cent of informal settlement residents had access to a sanitary latrine and only 5 per cent of households had a private latrine, with the others all sharing facilities. Fifty three per cent use water from tube wells despite it being polluted with naturally occurring arsenic and only 14 of the 22 informal settlements have any drainage provision (ibid., p. 221).

Source: Saha and Rahman (2006)

As micro-credit initiatives spread, there was an acknowledgement that savings facilities were often all that was required (Collins *et al.* 2009). Savings are important for multiple reasons. In the context of enterprise development, it enables the expansion of activities (or managing cash flow) without any interest charge or repayment obligation, and is therefore lower cost and lower risk than borrowing. For housing, as elaborated below, it is the major source of investment capital.

Shelter micro-finance. Loan finance for shelter-related investments is also rarely available through the formal commercial sector and savings are a major source of finance for incremental housing development. In India, more than 80 per cent of housing finance comes from private savings (Biswas 2003). In Bhopal, the capital of Madhya Pradesh (with around 1.8 million inhabitants), only 5 per cent of those moving from informal into formal settlements obtain formal housing finance (Lall *et al.* 2006, p. 1031). In Angola less than 2 per cent of a family's investment in housing comes from banks. Instead, most funding for housing is borrowed from the extended family (62 per cent) or from friends (27 per cent) (Cain 2007). There are numerous (and familiar) problems with formal financial institutions in terms of being able to borrow for housing in addition to those mentioned for more general financial services: loan products are not appropriate to the housing finance needs of low-income households (as well as being unaffordable), collateral requirements cannot be met, and relatively high administration costs are a further difficulty (Ferguson 1999). The Kenyan Banking and Building Societies Act, for example, explicitly forbids financial institutions making loans for plots of land with no or only partly constructed housing on them (Malhotra 2003, p. 225). Such resistance to incremental housing arises from the perceived risks of construction and the potential for contravention of building regulations, and the difficulty of selling partially constructed housing if the loanee fails to repay. Discrimination may make it impossible for women to secure formal housing finance (Datta 1999, pp. 192–193).

Recognising the importance of housing interventions for poverty reduction, a distinct set of programmes has been developed to provide micro-finance for shelter-related investments. When micro-finance lending first began, there was a reluctance to lend for housing due to the larger size of the loans – between US$500 and US$5,000 – and the belief that shelter investments are not productive and do not generate the income needed to ensure loan repayments. Over time that belief has been challenged. Most shelter micro-finance loans are made for terms of between one and eight years, though usually at or near the one-year end of that range (CGAP 2004). Security requirements may be similar to enterprise development loans, i.e. the loan contracts insist on group guarantees and co-signers, or may be based on minimal legal documentation declaring the property and other non-mortgage assets as collateral. Some shelter micro-finance lenders issue a conventional mortgage for loans at or near the high end of the US$500 to US$5,000 range. Chapter 3 describes how agencies have taken these programmes forward.

Most shelter micro-finance is awarded to individuals with some degree of tenure security. Loans are generally taken to build additional rooms, replace traditional with modern (permanent) building materials, improve roofs and floors and add kitchens and toilets. Such investments are highly popular: India's Self-Employed Women's Association (SEWA) estimates that almost 35 per cent of the housing loans from its bank go towards improving facilities such as a private water connection or a toilet (Biswas 2003, p. 51). The structure of this lending with its focus on the individual makes it difficult for lending practices to extend to collective activities; and it is the collective that is essential in negotiating for the acquisition of tenure (and often of land acquisition) and investment in basic infrastructure and services. There has been a long-standing interest in financing such investments: see, for example, how families in a low-income settlement in Dakar (Senegal) borrowed to install a water supply system and drainage channels with the investment paying for itself within a year due to savings in medical bills (see Gaye and Diallo (1997) for a discussion of this scheme; the cost data are from a field discussion with the authors). However, this has not been at the scale required (Mehta 2008).

Housing and enterprises: The benefits of housing improvements for home-based enterprises have long been recognised. Many urban households draw a significant proportion of their income from informal trading or productive activities such as selling prepared food, providing personal services such as hair-dressing, repairing cars and household goods, building material production and/or other small-scale manufacturing. Electricity provision and accessible water supplies (especially water piped into homes) bring many advantages to most forms of home-based enterprises. Home-based production may be more likely to involve women owing to the gendered division of household tasks and women's need to remain close to the dwelling (Kantor 2009). The women's bank in Sri Lanka extended their lending from micro-enterprises into housing because of these benefits: food producers could have increased space, improved ventilation and an enlarged counter space; garment producers could have space for private fittings, storage and production design; and traders and shop owners could have enlarged windows for trading, and slightly wider footpaths (Albee and Gamage 1996). Entrepreneurs find that custom increases as neighbourhoods are improved. Rental income is also an incentive – and home owners take loans for the construction of additional rooms and sometimes dwellings (Datta 1999; Lemanski 2009). Shenya (2007) argues that in Dar es Salaam a major factor holding up the expansion of rental rooms is a shortage of capital.

A common finding from studies of infrastructure improvements in informal settlements is that trade increases as local residents find it easier to walk to the shops (for example, when there are pavements rather than muddy pathways). The significance of transport links is illustrated by an example from Luanda (Angola) discussed above.

Non-credit-based market approaches: In addition to micro-finance, development interventions have sought to further the integration of the urban poor within the market through enhancing labour capabilities, strengthening better enterprise management, improving infrastructure to enable new markets to be reached, and strengthening local neighbourhood economies (Jones and Mielhbradt 2009).

Training may be provided alongside credit, especially when the micro-loans are provided by NGOs. For example, in Dhaka (Bangladesh), Practical Action provides training in conjunction with a revolving fund for loan finance. In this case there is an emphasis on food processing because the capital requirements are small, there are readily available raw materials, easily understood technology and accessible markets. Construction skills such as block making, bricklaying and carpentry may also be taught (Kinyanjui and Ngombalu 2006, p. 127). Training may also be offered in more complex areas such as production strategies, business planning and marketing. For example, an important component of the work programme of the Carvajal Foundation in Cali (Colombia) was to establish Productive Development Centres to provide local entrepreneurs with technical, administrative and commercial assistance (Cruz 1994).[10]

The potential of state-financed construction to support local livelihoods has been demonstrated through community construction contracts that have been popularised since their introduction in the Million Houses Programme in Sri Lanka. Such contracts enable local residents to come together in a group and bid for work within their neighbourhood. The federation of savings groups formed by women slum and pavement dwellers in India (*Mahila Milan*) have built or managed the construction (and running) of hundreds of toilet blocks (see Chapter 4 for more details). These seek to ensure that the low-income households themselves benefit when infrastructure investment takes place, and hence that local incomes grow (see the discussion in Chapter 3).

Collaboration between workers and/or entrepreneurs may be encouraged to improve the efficiency of their activities, or to enable greater bargaining power. As noted above, those using the rights-based approach may support vendors struggling to access the central city. There is also a more general recognition that enterprises can benefit from grouping together to strengthen their bargaining power. Hasan and Raza (2011) describe the motivation of Dr Akhtar Hameed Khan, founder of the Orangi Pilot Project, to establish worker cooperatives in Karachi to provide skills and hence better livelihoods. While a limited number of cooperatives were established, many struggled, and an alternative strategy of women work centres (managed by a single family) was adopted. These were successfully established and increased participation in the garment trade.

New kinds of relationships with suppliers and/or customers are possible when individual entrepreneurs come together. Waste recycling is one area in which there are evident commercial benefits from individual pickers being organised. The very limited returns that waste pickers get and the low prices they receive for the separated wastes from recycled material wholesalers or industries that use recycled materials has been documented for many decades (see e.g. Furedy 1992; Hardoy *et al.* 2001). In Brazil, networks of recyclers formed and challenged exploitative trading relationships; in some cases the income received from low unit prices was further exacerbated by fraudulent practices such as faulty scales. For example, in Salvador (Brazil) Wal-Mart stores in the metropolitan region have introduced recycling facilities for their clients (Fergutz *et al.* 2011). More than 200 waste pickers from the Cooperative of Ecological Agents and Waste Pickers from Canabrava collect 45 tonnes each month and transport them to the cooperative's centre for sorting, pressing and sale. A further example is the waste picker organisations in the Philippines who formed a federation to support local initiatives, negotiate with local authorities and become a legally constituted people's organisation (Vincentian Missionaries 1998).

Contribution to poverty reduction

The importance of access to financial services for low-income groups has been recognised. Savings in particular are a critical component of more robust household livelihoods (Collins *et al.* 2009). Without access to safe places to save, development is difficult. While studies have pointed out the difficulties of the lowest income households saving (see e.g. Kantor and Nair 2003), access to savings is important for all low-income groups. However, as discussed below, there are concerns about the expansion of financial services in respect of borrowing owing to the adverse consequences of debt.

The contribution of micro-enterprises to household income is significant particularly in low-income settlements with a high dependence on informal employment. For example, a study of 1,755 households living in Nairobi's informal settlements found that 30 per cent of households operated an enterprise and, encouragingly, that ownership of an enterprise is negatively correlated with income poverty (World Bank Kenya 2006). In this context, the greater availability of loan capital may be helpful in enabling the expansion of activities. At the same time, the significance of the broader context has been recognised as important to the prosperity of such activities (i.e. lack of micro-credit is not the only barrier to business growth), and this in part explains the development of the urban management approach to poverty reduction. It may be, for example, that enterprises in informal settlements are better served by electricity connections, cleaner water supplies or improved pathways to enable access for new customers than by access to credit.

A systematic review argues that despite extensive research, there is no clear evidence that micro-finance programmes have positive impacts on poverty reduction (Duvendack *et al.* 2011, p.14). In drawing this conclusion the authors stress that there is a wide variety of different programmes which may be orientated towards several different but related goals (economic, social and/or empowerment), and that there are genuine complexities involved in assessment. Johnson (2009, p. 296) agrees that some micro-finance-related promises with respect to poverty reduction may have been overstated. Whatever the reasoning, the understanding that micro-finance cannot be expected to reach and help the lowest income households is now widely acknowledged – and hence increasing emphasis is given to financial inclusion of those outside the formal banking system more generally rather than an attempt to make micro-finance reduce poverty in the lowest-income households (Johnson 2009, p. 294).

One attribute of loan finance is that higher income groups are more able to benefit than lower income groups because they can better manage the risks involved. This is true in the case of micro-enterprise support and also shelter micro-finance (although this may include some low-income households which have an asset in the form of housing to invest in). For example, an assessment of a micro-finance programme in the Copperbelt (Zambia) leads Copestake (2002) to conclude that while poverty was reduced, inequality was increased as the lowest income households struggled to participate. There is a risk, therefore, that local inequalities are exacerbated. Indeed Johnson (2009, p. 294) argues that the evidence suggests that the lowest income groups benefit more from the indirect effects of economic growth resulting from improved financial services than direct impacts.

In terms of the relationship between micro-credit and business success, there is now a greater recognition that some economic sectors may be more stable than others, and that encouraging many individual entrepreneurs to compete in the same sector may not be helpful. Greater emphasis has been put on aspects such as value-chain analysis to improve business performance through a much more sophisticated understanding of comparative advantage and competitiveness (Jones and Miehlbradt 2009, p. 310).

Finally and perhaps more fundamentally, the market approach to poverty reduction may have assumptions that are simplistic about the way in which the market operates and the difficulties that low-income households face. Johnson (2009, p. 293) summarises this critique: 'markets are (and have to be) constructed through a range of social, cultural and economic practices as well as being the subject of political action, while their rules, regulations and norms can be deeply entrenched and/or slow to change'. Moreover, market transactions are embedded within broad norms and values. For example, in Pikine (Senegal) women benefiting from a micro-credit programme appear to be investing in their social relations alongside financial accumulation and enterprise growth (Casier 2010, p. 616); and as discussed in the section on clientelism below and the work of Banks (2010) in Dhaka, the demands of community and political leaders present real difficulties for entrepreneurs and have to be managed alongside any loan repayments.

Approach 6: Social movement approaches

The scale of political exclusion and anti-poor rhetoric and practice is well established (Mitlin and Satterthwaite 2013). Massive neglect has resulted in around one in seven of the world's population living in informal settlements in urban areas – and for many cities to have between 30 to 60 per cent of their populations lacking adequate homes. The residents of informal settlements may lack the right to vote and will certainly be denied equal access to goods and services. Urban poverty has been under-estimated, ignored and otherwise denied by many national governments. This is acknowledged in some detail in the Poverty Reduction Strategy for Cambodia (Kingdom of Cambodia 2002, pp. 85–86), and what this document states could be applied to many other nations:

> Both the authorities and the better-off city dwellers tend to blame the [urban] poor for their wretched conditions and stigmatise the [urban] poor as socially undesirable, criminally inclined, even mentally defective. The usual response for middle-class people and from officials is that the urban poor should be sent back to the rural areas where they belong. Unlike the rural poor, who constitute the vast majority of the poor in Cambodia and who are considered to be innocent victims of poor administration and underdevelopment ... the urban poor are deemed to be responsible for their predicament. They are given a much lower priority for assistance because, on paper at least, they are much better off than their rural peers in terms of income, nutrition and proximity to basic services such as education and health. Even well-informed personnel in donor agencies ironically downplay the privations of the urban poor by using crudely conceived statistical comparisons with the rural poor to justify anti-urban grant and lending policies. By so doing, the donor community often reinforces prevailing local confusion and prejudices against the urban poor.

The historical contribution of social movements as mass-based agglomerations of organisations, networks and individuals challenging structural disadvantage and securing pro-poor political change is well established (Harvey 2012; Moore 2005; Tilly 2004). Moore elaborates on this perspective: 'it is one of the "givens" of political science that poor people in poor countries have few political resources and become politically effective only through collective action' (2005, p. 273). Hence it is not surprising that efforts have been made to support citizen agency towards improved political processes, such as more inclusive governance, better policies and pro-poor practices by state officials. Those supporting social movements believe that political parties and politicians are unlikely to secure inclusive and progressive development without sustained local activism.

Movements may be best viewed as politicised collective agglomerations of activities of *and for* the poor – and the concept extends beyond specific organisational form to refer to a process of mobilisation. Formal organisations can be *part of* social movements, but broadly speaking movements extend beyond formalised agencies to include the more nebulous, uncoordinated and cyclical forms of collective action, popular protest and networks that serve to link both organised and dispersed actors in processes of social mobilisation (Mitlin and Bebbington 2006). At the same time, while movements are processes of collective action that may be dispersed, they are also sustained across space and time; i.e. they are not single isolated events. NGO alliances with people's organisations have strengthened over the past ten years, in part due to the adoption of a rights-based approach to development (Hickey and Mitlin 2009). More generally, Korten (1990) identifies such alliances as a 'fourth-generation' NGO strategy as they moved beyond welfare (first generation), to programming and capacity development (second generation), and systemic integrated interventions (third generation) towards sustained pressure for pro-poor political change.

Despite this emphasis on processes that go beyond specific organisational forms, in terms of development approaches support for social movements means support for social movement organisations and/or social movement support organisations (Batliwala 2002; Mayo 2005). Such organisations are then expected to bring together more loosely formed coalitions and alliances. In part this work has emerged from earlier support to NGOs and an awareness that there was a need to move beyond this group into a much wider appreciation of the role of civil society (Edwards and Gaventa 2001).

In regard to the Global South, there have been several different types of movements and related activities. While much of the emphasis of documentation is placed on social movements in general, also discussed are urban movements where the focus of activities is on collective consumption goods such as housing, land and basic services, i.e. those required for urban living and livelihoods. Some movements were manifest in the struggle for democracy against colonisation and dictatorship: in such cases movements sought control of the state, and hence became or made alliances with political parties. While towns and cities may have been locations of struggle and sources of popular support, such movements are not particularly urban. With respect to the specific nature of urban space, struggles in towns and cities may be divided into five broad types related to: formal work, informal work, land (plus housing), basic services, and social discrimination (subdivided by gender, age, ethnicity and other distinctions).

The theory of change in each case has elements that are overlapping and those that are distinct; here we highlight some of those that are key. Overlapping elements centre on the strengthening of political capabilities (Whitehead and

Gray-Molina 2005) and the more general organisational development of social movements. Within these broader categorisations, there are considerable differences between agencies that encourage formalisation and those that prefer to work with more informal strategies. The significance of social movements for inclusive cities and effective poverty reduction means that the specificities of different modalities of support are raised throughout this volume. Already in this chapter we have discussed the rights-based approach to development which has resulted in NGOs developing these relationships and raising the priority of this work (Chapman *et al.* 2009). However, there are many other notable long-standing efforts to support social movements and related citizen activism.

Many of the social movements active in supporting the rights of the working class and challenging political exclusion and injustice in nineteenth-century Europe have emerged from within the trade union movement. Workers across the Global South continue to combine into movements to campaign for improved pay and employment terms and conditions. The primary relevance of trade unions is in formal sector workplaces, since informal workplaces are unlikely to permit unionisation – this limits the significance of trade unions for low-income urban dwellers, as most do not have formal jobs (Pearson 2004). For example, only 8 per cent of India's workforce works in a sector that is organised by trade unions (Sinha 2004). In the absence of formal union organising, other agencies have sought to support informal workers (Lindell 2010). Such initiatives seek to strengthen the ability of workers' organisations to defend their rights and advance their livelihood struggles. For street vendors, this includes campaigning to defend their access to well-located plots in central areas. Crossa (2009) discusses the strategies of two movement organisations in Mexico City, highlighting the contestation over alternative visions of city development and the multiple strategies followed to seek to protect the interests of traders, and Etemadi (2001) discusses similar issues in the context of Cebu City.

Notable movements have emerged to contest evictions and prevent the privatisation of basic services. Such movements are primarily defensive, as is also the case for those seeking to prevent the expulsion of traders from central areas. The strategies may involve physical resistance, legal challenges and the support of international agencies who criticise the evictions. The work of the Centre for Housing Rights and Evictions (COHRE 2006) exemplifies the ways in which international agencies document and publicise local events and in so doing seek to support local action.[11] In some cases the United Nations has been involved in such investigations, one of the most notable being their response to the eviction of 700,000 in Zimbabwe (Tibaijuka 2005). The work of the Centre for Housing Rights and Evictions since its foundation in 1994 is an example of the ways in which international agencies have sought to support local struggles, and global action is recognised as being an important contribution to local struggles for social justice (Edwards and Gaventa 2001).[12]

As theorised by Castells in his 1983 volume, *The City and the Grassroots*, movements to secure access to collective consumption goods and services (such as secure tenure and housing, and services such as water, sanitation and electricity) in towns and cities in North and South have both historical and ongoing significance. His analysis emphasises the essentially conflictual nature of urban residents' relations with capitalist economies, and he locates urban centres as a place where there are both practices of exploitation, dispossession and discrimination and, at the same time, places of resistance. The required public investment leads to the coming together of citizens to pressure the state to provide for these needs. Castells's 'grassroots' is not a simple reactive defensive body but rather an agency that takes multiple forms to contest the values of the city, seek cultural identities and search for decentralised forms of urban (self) governance (Castells 1983, pp. 319–320). Autonomy is criticial:

> when squatter movements break their relationships of dependency *vis-à-vis* the state, they may become potential agency of social change.
>
> (Castells 1983, p. 194)

Processes of social stratification and subsequent discrimination reccur in towns and cities across the Global South and numerous movements have emerged to challenge such processes. In the context of addressing poverty and inequality, there is an evident overlap with the absolute lack of income and assets. Box 2.6 provides more detail on one of the examples of citizen action mentioned above and describes the work of the women's nutrition movement in Peru, specifically the contribution to two networks of organisations to address the need for food in a context of acute and widespread need, and in addressing gender discrimination.

Box 2.6 Women's nutrition and citizen action in Peru

In Peru, the women's nutrition movement (*Organizaciones de Mujeres para la Alimentación*) centres on two distinct traditions: community kitchens (*Comedores Populares*) and the glass of milk committees (*Comités del Vaso de Leche*). Both provide members with food at a reduced cost offering free support to those in particular need. Women's organisations manage the provision using both community volunteers and state resources: extensive community involvement in food provision and cooking meals reduces processing and delivery costs. These groups also determine the beneficiaries.

The tradition of food relief managed by local women began between 1948 and 1956 when the government set up groups to distribute overseas food aid. During the 1970s and early 1980s, these local kitchens evolved to become an identifiable group of community kitchens with some securing independence from the state. Shortly afterwards, the glass of milk programme was catalysed by Lima's local government in 1984, at a time of acute need, and rapidly resulted in independent organisations. Just months after the programme was initiated, more than 25,000 mothers from 33 districts in Lima marched on Congress to demand official support; this campaign was followed by legislation in January 1985. Activities provide cereal and milk, primarily for children, pregnant women and breast-feeding mothers. The women's community kitchen organisations have also been involved in similar activities, negotiating for support from the Municipality of Lima as far back as 1986. The community kitchen organisations campaigned for legislation (Law 27307) with associated regulations to ensure that the state finances food-processing activities. Participatory planning committees mean that they sit alongside central and local government representatives – and that demonstrations are no longer needed.

The movement organisations believe they have broadly succeeded in establishing their entitlement and securing involvement in delivery – at the same time affirming their identity as citizens with both rights and management capacity, as opposed to 'poor people' with unsatisfied needs and few skills who require patronage and social welfare practices. They have also succeeded in challenging traditional male-led residents' associations and enabling women to play a role as community leaders and political activists.

Source: Barrig (1991) and Bebbington *et al.* (2011)

The social movement literature commonly distinguishes between old social movements and new social movements with the former being viewed to be those related to employment relations and the struggle for democracy, and the latter concerned with identity – for instance, the movement for gender equality, those concerned with sexual identities and those with committed environmentalists. However, that distinction does not work well in towns and cities in the Global South. As Box 2.6 illustrates, women's struggles involve both identity and financial redistribution; the same is true of movements of informal residents struggling for land, housing and basic services who may have particular ethnic affiliates, and who also have a strong identity based with their informal localities (Mohanty *et al.* 2011). The example of sex workers from Dhaka (see above) further exemplifies the overlap between identity and livelihoods.

Contribution to poverty reduction

This approach supports movements with the understanding that progressive social change has, in general, been secured through political changes supported by movements (if not led by them). However, this does not mean that all movement activities lead to social change.

As noted above, the increasing informalisation of the labour market makes labour market organising increasingly difficult. The consequences are exemplified by a report that documents how the number of industrial disputes in India fell from 3,500 in 1973 to less than 75 in 2001; the informally employed urban poor appear to be shifting their focus from employers (and employment conditions) towards the provision of basic services from the state (Agarwala 2006).

It has proved hard for movement activities to maintain their momentum and frequently intensive activity takes place but is not sustained. This is true following both success and failure (see Dávila (1990) for an account by Pedro Moctezuma of how the strong social movement in Mexico struggled against the control of the dominant political party, PRI, but weakened or dispersed when they felt they had achieved their goal).

The reduced level of mobilisation makes it hard for the gains to be maintained and advances may be withdrawn (Mageli 2004; Racelis 2003). In the case of anti-evictions and campaigns against privatisation, these are defensive battles and often the best that might be secured is the status quo (which frequently means continuing tenure insecurity and very partial access to services for most of the population). As described by Perreault (2006) in the context of Cochabamba (Bolivia), many of the urban poor did not receive water under the system that the movement lobbied to retain. In other cases movements are more pro-active but politicians find promises are easy to make and equally easy to ignore.

There are concerns that movements are influenced by a range of other agencies including political parties and NGOs. Crossa's (2009) account of the struggles of movement organisations in Mexico City describes the impacts of such affiliations upon movement tactics and successes. Some of the urban popular movements in Mexico resisted such affiliations in the 1980s due to their wish to maintain their ability to make alliances according to their political interests (Davila 1990). Castells (1983) describes the compromises made by squatter movements in Latin America who traded political support for access to land. These themes are returned to in Chapter 5.

The concern that movement organisations may be captured by elites is long standing. In 1911, Robert Michels wrote about the domination of trade union movements by labour elites. Such practices continue to this day. Some of these concerns have been touched on more generally in the discussion about participatory governance. Cabanas Díaz *et al.* (2000, p. 87) illustrate the consequences when they describe the outcome of an organised land invasion in Guatemala City:

> The sites that were occupied and the form of their occupation reflect different economic situations and levels of organisation. The poorest households ended up on the steep slopes of the ravines and in the areas around the sewage and wastewater outlets.

Chapter 4 describes a number of programme interventions that work with urban movements to improve access to secure tenure, basic services and, in some countries, housing improvements. A central goal is social and political inclusion through city-wide development strategies. Their strategies build on experiences to date, and Chapter 5 elaborates the underlying analysis and conceptual frameworks that are used. The goal is to find effective mechanisms to support a grounded set of collective activities that incorporates mass involvement from residents, addresses immediate needs and enables a sustained political engagement that secures both policy reform and changed relations between the state and local citizens.

Approach 7: Aided self-help and community development

There are two more approaches that emphasise the importance of the ways in which the urban poor themselves engage with urban realities to create development and reduce poverty. The first of these is aided self-help, or recognition of what communities do for themselves which may include their attempts to use clientelist networks but which also recognises practical activities at both the household and community levels (see Bredenoord and van Lindert (2010) for a recent elaboration of the importance of this approach).

In many cities and towns across the Global South, individuals, households and communities have little government support to address their basic needs for shelter including secure tenure, basic services and housing. As described in the introduction (Chapter 1), they face a considerable lack of investment in the planning and infrastructure needed for safe and secure homes. They may also face few opportunities in labour markets that offer little pay and difficult and sometimes dangerous working conditions. In the absence of alternatives, people have to help themselves. Moser (2009), who draws from her research in an informal settlement in Guayaquil over 30 years, outlines the multiplicity of strategies used by low-income women to address collective consumption needs (see Box 3.3 in Chapter 3). She explains how the women undertake self-help, negotiate with local politicians and an emerging political party, are active in protest movements, participate in national government programmes to address local needs, and collaborate with an international NGO and international development agency. Such self-help may include both individual and collective efforts to address development needs. At the individual level, households make multiple efforts to invest in human capital, including education as well as other learning opportunities. Households may also seek diversification of their income sources to reduce the risk of poverty; see, for example, the discussions in Moser and McIlwaine (1997, p. 52) and Moser (1997, p. 58). However, this approach in an urban context focuses in particular on the provision of support for the self-help of land acquisition and development, and housing. Such support seeks to replicate the provision that people are otherwise arranging on their own.

The conceptual framework to support this approach emerges from these long-recognised practices of self-help community activities in respect of public goods such as water, security and waste management, as well as housing. Such self-help strategies support incremental development because people cannot afford to raise the capital to develop their homes and neighbourhoods with a single investment. Through such self-help, low-income communities provide themselves with the goods and services that neither the market nor the state is offering (Devas 2004). Turner (1976) illustrated the development of low-income settlements in Peru and the considerable investment that households make over time, even though tenure is not formal and may not be secure, and services not provided. Barrig (1991, p. 66), writing 15 years later, describes a similar situation in Lima (Peru) thus:

> The squatters, organised in self-help groups, build their own housing with extremely few resources, investing money and manual labour in developing road networks, constructing medical facilities and linking up with the city's water and electrical systems. Such activities are significant in size, as more than 30 per cent of the population of Metropolitan Lima lives in shantytowns, which have followed this pattern of development.

Peattie (1990) describes in detail the careful organisation required to mount a large-scale invasion of land in Lima that then served as the basis for negotiating land.

As a result of John F.C. Turner's work and that of others such as Matos Mar and William Mangin (that go back to the 1960s), urban professionals (architects and planners) have sought to build on grassroots energy and creativity with programmes to enable such self-help. These professionals helped legitimate more affordable and less exclusionary housing policies and associated programmes, including both serviced site programmes and upgrading programmes. The goals are both to replicate the strategies that people use to get started, and to provide both these people and those in existing informal settlements with the support they need to speed up the processes of improvement. For those without homes, helping people get started has primarily centred on the provision of land with provision for minimal infrastructure and services that can be upgraded over time. The World Bank began funding for sites and services programmes in the early 1970s (Buckley and Kalarickal 2006; Cohen 1983). However, even then the concept of sites and services progammes had been used in Peru from the 1960s and was already underway in a number of countries including Malawi (Manda 2007; Riofrio 1996). As regulatory standards increased in middle-income countries in the Global South, shell houses were also provided in countries such as the Philippines and India as well as in El Salvador (Mukhija 2004). At the same time, the informal and erratic upgrading process described above in the context of Peru is addressed through subsidised local government or utility support, sometimes supplemented by loan programmes to enable households to finance their own improvements more easily. Recognition of the quality of life that has developed in areas that were formally seen as squatter sites helps persuade local government that upgrading is a viable strategy for urban development. At the same time, and as elaborated in Chapter 4 for the case of the Orangi Pilot Project, a further approach has been to identify improved modalities of infrastructure investment to enable household efforts (including time and money) and state funds to be used more effectively. Relevant to both upgrading and new build have been self-help financial arrangements. Credit unions and housing cooperatives are examples of more complex self-help strategies that have emerged.

The underlying assumption is that support will speed up the processes by which households secure shelter with tenure and infrastructure and basic services. Moreover, not only will it speed up the processes that people are undertaking anyway (i.e. responding to local demand and therefore providing an appropriate product), it will also enable scarce resources to go further in reaching a significant number of people because international and/or national support is added to people's own efforts and investment finance. Rather than encouraging people to adopt the approach of the professionals, aided self-help and community development is all about responding to what the people are doing anyway.

This approach is particularly focused on the needs of women who bear the consequences of gendered roles and responsibilities that require them to manage reproductive needs within the households including the care of the elderly and young. Expensive formal solutions are out of touch with their realities, and the pragmatic focus of aided self-help may be more likely to assist them in addressing their needs for water, sanitation and safe neighbourhoods. In some cases, particular efforts are made to draw women into local management processes.

Contribution to poverty reduction

Such strategies are recognised to be supporting residents to address their core needs but it is also recognised that there are limitations on what can be achieved. A first substantive critique is that the approach detracts from the core business of ensuring that the state makes finance available to support the housing needs of low-income groups. In working with people's own strategies, it implicitly validates the idea that people alone can address their housing needs, or at least the state has to do no more than a minimum (Burgess 1978). This kind of neighbourhood construction process adds considerably to the burden of work as people have to struggle to make a living, and to secure (and build) their own homes and often also infrastructure. A consequence of the lack of redistributive finance is that the lowest income groups often cannot afford to participate in the emerging solutions. Shelter micro-finance, housing cooperatives and credit unions may address the needs of the better-off among the urban poor, while upgrading programmes may displace tenants who move elsewhere.

A second concern is that government support, when available, has been too small in scale to make a difference. Consequently it has been captured by higher income groups which are better placed to secure the scarce resources (Manda 2007; Rakodi 2006a). Due to the low levels of investment progress at scale is difficult to achieve. Even by 1993, Riofrio (1996) points out that only 63.6 per cent of Lima's population enjoyed piped water in the home and only 60.2 per cent of households were linked into the main sewer system. The low level of government investment in part reflects the ambivalence of the government to this solution (particularly on the side of upgrading), as they prefer a more 'modern' city to fit with the most commonly promoted global vision. One site and service project close to the centre of Lilongwe was demolished due to the fact that it did not fit with the vision of the city held by the autocratic 'president for life' Hastings Banda (Potts, quoted in Myers 2003, p. 154). This highlights that even if the solution works for the lowest income groups, if it does not also result in required political change (both in attitude and policy) it is unlikely to move forward.

An assessment of the contribution of aided self-help to poverty reduction is difficult in that so much of it is not documented or counted – and it is also difficult to gauge its influence on policies and practices. Of course there are self-help groups that are neither inclusive nor progressive – for instance, some of those based around particular political, ethnic or religious factions. There are also the settlements and households where aided self-help can provide no solution – for instance, for tenants in many informal settlements in Nairobi (and the power of landlords and their organisations – see Weru 2004). At the settlement and city level, Simone (2006) talks about pirate towns to emphasise the lack of order and governance (even self-governance), and Rodgers (2009) discusses the ways in which drug trade-related gangs create substantive difficulties for any development intervention in local neighbourhoods.

There are also those within informal settlements who have difficulties getting benefits from aided self-help initiatives. Hanchett and colleagues (2003, p. 48) highlight the difficulties that the lowest income households face in making the regular monthly payments to access community-managed water services in Dhaka, hence their low levels of participation in water committees. With a changed payment structure, their needs could be met more easily. Many financial cooperatives (both credit unions and ROSCAs) have minimum monthly savings requirements that effectively eliminate the participation of lower-income residents. These examples point to real limitations of such self-help solutions in providing a substantive solution to the level of poverty experienced by the lowest-income urban dwellers. But in many cities aided self-help has improved housing conditions and helped secure tenure and better provision for infrastructure and services for large numbers of low-income households. Perhaps as importantly, it has also helped change perceptions of politicians and civil servants, as it shows the capacities of low-income households to build or improve their homes, and has helped influence the much wider official acceptance of 'slum and squatter upgrading'. As discussed in Chapter 4, aided self-help is an important component of the strategies of federations of slum and shack dwellers – although this is within a larger set of goals and activities that seek political change, especially in relations with local governments. This is also true of other initiatives discussed in this chapter.

Approach 8: Clientelism

Clientelism, or the use by the urban poor of the patron–client networks that link them to more powerful social groups to secure advantages (particularly access to tenure and basic services), is broadly accepted to be ubiquitous in the Global South (Kabeer 2002; Wood 2003). Clientelist relations have been described and widely critiqued in respect of their ability to catalyse and support inclusive and pro-poor urban neighbourhoods, and towns and cities (Bawa 2011; Peattie 1990; Pornchokchai 1992; Scheper-Hughes 1992; Valença and Bonates 2010; van der Linden 1997). While there are different kinds of clientelism, the major focus of these authors is on the exchange of partial public services (such as water and roads) and sometimes jobs or wages[13] (for community leaders) for votes and other manifestations of political support.[14]

The primary concerns are that these impose highly stratified social relations on to low-income communities, result in the partial provision of infrastructure (usually with little regard to quality) and services, and hence maintain their powerlessness and disadvantage.

Despite such evidence, this approach may be viewed through an alternative perspective. Clientelism is unusual in that it does not come from particular actions proposed by external groups seeking to reduce poverty. Rather it comes from existing political relations – and an alternative approach is to recognise that the present system offers something to the urban poor (Auyero 1999, 2000), and that many more ambitious interventions are at best ineffective or counter-productive. In part, they are ineffective because they fail to recognise the way in which power functions, and the depth of entrenched advantage that lies behind such power relations. This is recognised by local culture as well as academics – van der Linden (1997), in his discussion of patrons and grassroots organisations in a sites and services project in Pakistan, quotes a local proverb which says that if you live in the river it is better to stay friends with the crocodile (p. 81). For example, in Brazil housing investment has increased significantly with millions of dwellings being built but 97 per cent of units within one of the larger new initiatives are managed by private housing construction companies with the urban poor themselves not having the opportunity to self-construct (Valença and Bonates 2010, p. 172). This approach emphasises that it is better to recognise what exists and how it functions. It may also be the case that organisations cannot be extracted from their local political context – and hence clientelist forms of political relations may reproduce themselves regardless of the preferences of external agencies (Devine 2007, p. 309). Benjamin (2004) develops this position when he argues that the professional critique of clientelism is used by national political elites because they are threatened by this local clientelist political process that introduces uncertainty and reduces their control over urban development. The 'problem' that this approach seeks to deal with is the naïve challenges to clientelism that may satisfy the professionals who design alternatives but which offer little to the urban poor themselves because they cannot be realised given the existing configuration of power.

This approach recognises that making use of clientelist relationships is very much a part of the strategies used by the urban poor (Robins *et al.* 2008, pp. 1078–1079). In the absence of adequate land for housing, services and employment, households develop strategies to secure patrons; building relationships with powerful individuals who help them secure access to needed goods and services. Wust and colleagues (2002, p. 216) highlight the significance of such relationships in Ho Chi Minh City (Vietnam):

> households enter into vertical protection networks operating according to Mafia-like (clientelist) logic. To find a job or obtain credit or an administrative favour, people look for backing by a protector, an influential person

> able to defend their interests and get them what they need. These are often small entrepreneurs or local political or administrative leaders. Generally, the various social relations that the households establish in their neighbourhood aim to ensure their integration in the urban environment. They are often of paramount importance for the survival of the poorest families.

Although this quote identifies multiple goods and services secured through these relations, the approach has a particular focus on the collective consumption goods that are essential to neighbourhood development and begins by recognising what is provided. Such goods and services mean that the relationships of concern are those to do with the interface between the residents of informal settlements and the political system. In this context, patrons help residents secure access to the goods and services they need and a typical return gesture is that residents commit their votes to patrons or to those that the patrons represent. Patrons work within state structures that are broadly supportive of their role. The prevalence of clientelism reflects the benefits secured by local political elites through using this means to allocate scarce resources and secure their vote banks. In some cases this has developed a national element through very populist parties but this pattern does not emerge everywhere. The state does not have sufficient resources to provide essential infrastructure and services. In the absence of scale, it uses personalised relations to manage protest as it buys off, co-opts and absorbs pressure and protest from the urban poor.

Clientelist relations enable an engagement between the urban poor and the policy-making process including richer, more powerful and better-placed decision-makers (Bénit-Gbaffou 2011, p. 456; Benjamin 2000; Robins *et al.* 2008, p. 1075). Benjamin and Bhuvaneswari (2001, pp. 2 and 35) use the term 'porous bureaucracy' to encapsulate the findings of their research in Bangalore and the process of lobbying and response that they observed. They document the potential of the poor to advance their interests to 'vote bank politics' which opens up the possibility for community leaders to play a complex political strategy with councillors and higher-level politicians within the city (Benjamin 2000, p. 44). Rather than clients being passive observers, these relations offer a potential for community leaders to develop an active strategy to advance their interests (Amis 2002, p. 7). Benjamin and Bhuvaneswari (2001, pp. 73–74) also argue that such networks tend to be inclusive rather than exclusive, as the power of the local community leader is in part related to their capacity to create an agreed position within the settlement and ensure that all the local residents are willing to support that position. The ways in which 'patrons' deliver benefits which help local people in multiple ways such as finding employment, securing essential medicines, and simply enjoying themselves at rallies is elaborated in Auyero (1999).

While Amis (2002) emphasises clientelism in the context of democracy, Castells (1983, p. 193) notes that even within authoritarian regimes, the organised urban poor have been able to negotiate for concessions and may have been able to secure improved policies and approaches by the state. This engagement reflects a level of interdependency between the ruler and the ruled (Robins *et al.* 2008).

While the urban poor need access to state resources, the political system requires some level of acquiescence from those being so ruled, and this gives space for some negotiation and redistribution.

Within these political regimes, there is space for negotiation and potentially a reclaiming of autonomy (albeit within constraints). Bayat (2000, p. 546) describes how hundreds of informal neighbourhoods formed in Tehran and Cairo despite state opposition as a result of such informal negotiations. He also describes the non-payment of utility bills which takes place in many informal settlements, alongside illegal tapping of lines. The argument is that these are at some level permitted by local power brokers as part of an unwritten agreement. Bayat (2000, p. 547) discusses the potential for these to be acts of contention or challenges to the status quo but at the same time refers to them as acts of quiet encroachment, the testing of boundaries that may or may not be negotiable. To have a political impact such processes need to be collective – and the collectivity of clientelism is highly local. Such processes are not entirely individual as they are about the acquisition of neighbourhood goods and services as well as individual leader benefits; but the collective element remains specific and limited as alternative vote banks compete with each other to attract benefits in return for loyalty to the political elite.

Contribution to poverty reduction

Three issues need to be highlighted. The first is that the gains may be real but they are limited and come at some cost. The clientelist state pre-empts a political collective response by creating and reinforcing vertical relationships between leaders and the state, and between leaders and residents in informal settlements. There are many illustrations of the ways in which local protest is controlled and a local clientelist community leadership consolidated (Garrett and Ahmed 2004; Henry-Lee 2005; Hossain 2012; Perlman 2010; Scheper-Hughes 1992; Thorbek 1991). An example from Sri Lanka illustrates the constraints they may face; when community associations begin to negotiate for more autonomy, even though they are not yet as politicised as social movement organisations, political elites begin to undermine the local process (Russell and Vidler 2000). There may also by self-policing by residents who avoid being too involved in negotiations with the state in case they are seen as threatening the role of community leaders (see Desai and Howes (1995, p. 234) for an example from Mumbai). In other cases, as described by Auyero (1999, p. 314), elites simply reinforce dependency with personalised relationships of gratitude, and benefits are necessarily limited (ibid., p. 326).

The second issue is around coercion. The common practice of coercion by local leaders who manage such vertical relations of power can create a framework of violence in low-income settlements. Henry-Lee (2005) discusses the situation in some Jamaican urban neighbourhoods where political parties control electoral politics in low-income communities with high levels of gangsterism and crime. In such cases, the links are very evident as politics and crime combine to advantage a small elite. Links are also evident (although very different) in Bénit-Gbaffou's

(2012) account of political decision-making in one Johannesburg settlement when she specifically connects the desire of ANC activists to dominate political space with the physical violence they use to control their political opposition (see Chapter 5). However, less explicit links may be more commonplace and patron–client relationships may result in ongoing insecurity and fear with a relatively small number of leaders able to exploit others to their own advantage using their political position to secure gains. Garrett and Ahmed (2004) discuss a survey of 585 households in informal settlements in Dinajpur (Bangladesh) to understand issues related to crime violence and insecurity. One in six households experienced a crime and of these, 25 per cent were severe beatings. Twenty-six per cent of crimes were attributed by interviewees to be perpetrated by unofficial and unelected community leaders (*mastaans*), and only 8 per cent of crimes were reported to the police in a context in which formal security is weak. Banks (2010), in a study of households' own strategies to avoid poverty in Dhaka, highlights the problems faced by businesses from *bhaki khay* or 'to eat without paying' as *mastaans* require access to free goods and services: as a result, some traders prefer to operate outside of the settlement despite police harassment. However, we should be cautious against generalisation. In both Ecuador and Nicaragua, the levels of violence from gangs trading drugs appear to be significantly higher than those under previous clientelist politics (Moser 2009; Rodgers 2009)

The third issue is that if benefits are secured these may not be available to the lowest income and most disadvantaged groups. Benjamin (personal communication) accepts that the benefits that are secured through informal negotiations may not reach down to all of the residents. Roy (2004, pp. 160–163) discusses the gender implications of this kind of politics as it is manifest in Kolkata and the ways in which women are excluded from what in this city is party-managed patronage. These groups are recognised to reinforce patriarchy and women's subservience (Roy 2004; Scheper-Hughes 1992). The point is that a clientelist process relies on scarcity and partiality – all cannot benefit. Bénet-Gbaffou (2011) describes the history of a group of party loyalists threatened with eviction in Johannesburg who negotiated a private deal with the local authority enabling them to stay on the contested land at a time when other evictions of 'less organised and less politically-resourced groups of residents' were taking place. Any move to universal provision is likely to require recognition of the importance of solidarity.

The schizophrenic nature of relations between local community groups (and wider networks of such groups) and clientelist political processes is illustrated by Castells (1983, p. 193) when he argues that the squatters in Lima were 'deeply realistic' in what could be achieved in the 1970s but appeared to have behaved as a 'manipulated mob' as they sought to negotiate between competing interests, 'walking a line' between survival and integrity that did not always exist. On the one hand, he recognises that they secured land for their members; on the other, he suggests their alignment to the political strategies of the military state made the movement 'an instrument of social subordination to the existing political order instead of an agent of social change' (Castells 1983, p. 194).

Conclusions

The eight approaches to poverty reduction discussed in this chapter, all of which have been tested over the past few decades, provide a framework through which we can understand the ways in which urban poverty reduction has been sought. Each is somewhat distinct. Each includes a conceptual understanding of the ways in which structures limit and constrain possibilities. Each also offers an understanding of how activities link together in chains of causal relationships that offer (it is believed by at least some) development potential. We believe that the eight approaches described here underpin most of the programmes that can be seen today within both national and local governments, and international development assistance agencies. Hirschman (1970) suggested that the choices facing consumers are exit (i.e. to leave), voice (i.e. to participate through argument) and loyalty (i.e. to attach to vertical relations and wait for reward). The discussion here suggests that there are more options for those designing programmes and participating in processes of social change. In particular, the diversity of options is illustrative of the fact that in the modern capitalist society power may be concentrated but it does not have a single source and it may be economic or political, and belong to elites or to a more diffuse and contested leadership. Political directions may be led by self-interest or ideological, while strategies may be related to the here and now or to visionary goals.

Each of the approaches outlined in this chapter responds to a particular observed need or opportunity, and/or has responded to an understanding of the underlying problems. In general they are distinctive in their approach and some are particularly associated with ideological positions and political affiliations. Hence the market approach may be linked to a neoliberal political position by those who are dubious about either the commitment or capability of the state (and often both) but dependence on state-led development is attractive to those pursuing liberal social democracy and who believe it is possible to elect a social democratic government that redistributes to those in need. Managerialism offers a solution to those who despair of the self-interest of politicians and have decided that the best to be expected is a somewhat functional state in key sectors. However, this is also too simplistic and few of those planning interventions subscribe to one approach alone. Both aided self-help and clientelism, it may be argued, represent pragmatic advances that are sometimes acknowledged by their supporters to be far from ideal but are the best that might be expected. In addition, it is perfectly possible to support market-based approaches in many aspects but to believe that the state should be reasonably efficient in the delivery of the services that it does provide; also, to support co-production as a mode of service delivery applicable to a temporary strategy leading to the development of comprehensive welfare state provision as the state develops greater capacity and capability (Joshi and Moore 2004).

However, whatever the ideological or pragmatic position taken, experience shows the limited outcomes that have been achieved and the analysis here suggests that substantive problems remain. Market opportunities are difficult for those who have few assets and lack political connections. The welfare state is under-financed and rarely provides comprehensive assistance that is flexible according to need or universal provision. Social movements have proposed radical alternatives to address injustice, exploitation and repression, although it has long been acknowledged that such organisations may be dominated by a few individuals or groups who may not represent the interests of particularly disadvantaged groups. Hence few of the proposed solutions derived from these eight approaches are, in themselves, likely to result in long-term reductions in poverty for all of those with low incomes and/or who are otherwise disadvantaged, even though they may help particular groups. The difficulties in realising success may relate in part to the institutional complexity within the urban context and the need to blend responses to particular contexts to achieve the variety of reforms required for securing social progress.

Table 2.1 summarises how the approaches described above translate into particular policy processes, and some of the more common problems that each faces. Chapter 3 explores how these ideas are taken up and transformed into more coherent programmes of intervention to address urban poverty.

Before leaving this discussion about approaches to development and their implications for efforts to tackle urban poverty, it may be worth registering the different underlying power relations or authority and hence the legitimacy of action associated with each approach. Our argument in elaborating this is that the way in which power is legitimised is important for the nature of challenges to existing power distributions. It is also important for the ways in which existing agglomerations of power are replaced by alternatives, and whether those alternatives result in new distributions of power or the same distribution with an alternative individual or group at the 'top'. Of the eight approaches discussed here, all but two (social movements and self-help) encourage vertical types of authority. In government-led poverty reduction, the state is the legitimate authority, and in market approaches the more powerful in the market (larger companies, those with monopoly control, employers able to choose from many seeking work, large-scale landlords) have an advantage and also represent a vertical system of authority. Participatory governance is an attempt to reduce the vertical authorities within the state and replace them with a more equal negotiated outcome – but it is placed under representative democratic structures (i.e. local authorities or national governments).

There are exceptions in the market if it is highly competitive and in participatory governance if it is genuine and substantive, as well as, arguably, in some forms of the rights-based approach which stress the right to participate in decision-making rather than the enactment of formal rights that are realised and protected by the state and/or judiciary. In these contexts, it may be argued that the nature of power is more diffuse and that there is less opportunity for powerful groups and individuals to dominate. To elaborate, there is a strong sense that the authority of the more powerful is restricted because the structures that nurture participatory governance hold the state to account for its actions. In the case of participatory budgeting, for example, it has been

Table 2.1 Summary of the eight approaches

Approach	*Primary concern that approach addresses*	*Theory of change*	*Major concerns*
Welfare	Assistance to those lacking the resources and access to services to meet their basic needs.	Establish the ability to provide cash or in-kind goods or services to alleviate immediate needs. Possibly make this conditional to change behaviour in favour of keeping children in school and attending health care. May be part-funded by compulsory individual and/or collective savings that helps to prepare for life-cycle needs and reduces the role of the state.	Achieving scale is critical but this is expensive and hence has to be a political priority. Programmes may tend to be top-down in management, dividing groups into deserving and non-deserving poor with discrimination against some groups. Some modes of delivery encourage the individualisation of citizen–state relations, preventing the consolidation of social movements.
Urban management	Lack of planning, basic infrastructure and services for urban well-being and prosperity.	Investment in infrastructure and services will increase income generation opportunities and support enterprise development. Those investments need to be located at the local government level to be effective. Evident need to manage land use and land-use changes guided by a city plan.	Emphasis on the management of urban centres for economic growth may lead to models of urban development that exclude low-income groups from the city centres and other prime locations. Modern urban management models may be expensive and unlikely to be an efficient use of scarce resources. Professional designs may be less effective for inclusive pro-poor cities than alternative approaches. The focus is the city, excluding consideration of the nesting of city economics within the macro-economy and social links at the household and other levels.
Participatory governance	Need for improved processes of democratic local government to ensure that it is more responsive to the needs and interests of its low-income and disadvantaged citizens.	Creating institutions of participatory governance to ensure that democracy becomes more pro-poor. This can be achieved through a diversity of strategies offering citizens and community organisations greater inclusion and influence	Participatory forums can be limited in their decision-making role. They may also not be inclusive or pro-poor. Participatory opportunities may also be dominated by non-poor groups, or include only some of those who were previously excluded.

Table 2.1 (continued)

Approach	*Primary concern that approach addresses*	*Theory of change*	*Major concerns*
		in political decision-making and state action. May extend to co-production.	Government policies can be influenced by elites whatever the intentions of politicians. Clarity between models of representative and participatory democracy needs to be in place.
Rights-based	Failure of state to treat all nationals as citizens with equal rights. Failure of state to meet its duties and obligations.	Extending rights and entitlements will protect low-income and disadvantaged groups and individuals. The emphasis on rights rather than needs reinforces a broader understanding of social justice.	Rights can be difficult to achieve by groups that have little power. Legal processes to claim rights can be complex and formal, and hence exclude low-income households. Rights-based approaches strengthen the power and legitimacy of the state, which may be more concerned with property rights than the urban poor.
Market	Lack of access of low-income individuals and households to private services that have to be paid for, to market opportunities required to provide needed incomes over the life cycle and for enterprise development.	Improved access to financial markets will enable scarce cash to be used better to address needs and generate further income. The market encourages improved access to a range of goods and services. Often anticipated that an emphasis on markets will provide livelihoods opportunities.	Not all people are able to enter the market, and/or withstand the competition. Market approaches favour those who are already relatively better-off. Market approaches may increase vulnerabilities for households unable to manage debt. Does very little to address adversity in difficult macro-economic conditions.
Social and urban movements	Without strong mass organisations and associated processes to represent their political interests, the urban poor will be disadvantaged and will be excluded from political decisions and infrastructure and service provision.	Strong and capable urban poor organisations will be able to develop effective strategies and realise them. This includes making alliances with each other, building relations with a range of professional organisations, and negotiating with the state.	Movements respond to both immediate and long-term difficulties that the urban poor face. However, movement activities may be short-lived and with a focus on making demands on the state, and so may not sustain the pressure needed for substantive change. Movements may also be manipulated or

Table 2.1 (continued)

Approach	*Primary concern that approach addresses*	*Theory of change*	*Major concerns*
			co-opted by political interests. Movements may not represent the interests of the lowest-income members.
Aided self/ help	The urban poor have to provide themselves with housing, basic services, infrastructure and land acquisition. They can do this better with the support of local government and access to financial services.	Individuals and groups make substantive investments in addressing their own needs for housing, infrastructure and basic services. With government provision of bulk supplies, trunk infrastructure, technical assistance, loans and with appropriate regulatory regimes, much more can be achieved.	Without external resources and subsidies, the lowest-income households may not be included in the solutions promoted by aided self-help. Local solutions for improved basic infrastructure and services are unlikely to be fully effective without access to trunk infrastructure networks which requires considerable state investment.
Clientelism plus	Lack of appreciation that clientelism provides an avenue for low-income disadvantaged citizens to access the state, albeit within vertical relationships that are often exploitative and which provide limited resources to some.	Clientelism does a little for more of the urban poor than many more formal interventions. Recognising the value of clientelist politics in assistance and access to land and basic services is important in developing pragmatic strategies for a more inclusive politics. *Generally not an active intervention.*	Glass half-full and glass half-empty debates may not recognise the insufficiency of the half-full glass. Hence while clientelism does something to help some access essential infrastructure and services, it does not result in adequate services or adequate coverage. These practices reinforce vertical relations of authority and may be associated with violence and fear. The lowest income groups may not benefit from the gains.

acknowledged that the annual reviews in which citizens consider what has been achieved as a result of the change that they made the previous year have been important in holding the state to account. However, in the case of market approaches, the 'invisible hand' that regulates markets in the interests of all as described by Adam Smith ([1776] 1982) has been recognised to relate only to a very specific set of circumstances and very often the larger companies and commercial interests are able to dominate trading, securing their own interests at the expense of customers and smaller companies.

In general, many of the approaches appear to reinforce social relations that are at best hierarchical and disempowering, and at worst patriarchal and violent. Such relations may help what those in power consider to be the 'deserving poor' but they may also demand a high level of psychological dependence as the price of assistance (see above). In some cases, this is an engagement between the individual and the state (e.g. the majority of cash transfers). In other cases, the authority is exercised over a group. Clientelism, for example, may impose control over the majority of people in the settlement, and the familial and exploitative relations between political and community leaders may be reproduced by relations between individual residents and the community leadership undermining collective support and associated activities.

The next chapter continues to explore poverty reduction efforts with a focus on the agencies that are active in southern towns and cities. Chapters 4 and 5 then explore a small number of interventions that have sought to change power relations in greater depth to learn from their strategies and experiences.

Notes

1 Donor agencies frequently ask for the theory of change to be elaborated in applications for development assistance.

2 See the Dublin Statement on Water and Sustainable Development: http://www.wmo.int/pages/prog/hwrp/documents/english/icwedece.html#p1.

3 Britto (2005) notes that Brazilian municipalities introduced conditional cash transfers in 1995 prior to the first nationwide programme in Mexico (*Progresa*) in 1997.

4 See SPARC (1985); in this instance, the census of pavement dwellers undertaken by the pavement dwellers with support from the Mumbai-based NGO SPARC produced maps and addresses that allowed them to negotiate for and obtain ration cards.

5 Baskin and Miji (2006, pp. 56–58) discuss the outcomes of bridge construction in Luanda. Before the bridge was constructed, there were ten enterprises along the main road. In two years of improved access, 44 new income-generating activities had appeared along the same road, producing an additional income of US$124,080 a year. That yearly gain was nearly equal to the total cost of the improvements.

6 See 'Towards participatory urban management in Latin American and Caribbean cities: a profile of the Urban Management Programme for Latin America and the Caribbean', an Institutional Profile in *Environment and Urbanization*, Vol 13, No 2, 2001, pp. 175–178.

7 There are interesting parallels here in the support of the police for new police stations in informal settlements in Mumbai, served and supported by elected citizen representatives; these were formed through an agreement between slum dweller federations and the police (see Roy *et al.* 2004).

8 The Shack Dwellers' Federation of Namibia's group in the town of Gobabis noted the importance of having meetings with their local authority within their own meeting room in the informal settlement and not in the town hall. More members attended and those who did attend found it easier to express their views in a familiar place (Muller and Mitlin 2007).

9 This section draws on Hickey and Mitlin (2009).

10 See http://www.fundacioncarvajal.org.co/sitio/ for details of its current work on this and in other areas, including housing, education and social development. The current mayor of Cali, Rodrigo Guerrero Velasco, who was also mayor of Cali between 1992 and 1994, was previously director of the Carvajal Foundation.

11 http://www.cohre.org/.

12 This Centre no longer seems to be active; in its website the most recent annual report available is for 2010 and the most recent press release is dated August 2011.

13 Community leaders may be included on lists of municipal employees but not have to work, even though they get the income. In Argentina, they come to collect their wages at the end of the month and are known as gnocchis, since there is a tradition of eating gnocchi at the end of the month when incomes are running low because it is a cheap meal (see Hardoy *et al.* 1990).

14 Kitschelt and Wilkinson (2007, p. 2) define clientelism as a particular form of party–voter linkage; it is a 'transaction, the direct exchange of a citizen's vote in return for direct payments or continuing access to employment, goods, and services'. We define it more broadly to include other forms of political support beyond voting, since this is more consistent with the literature on this topic.

3 The work of local, national and international agencies

This chapter analyses the work of official agencies active in urban poverty reduction from local urban governments and civil society organisations through regional/provincial/state governments to national governments and international agencies. In so doing it builds on the analysis of the eight conceptual approaches in Chapter 2. The discussion here recognises that in practice urban programming draws from across a number of approaches as agencies have sought effective interventions. Different sections highlight the evolution and activities of agencies working at the national and local government level as well as those of international agencies.

The analysis here combines with the discussion in Chapter 2 to provide a summary of learning to date: we explore both successes and failures, and highlight some of the critical experiences that have advanced our understanding of effective work in urban poverty reduction. The conclusions to the chapter highlight the need to build a political capability among the organisations of low-income and otherwise disadvantaged groups. Working together, citizens need to be able to manage electoral politics and volatile state support for urban poverty reduction programmes. These organisations also need to make sure that citizens are activity engaged in and 'own' the programmes while minimising the associated exclusion for the lowest-income residents, and the negative effects of the clientelism present in many citizen–state relations. If there is one almost universal failing in the programmes and interventions described in this chapter, it is the failure to engage low-income groups and their organisations sufficiently – both in prioritising what should be done and in actually doing it and supporting these organisations to do so.

Introduction

Chapter 2 outlined eight different conceptual approaches that underpin development efforts – four that that are state-directed, two that work within the status quo and then two further approaches – market-based and depending on social movement action. In practice, development actions, from government, civil society and/or development assistance organisations, are realised through agencies. In most aspects of poverty reduction, higher levels of government and international agencies are only as effective as the local organisations/institutions they support (and help fund). In practice it is unusual that a single one of the eight approaches is followed for reasons of ideology and the complexity of urban poverty (where there is rarely a single unambiguous cause). Rather, these approaches are taken up, and blended together through programmes and projects that attempt to address specific contexts and the associated needs and interests of the urban poor.

Further blending of approaches arises because, in the urban context, there are many institutions with overlapping agendas and responsibilities, and hence the design and realisation of urban programmes almost inevitably involves a high degree of negotiation between agencies. While small NGO projects may be able to take place with little interaction with other organisations, these are the exception in urban contexts. Indeed, the head of one country programme for a well-known international NGO explained their lack of attention to urban areas not by the lack of need but by the complications of having to work with different agencies within local governments and to get official approval for what needed to be done.

Hence, urban poverty reduction programmes and projects are generally realised through the combined efforts of different agencies that may include residents' associations and/or livelihood groups, local government officials and politicians, national government departments, and in some cases other professional civil society groups including NGOs and academic institutions. In many nations, state or provincial governments also have importance. There are often problems with overlapping responsibilities for particular investments or services within the area – and contestation over relative roles and responsibilities. In this chapter, we describe some of the more substantive directions and associated programmes followed by the agencies active in this area of work. This first introductory section introduces the types of agencies involved and offers a flavour of their work, prior to a more in-depth elaboration of their roles and activities.

The division of responsibilities between local, national and international agencies

Much of what is needed to reduce urban poverty lies within the regular activities and responsibilities of local (urban) governments and local offices of higher levels of government, although it is equally true that in most cases they fail to address these responsibilities at a substantive scale. A frequent assumption among those critical of the decline of state activities as a result of structural adjustment programmes is that services were functional prior to this. However, there is much evidence to suggest that this was not the case in most nations, and that national and local governments and international agencies have long neglected to address the needs of informal settlement residents (Mitlin and Satterthwaite 2013). Despite this neglect, some provision continued to be made. Standpipes in informal settlements, site and services programmes that provided some households with access to land, mother and child centres that provide daycare and help to improve nutrition, equipment loaned to communities to help lay water lines, communal waste bins, pathway improvements, public toilets, land fill for settlements on waterlogged sites – these are just some of the supports that might be offered.

Although the historic record of national governments in urban poverty reduction is not impressive, national government action is critical to effective poverty reduction. National governments are responsible for both economic policy (that influences prosperity and who prospers) and redistribution between citizens (which is meant to ensure that each individual has enough to meet basic needs). Despite the constraints of globalisation and long-standing vulnerabilities in a highly stratified global economy, the significance of the nation-state for development remains widely recognised. In recent decades, national government programmes on poverty reduction have concentrated on several particular lines of action. In an urban context, the most significant measures to address poverty and inequality are cash transfer programmes, national housing programmes using direct-demand subsidies, upgrading programmes in informal settlements, structural reforms promoting market systems of resource allocation, and the decentralisation of government functions along with the strengthening of forms of public accountability. One consistent theme is that the role of the public sector has been developed more to influence the choices of individuals and agencies, rather than predetermining outcomes, i.e. away from supply-side provision to support demand-led improvements. This is evident in both of the programme areas that we consider below, namely housing support and cash transfers.

However, in urban areas, most of the responsibilities for providing infrastructure and services and for ensuring that health and safety standards are met within homes and workplaces fall within the responsibilities of city or municipal governments. These organisations are also responsible for implementing many initiatives that contribute or should contribute to poverty reduction, although the form of these and the availability of external finance to support them is determined by higher levels of government. Table 3.1 lists the many forms of infrastructure and services, housing responsibilities and other responsibilities that influence service provision or poverty reduction that fall to urban (city and municipal) governments. Of course, the nature and extent of local government involvement in this long list varies greatly between nations. Responsibilities for provision are also often shared between different levels of government – or higher levels of government set policies – and define and manage what funding is available to address some of these. The extent to which local governments fulfil many of these responsibilities also varies greatly; what is often most notable is how few local governments fulfil most of these responsibilities – or indeed have the capacity to do so. But Table 3.1 is also a reminder of how much local government contributes to the provision of housing, infrastructure, services and other underpinnings of good living standards and healthy working conditions, if it meets its responsibilities.

With regard to housing, local governments usually have responsibility for the management of land within their boundaries including the application of

Table 3.1 The different local public infrastructure and services in which city/municipal governments have roles (as providers, supervisors or managers)

Infrastructure needed for service provision		*Role of local government*
Water supply	Piped water supplies and water abstraction and treatment; other water sources provided or supervised.	In many nations, local government as the provider of these. In some, as the supervisor of private provision.
Sanitation	Provision for sewers and other services relating to sanitation or liquid waste disposal.	In most nations a municipal or city authority responsibility.
Drainage	Provision for storm and surface drains.	In most nations a municipal or city authority responsibility.
Ports and airports		Often shared responsibilities between local and higher level of government.
Roads, bridges, pavements	Provision to link informal settlements to the formal city. Also important for peri-urban settlements.	Usually divided between local and higher levels of government (often on the basis of hierarchy of roads).
Solid waste disposal facilities	Landfills, incinerators, dumps.	These are usually the responsibility of local government.
Electricity supply	Street lighting, household connections.	Frequently private sector provision or national agency. Local government may have a role in expanding connections.
Parks, squares, plazas, other public spaces		Almost always local government responsibility for provision and maintenance.
Waste water treatment		Usually a local government responsibility although so often not fulfilled.
Services		
Fire brigade and fire protection services		Usually local government responsibility.
Public order/police/ delivery of early warning for disasters		Police usually a national or state government responsibility. Disasters also a national responsibility although local government may also play a role.
Solid waste collection for homes and businesses	May be household connections, or communal bins.	Local government responsibility, although may be contracted out.
Child care, schools, libraries		Some under local government, some under local offices of higher levels of government.

Table 3.1 (continued)

Infrastructure needed for service provision		*Role of local government*
Public transport – road, rail	State companies, regulation of private providers, investment in the infrastructure for trains.	Public road transport – usually under local government – although much provision contracted out. Suburban railways and metros may be under local authority.
Health care/public health	Provision from primary health care through different levels.	Primary health care services often under local governments; higher level services often under higher levels of government.
Environmental health		Usually a local government responsibility. May include licensing of certain enterprises and markets.
Occupational health and safety		Usually a local government responsibility.
Pollution control and framework for hazardous waste management		Usually with standards set by national governments and implementation by local governments.
Ambulance service		Usually a local government responsibility (although wealthier groups may contract for private provision).
Public toilets		Local government responsibility.
Social welfare (includes provision for childcare and old-age care)		Mostly national although local government offices may play roles in determining eligibility and providing payments.
Cleaning of streets, squares and other public spaces		Local government responsibility. May be passed to local communities at the neighbourhood level.
Disaster response	Range of measures from disaster preparedness to response.	Much responsibility for this within local government although often not addressed.
Registration of births and deaths		Usually a local government responsibility.
Responsibilities for housing		
Public provision and/or maintenance of housing		Often a local government responsibility – including rent collection.
Public provision of access to housing		Depends on which level of government was responsible for housing provision. Local government frequently maintains lists of those in housing need but may not be responsible for all supply.
Regulations for rental housing		

Table 3.1 (continued)

Infrastructure needed for service provision	*Role of local government*
Other local government responsibilities that influence service provision or poverty reduction	
Urban planning	Local government responsibility. Should play major role in defining infrastructure provision for expanding urban area.
Building regulations	Local government responsibility for enforcement; most regulations from higher levels of government.
Land-use controls	Local government responsibility for enforcement; some regulations set by higher levels of government.
Site clearance for infrastructure and resettlement	Usually local government responsibility.
Raising of local revenues	Local government responsibility.
Provisions for public employees	Local government with responsibilities defined by higher levels of government
Provisions for disabled persons	Local government with responsibilities defined by higher level of government.

land-use/zoning and building regulations. In many cases, they are responsible for the utilities that are meant to service all residential (and other) buildings; in other cities, some provision for these services falls to specialist agencies – for instance, for water, sanitation and electricity – although often with supervision falling to local governments. Municipalities can influence access to secure and adequate housing (or not) through their willingness and ability to identify and service land suitable for settlement, the appropriateness of local regulations and their enforcement, and their willingness and ability to upgrade informal neighbourhoods that have developed outside the regulatory system. The extent to which they are interested in collaborating with the urban poor through co-productive strategies to improve local facilities influences the costs as well as the scale, multiplicity and nature of improvements. Local governments influence employment opportunities through their strategies for economic development and through the regulation of trading and management of market areas (including informal markets). Either through their own jurisdiction or through collaboration with other state agencies, they influence the transport network (and public transport) and the ability of the local labour force to reach industrial or business areas offering employment. In most urban centres, they also influence the quality and accessibility of essential health and education services, and in some are the key providers of these services. In some countries, regional government and/or specialist development agencies may also be important in allocating resources, raising funds and undertaking direct activities. The discussion of local government below focuses on three areas of their work. The first is the provision of serviced sites and the upgrading of informal settlements, the second is the provision of basic services, and the third is participatory governance.

After this, the contribution of development assistance agencies to urban poverty reduction needs consideration, even if this is to point to the very limited support given to this by many such agencies. Development assistance includes loans that have to be repaid in full (often termed non-concessional loans), concessional loans (that have a grant element) and grants. Some development assistance agencies have sought to address aspects of urban poverty through activities such as upgrading, site and services, housing support, basic services, cash transfers and micro-finance. Some include support for greater private sector roles in infrastructure and service provision. There have also been measures to increase accountabilities of service providers, and citizen consumers or 'clients'. In the last decade, some agencies have also promoted an urban-focused pro-economic growth agenda. While it is possible to point to many specific projects or initiatives that contribute to reduce one or more aspect of urban poverty, what is also evident among the long list of multilateral and bilateral agencies is the lack of any coherent urban policy focused on poverty reduction. The official multilateral and bilateral agencies inevitably work through national governments and so often have little knowledge of or contact with urban governments. They also have little capacity to support the kinds of local government reforms and processes that bring real benefits to low-income urban dwellers. In recent years, many have supported cash transfers and at least these benefit a proportion of low-income urban dwellers. However, they do nothing to address the lack of infrastructure and services that penalise low-income households living in dense areas, nor do they support greater voice and influence for low-income groups.

Following the discussion of official development assistance, we turn to the contribution of NGOs and other civil society organisations. We discuss the evolution of NGO strategies in recent decades and their use of both rights-based strategies and new innovative approaches to components of urban development. Over time, there has been an increasing emphasis on alliances between professional agencies and citizen groups. This may be understood as a response to the lack of political influence that NGOs alone have been able to secure and hence the limited scale of much of their work. At the same time as NGOs have put an increased focus on alliance building, there has been greater recognition of the role of a range of other civil society organisations, including organised grassroot groups, by many other agencies working in development, and our discussion reflects this.

The conclusions to the chapter discuss the problems that have emerged as agencies design and realise their programmes. The discussion highlights the need to support greater political capability among the organisations of low-income and otherwise disadvantaged groups to work and organise together. Challenges include the need to manage electoral politics and bureaucratic state processes within urban poverty reduction programmes. To achieve an improved set of development options, residents need to find ways to work with the state – in co-productive approaches – and in most cases this will include the need to blend their own investment finance with the monies of the state. Core challenges are to ensure that the state collaborates with the organised urban poor, and that the lowest-income and most disadvantaged residents are included in the options that emerge. Residents have to manage considerable political complexity, including minimising the negative effects of the clientelism present in citizen–state relations.

National governments

National government policies impact on the lives of the urban poor in many ways. We include here both on policies that have a specifically urban component, and those that are more general in their impact. We discuss national governments' shelter policies and the growing importance of cash transfer programmes. We use the term 'shelter' to include land and housing tenure, basic services and infrastructure as well as the dwelling. We have chosen this focus because, as summarised in Chapter 1, shelter is a particular challenge in urban areas in which land is a valuable commodity and high densities and large population concentrations mean that access to infrastructure is critical for health and well-being. The capability and commitment of the state in the financing and other aspects of making available infrastructure and services, serviced land and housing development has a major influence on the possibilities for low-income families to be adequately housed. Equally, the measures the state takes to support those on low incomes are also important in addressing immediate needs and in enabling households to move forward from crises and secure their development goals.

Housing policies and programmes

There has been a long-standing recognition of the need to intervene in shelter provision in urban areas.[1] The rationales for intervention include the necessity to provide bulk infrastructure to enable basic needs to be met in a context of high population densities and the high costs of housing, particularly formal housing, and hence the requirement to raise a significant amount of capital at one time with repayment over a longer period. In a modern urban context, housing is (at least in theory) 'planned' with regulatory interventions both in location (for example, zoning of land uses and requirements for obtaining government approval for new developments) and in housing standards (for example, the size of the unit relative to the stand or plot, the minimum size of the stand, infrastructure and building standards). However, the inability of both local and national government to plan adequately for urban growth (and growth in households) has underpinned the expansion of informal settlements. If the formal market – for land, for housing, for services – does not work for a significant section of the population, informal markets develop to serve their needs, however inadequate these are in regard to tenure, housing size and quality and infrastructure and service provision. In some cases, governments have responded to this situation with housing programmes. A further rationale for shelter interventions in both formal and informal settlements is health risks from high population densities if good-quality infrastructure and services are lacking, both for the immediate population and for others who may also be at risk.

Government programmes to address housing needs have been varied, reflecting both the multiple rationales for such interventions and the experiences with different solutions that have been attempted. In Africa and Asia during the 1950s and 1960s, newly independent states built formal housing, often through institutions that had previously built these for the colonial government key workers (Hardoy and Satterthwaite 1981). With high unit costs and with little cost recovery – whether from rent or from those that got tenure of the units allocated – this solution was insignificant due to an inability to increase its scale. If the units built required large subsidies, relatively few got built. Even when the units built had smaller subsidies so that available government funding could build more units, these units could not be afforded by low-income groups (see Hardoy and Satterthwaite 1981). It was also common for the (often highly) subsidised units to be allocated to non-poor groups (for instance, civil servants, the military or police officers) or to end up with non-poor groups (as the households allotted the units sold these or rented them out). Many governments were still engaged in large public housing programmes or housing finance systems that supported such housing in the mid-1970s when the UN Conference on Human Settlements recommended changes in approach – towards upgrading and serviced site schemes. There are also many examples from the late 1970s and early 1980s of national governments still funding public housing programmes, despite the evidence of their high cost and limited impacts (Hardoy and Satterthwaite 1989).[2]

There are also recent examples of this. For example, although 27 per cent of the population in Niamey (Niger's capital city) are below the poverty line,[3] the Poverty Reduction Strategy Paper for Niger (2008, p. 43) reports on formal housing construction:

> [H]ouses have been constructed by the Government and its related services in occasional operations in urban areas; from 1960 to date, the total number of houses is below 1,500, in addition to the construction of 551 housing units in Niamey during the 5th Francophone Games in December 2005.

However, the Strategy also recognises the lack of functionality of formal housing finance in terms of reaching the urban poor: 'The housing loan granted by Credit du Niger (CDN), which is the sole financing institution in this sector since 1966 is low and targets only wage earners' (ibid., p. 43).

In India, while the official discourse on urban poverty reduction and housing has shifted significantly to 'upgrading' and 'community-driven' solutions, much of the government provision for housing is still funding private companies to build small apartments to 'rehabilitate' slum dwellers (Patel 2013). Here the same issues that dogged the public housing programmes of earlier decades are still evident. These include the building of these units on land sites far from where low-income groups want and need to be in relation to income-earning opportunities and services – although the Government of India's Basic Services for the Urban Poor programme does allow for 'in situ' redevelopment of informal settlements. Other problems are those of poor-quality construction, very small units, poorly designed buildings and lack of provision for maintenance. Although this programme was meant to support in situ upgrading and participation, the projects it supports on the ground have often involved wholesale clearance of the site (that may include the destruction of good-quality homes), little or no provision for temporary accommodation while the site is redeveloped – and only a proportion of those cleared from the site gaining (or being able to afford) access to the newly built units. A similar situation may be seen in the re-development of informal settlements in South Africa where shacks are replaced by dwellings financed by the capital housing subsidy (see below). Perhaps it is only when we consider the profits made by the companies that get the tenders to build these (and the 'backhanders' to local government officials) that we understand how an approach that has been shown to be so ineffective continues to be supported.

In most cases governments have realised that such approaches based on addressing housing needs through construction are unlikely to address the housing problem on a sufficient scale and have looked for alternative strategies. Many governments, particularly in Asia and Latin America, have sought to extend mortgage finance to enable households to access long-term loans (UN-Habitat 2005). However, this approach offers little to most low-income households because it is typically available only for completed legal dwellings that are unaffordable to the urban poor. The Kenyan Banking and Building Societies Act, for example, explicitly forbids financial institutions making loans for plots of land with no or only partly constructed housing on them (Malhotra 2003, p. 225). Equally constraining is the loan requirement that employment is formal, with the loan repayment being deducted from the salary (Buckley and Kalarickal 2004; Calderón 2004). While incomes have increased for some in nations in Latin America and Asia, in most cases less than half the population can afford formal loans for conventional housing. In the Philippines, for example, this falls to less than 30 per cent, although a government scheme offering mortgages at subsidised interest rates helps low-income formal sector workers such as nurses and police officers access small dwellings (Llanto 2007). Even among more affluent countries such as Mexico and Colombia, this proportion only rises to 60 per cent (UN-Habitat 2005). Further problems are those mentioned in Chapter 2 and include the reluctance of formal financial institutions to address the needs of the urban poor as shown by their choice of location for banking outlets and minimum deposit requirements for a bank account.

As the population of many urban centres grew rapidly with a high proportion of households unable to afford formal housing, so an increasing number (and often proportion) of the urban population came to be housed in informal settlements. These included many who illegally occupied land belonging to government agencies or institutions and private landowners, although it was common for such informal settlers not to settle on the best land sites because they knew they had little chance of avoiding eviction. Sometimes they invaded land sites that they knew they would be evicted from as a strategy to engage with the government and negotiate for another site (see Peattie 1990).

The eviction of low-income groups from urban land sites that more powerful groups want (including national and local governments) may be as old as the cities themselves, and may be inherent to the competition by different groups to secure access to the best locations. Certainly large-scale evictions of 'slums' have been common throughout the nineteenth and early twentieth century in cities all around the world. The documentation of evictions in the development literature goes back to the 1960s (see Abrams 1964; Marris 1979; Portes 1979) as do the recommendations for alternative approaches (Mangin 1967; Turner 1968) and continues up until the present (see e.g. Bhan 2009). However, the lack of affordability, lack of state investments and continuing need for low-paid labour has helped shift the official perspective on informal settlements. In some nations, especially in Latin America, the return to democracy and strong civil society pressure has also helped to change policies, programmes and attitudes. In some informal settlements where the residents have paid the landowner for their plots, there is less pressure for evictions – even as these are still illegal in regard to zoning, building and planning regulations.

As the need for far more low-cost accommodation was increasingly recognised from the 1960s, sites and services programmes, along with slum and squatter upgrading programmes, developed as alternative interventions. These have been used by development agencies at all levels (including national government, official development assistance agencies and local government). Such programmes were introduced in Chapter 2 within our discussion of the aided self-help approach to urban poverty reduction.

For national governments, sites and services programmes are attractive because they are significantly cheaper than public housing programmes, as no house is built. Legal tenure is provided for a residential plot together with some infrastructure (for instance, a piped water supply to the plot) and this can greatly reduce or may even remove the need for subsidies. They provide a partial response to housing need but one that is potentially valuable for low-income households if they can get a serviced site on which to build at a price they can afford. Unit costs may be brought down by keeping plot sizes small, providing only limited access to services and making an initial investment in infrastructure. The expectation is that households will build and over time improve their homes, and that infrastructure and services will be upgraded. What distinguishes sites and services from the provision of serviced sites for higher-income families is minimal infrastructure, official acceptance of delayed or incremental housing construction and building standards being modified to ensure affordability for the lowest income households. Sites and services may also have smaller plot sizes. While mortgage finance in many cases seeks to assist households to enter the commercial lending market (some such finance is subsidised but this is not always the case), sites and services owe more to the ways in which communities help themselves to address their housing needs. The popularity of the approach has declined as some problems became evident, including the lack of serviced sites for housing for other income groups and hence the capturing of the benefit by higher-income groups. Their costs were often too high for low-income households, especially where no long-term finance was available. Many serviced site schemes were also in areas where land was cheap because it was far from income-earning opportunities and services – just like many public housing programmes. Serviced site schemes were also viewed less favourably by politicians as the photos of them handing over serviced sites look less impressive than handing over public housing units (Simelane 2012). Most of these schemes also faced a continued need for a subsidy (Rakodi 2006b; see also Manda (2007) for an example from Malawi). However, as discussed in the section on local government, there are examples of serviced site schemes that benefited low-income groups that were designed and implemented by local governments.

Rather than simply focusing on one solution, some national governments also recognised the need to develop programmes able to provide multiple types of housing support at scale. One significant example was the Mexican National Popular Housing Fund (FONHAPO) which was created in 1981 to improve housing through providing a range of grants and loans for low-income households (Connolly 2004). Between 1982 and 1998, FONHAPO received 4 per cent of the public housing monies and provided 23 per cent of all housing financed through these funds. The programme targeted people earning less than 2.5 times the minimum wage[4] and offered sequential loans to agencies (including state organisations, local government, civil society organisations, including cooperatives) for different phases of housing development from preparation through to self-build housing construction on serviced plots. The types of project supported included sites and services, incremental housing, home improvements, finished dwellings, and the construction and distribution of building materials. A further significant national government contribution has been the Million Houses Programme in Sri Lanka whose work is described in Box 3.1.

Box 3.1 Housing innovations in Sri Lanka

In Sri Lanka, the implementation of the Hundred Thousand Houses Programme (1978 to 1983), the Million Houses Programme (1984 to 1989) and the 1.5 Million Houses Programme (1990 to 1994) helped low-income households access affordable houses, with around 60 to 70 per cent of residents in under-serviced settlements in Colombo benefiting in terms of tenure rights and investments in shelter improvement. The experiences built on the earlier UNICEF-funded Urban Basic Services Improvement Programme (1978 to 1986) that was implemented in Colombo.

The Housing Minister (who was also the Prime Minister), Ranasinghe Premadasa, introduced the Hundred Thousand Houses Programme in 1978. Political support, public resources and a new legal framework resulted in 50,000 houses built in rural areas through aided self-help; 30,000 in urban areas through direct construction by the private sector; and 35,000 through informal settlement upgrading in Colombo. A programme assessment recognised that progress was limited owing to high construction costs, high standards, poor-quality commercial construction and a lack of finance. Nevertheless, the merit in the design of the programme was recognised and a further phase, the Million Houses Programme, was introduced in 1984. This phase had a greater level of decentralisation and included new house construction, the upgrading of existing houses, and sites and services projects in both urban and rural areas. The realisation of the programme in urban areas included 12 Municipal Councils and 39 Urban Councils with provision for sites and services projects for residents who had to be relocated, settlement regularisation and on-site upgrading, basic services provision in relocation areas and support for individual housing improvements and/repairs. Loan finance was available for housing improvements; for most loans the interest rate was 6 per cent but the interest rate rose to 10 per cent for higher-value loans.

The implementation of the programme was assisted by a training and information component for both central government and local government staff. Resident participation was enhanced through the Community Development Councils (CDCs) that were established at the neighbourhood level to provide a community forum for the monthly monitoring of progress. Community action planning was developed to enable local residents to lead the process through designing their own layouts, formulating community-specific building codes, introducing the loan programme, familiarising the community with procedures for their involvement in minor infrastructure works, and initiating group credit programmes for income-generating activities. In 1986, a contract was awarded to a group of local residents after a community expressed its dissatisfaction with the quality of construction of a well built by a contractor through the conventional tendering process. This contract developed into a new system offering opportunities for local residents to be involved in construction through contracts with their Community Development Councils.

When the Million Houses Programme ended in 1989, 258,762 families had been reached in rural areas and 38,125 families in urban areas. It was followed in 1990 by the 1.5 Million Houses Programme, which reached 57 per cent of the target (nearly 859,000 families) by 1994. Then, after being in power for 17 years, the United National Party lost the national elections, and the policy and programmes were abandoned. The government once again took on the role of a housing provider, and local and national government ended this intervention in informal settlements.

Source: Joshi and Khan (2010)

The strengths and weaknesses of these housing interventions in Sri Lanka have been analysed, and housing activists in Asia have recognised that the programme depended too much on the commitment of particular politicians. Over time, the political interest in the programme dwindled in part because it was associated with a specific political party. Moreover, Russell and Vidler (2000) discuss the experiences of the Community Development Councils and their difficulty in managing relations with local politicians given the predominance of clientelist politics and their lack of financial autonomy. Community leaders now recognise that they need to develop their own autonomous process – not to go it alone but to engage government more effectively (ACHR 2011b, p. 36). Although the Community Development Councils had an effective role in implementation, many collapsed or did not function thereafter because their participants relied on the government. A survey of 1,614 settlements in Colombo in 1999 revealed that of the 625 registered Councils, only 126 were functioning effectively and 100 were functioning irregularly (Sevanatha 1999, quoted in Joshi and Khan 2010, p. 310). Nevertheless, this experience has been important in the development of further strategies by Asian civil society and governments seeking to address shelter need.

In the Philippines, a new alternative for those living in informal settlements emerged with democratisation. The Community Mortgage Program was launched in 1986 and drew directly on NGO experimentation with collective loans for low-income households to improve housing conditions (Porio *et al.* 2004). The Program (which still continues) provides low-interest loans that allow informal settlers at risk of eviction to acquire an undivided tract of land to be purchased through a community mortgage. Projects may be either 'off-site' or 'on-site' with on-site projects allowing the settlers to formalise their claim to the land they occupy already by buying it from the owner, and off-site projects securing finance for relocation. Loans may be used for infrastructure and housing development but this is relatively rare and high land prices mean that most of the loan is required for land purchase. With title, communities then negotiate with local government officials and politicians (generally at the lowest level or *baranguay*) for incremental improvements to services.

In 1992 interventions within Asia took a further step forward with the establishment of the Urban Community Development Office in Thailand which was set up to address the remaining problems of urban poverty despite considerable economic growth taking place in the country (Boonyabancha 2005). The work of this Office and its successor, the Community Organization Development Institute, is described in Chapter 4. Here, existing experiences with savings and loan finance were used to design an approach with neighbourhood savings schemes, low-interest loans and, as the programme developed, government infrastructure subsidies. Support is provided to networks of community organisations formed by the urban poor, to allow them to work with municipal authorities and other local actors and with national agencies on urban city-wide upgrading programmes.

A further (more recent) strategy to secure improvements in housing conditions has been the 'direct-demand' subsidies originating in Latin America which offer combined financing of subsidies, savings and loans for the purchase of completed dwellings. Although originally conceived in Chile as a strategy that would enable houses to be bought from existing providers (Gilbert 2002a, p. 309), in practice the private sector has been reluctant to lend and to build, particularly in the case of the lowest-income groups (ibid., p. 315). Hence, this has in part developed as a financing route for the private sector to supply to the government for groups of households that are recognised to have an entitlement within the programme. As a result, the programme has provided high levels of state support for construction, and between 1980 and 2002 average annual construction part-financed by state subsidies was between 50,000 to 80,000 units (Arrieta 1999; Posner 2012; Tironi 2009). Tironi (2009, p. 976) summarises the impact of the programme thus: 'In Santiago, where 40% of Chile's population is concentrated and 21% percent of the population lived in slums by 1987[...] 67% of all low-income households today live in subsidized housing units built between 1980 and 2002.' There are a number of sub-programmes in Chile which blend three financing components: beneficiaries' savings, government subsidy and (except for the lowest-income households) loans – the cheaper the housing, the higher the proportion of its costs provided by the subsidy. All sub-programmes require the families (even those with very low incomes) to make a savings contribution to reduce dependency on the government, and increase a sense of ownership (Gilbert 2002a, p. 310). Most sub-programmes have included a loan but there are an increasing number of subsidies given without loans to reach those on very low incomes and this has resulted in a shift in beneficiaries in favour of low-income groups (Posner 2012).[5] People apply for assistance through the regional office of the Ministry of Housing or through the local government. The process of applicant selection has clear rules, common assessment according to defined criteria and transparent procedures (results are published in a local and/or a national newspaper).

In Chile, the programme was introduced in response to political pressure. At the time Chile had a military dictatorship but even this government required a level of popular support. Gilbert (2002a, p. 310) argues that the Chilean government chose to put aside at least part of its neo-liberal agenda to provide housing subsidies. Posner (2012) in a more recent analysis suggests that this was a coherent strategy of the government to shift welfare systems towards programmes that individualise relations and reduce solidarity between low-income groups, and hence reduce the possibility of cross-class alliances with those better able to negotiate for higher subsidises. Whatever the longer term political implications, the use of direct demand subsidies has proved popular and variations of this have been replicated, particularly in Latin America. Such subsidies were introduced in Costa Rica in 1986 and almost immediately replicated in Colombia, El Salvador, Paraguay and Uruguay in 1991 (Mayo 1999, p. 36). These programmes are not exact replications of each other and in some countries there has been a willingness to offer support for incremental housing. For example, the subsidy, savings and loan programme in Mexico (*Tu Casa*) includes an option for incremental improvements, though it should be noted that 92 per cent of the programme's funds (about 60 per cent of its loans) are earmarked for buying completed, albeit minimal, housing.

Despite the emphasis on savings and loans within this direct-demand model, in South Africa, the African National Congress (ANC) government that took up office in 1994 favoured a complete capital subsidy. The provision of housing was declared to be a priority with a target of one million dwellings within five years. While the focus on housing reflected political priorities and social needs, the specific strategy of a capital subsidy for addressing housing need emerged from the business representatives and consultants who dominated the multi-stakeholder National Housing Forum between 1992 and 1994 (Baumann 2003; Gilbert 2004; Huchzermeyer 2003). The capital subsidy provided money for land purchase, infrastructure and housing development, and appeared to offer the government a win-win-win option, simultaneously addressing the needs of low-income households without adequate housing, providing reassurance to a struggling construction sector, and catalysing a lead sector for economic regeneration. By 2008, 2.8 million subsidised housing units had been provided or were being constructed.[6] However, neither the scale nor subsequent modifications to the programme resulted in unambiguous success. A number of concerns have been raised and minor modifications made. For those who have secured a subsidy financed house, there are notable shortcomings. The quality of housing constructed in the early years of the programme remains problematic (Bradlow *et al.* 2011, p. 269). Dependence on construction companies has resulted in adverse locations with increasing spatial disadvantage (Oldfield 2004; Pieterse 2006). Regardless of the policy and intent of these providers, participation is superficial due in part to the lack of interest by providers, and the lack of organisation among many households (Miraftab 2003). Efforts to introduce more community-led approaches have proved difficult to replicate at scale (Bradlow *et al.* 2011, p. 270). As the state responds, with increasing frustration at its lack of success in housing provision, by restricting the supply of informal solutions through controlling informal settlements, the continuing need for housing combined with a lack of alternatives simply means the growth of informal rental solutions in formal housing areas (SAIRR 2008). These households are significantly disadvantaged: not only do they face insecurity and continuing rental payments, they also do not benefit from the subsidies offered by the state to improve access to water and electricity. Need was and has remained considerable. By 2007, the housing backlog was bigger in absolute terms than in 1994 (Baumann 2007). One problem is a lack of affordability; in 1996, 80 per cent of South Africans were eligible for the housing subsidy as they earned R3,500 or less a month. By 2000, this had grown to 85.4 per cent of the population (Department of Housing 2003, p. 9).

Contribution to poverty reduction

In summary, despite being the preferred solution by governments wanting to show their capacity to modernise, public provision of formal housing and of finance to allow low-income groups to obtain housing are too expensive to address the scale of need. Alternative solutions are required. To find ways to resource the gap between what is needed for good-quality housing and what can be afforded, national governments draw both on the experiences with aided self-help (to reduce costs and unit subsidies) and the market-based approach to development (so households have to contribute more), particularly the emphasis on more accessible financial markets. However, substantive problems of affordability still remain. As argued by Porio *et al.* (2004) for the Philippines, the lowest income groups are not likely to avail themselves even of the Community Mortgage Program because of the costs involved and their more immediate basic needs like food, health and education. Similar conclusions emerge from UCDO's work in Thailand (Boonyabancha 2004). Even with loans available with an interest rate of 3 per cent, the land development process was unaffordable. However, despite a lack of affordability for the lowest-income households, efforts to 'downmarket' mortgage finance and provide new housing opportunities for lower-middle-income households have been very important for improving housing options in some countries, particularly those that have achieved significant economic growth.

Larger scale programmes to address the needs of the lowest income households either through programmes that enable the upgrading of informal settlements and through the sites and services programmes have been an important contribution. In addition to the programmes discussed above, further examples from Central America are provided by Stein and Vance (2007).

As described above, in Chile and some other countries, subsidised housing provision has made a significant impact, but the experiences suggest that there are difficulties. A major challenge has been maintaining the government's commitment to programme financing, and programmes do not outlast the particular administrations that support them, as Connolly (2004) illustrates for the case of Mexico. In some cases innovative housing programmes shift to being increasingly clientelist, reinforcing vertical relations between citizens and the state, and maintaining a high level of dependency as resources are delivered to those that are prepared to support an inequitable political system (see e.g. Valença's analysis of Brazil; Valença 2007; Valença and Bonates 2010). Connolly (2004) also explains how FONHAPO became used to reinforcing clientelist relations in Mexico. Some other consistent lessons emerge. Relocation has costs for the urban poor, including the disruption of social networks and associated support systems and often livelihoods, and hence in situ developments are usually preferred if the benefits for poverty reduction are to be maximised. The large financing allocations for government housing programmes are attractive for the formal construction sector which gets contracts to build the units, but with weak government oversight construction quality is poor and in many cases such housing is badly located with infrastructure costs passed onto local government (see Rodríguez and Sugranyes (2007) and Posner (2012) for a discussion of such problems in Chile). Posner also argues that the design of the Chilean programme has exacerbated competition between low-income residents and higher groups and within low-income communities, resulting in reduced political pressure for reform (ibid., p. 67). At the same time, some of these processes have sought to engage local citizens in a meaningful way in project design and implementation. But while national governments may be keen to support greater participation at the local level, these commitments often falter as these programmes are rolled out because they do not fit easily within the profit-making orientation of the project managers when construction is contracted to the private sector (see Miraftab (2003) and Oldfield (2008) for a discussion of this problem in South Africa).

As a result of such shortcomings the scale of need continues to be significant, even in nations where major commitments have been made to improve housing. Needs are exacerbated in a context in which, as described in Mitlin and Satterthwaite (2013), a significant proportion of the urban population (or any city's population) have very low incomes. Housing problems are further exacerbated in many cities with rapidly growing economies as there is pressure for the lowest-income households to be evicted from inner city areas due to both land speculation and public investments in infrastructure that are more likely to address the needs of middle- and higher-income households than those of the urban poor.

Cash transfers[7]

As noted in Chapter 2, the importance of governments (and others) in assisting citizens in need through a number of measures broadly grouped under the heading of social protection came to be recognised during the 1990s. In one sense, this was peculiar in that this was also a decade where emphasis continued to be placed on market solutions and an increasing role for the private sector. But perhaps the scale of need, some early successes and the very modest level of support these required per person reached overcame this. This was also a decade in which some of the weaknesses with respect to the market approach were becoming more evident as financial and economic crises resulted in growing numbers living in poverty – including in Latin America where there was continuing pressure for democratic and accountable government and from where these new approaches emerged.

During the 1990s, there is some evidence of a widening of approaches to poverty reduction. Governments' support for housing, for example, has been viewed in the context of support to enhance household assets as a way to reduce risk and vulnerability (see Lemanski (2011) for South Africa). However, the main effort in this regard has been social assistance (to address immediate problems related to low incomes where there is no alternative means of support) and social insurance (to enable households to prepare for old age, periods of unemployment or ill-health and an associated inability to earn incomes). Most recently, as noted in Chapter 2, cash transfer programmes from governments to low-income households have been expanded to address acute needs and to provide the basis on which to secure improved livelihoods.

By 2010, new forms of cash transfers have been estimated to reach 150 million households worldwide (Barrientos and Hulme 2008a, p. 3) or 850 million people in 173 million households across the global South (Niño-Zarazúa 2010, p. 10);

and the literature focusing on poverty reduction in general suggests that they are now the principal instrument used by governments. A significant number of those reached come within several large programmes, including *Oportunidades* (Mexico) – five million households, *Bolsa Família* (Brazil) – 12 million households, the Minimum Living Standards Scheme (China) – 22.4 million households, and Indonesia's Safety Net Scheme – 15 million households (Barrientos *et al.* 2010; Köhler *et al.* 2009; Niño-Zarazúa 2010). This figure includes all cash transfers such as pensions, child support, workfare, and payments from unconditional and conditional cash transfer programmes. Their significance varies considerably – in Latin America, transfer reach ranges from 6 per cent in Brazil to nearly 20 per cent in Mexico (Grosh *et al.*, quoted in Niño-Zarazúa 2010, p. 14). In some countries amounts are just a few dollars a month and it is difficult to see that these funds will have a substantive impact on household poverty (Barrientos *et al.* 2010). In terms of understanding their impact on poverty, it is also important to distinguish between transfers such as pensions and child support that are designed to be provided at particular points in the life cycle and at times of acute need due, for example, to unemployment (social insurance) rather than funds provided to households in need of help to move out of poverty (social assistance). Some of these programmes are associated with conditionalities, introduced by governments to improve the effectiveness of such transfers in tackling intergenerational poverty and improve human capabilities: these programmes seek to provide for long-term poverty reduction and growth. Both Mexican and Brazilian governments conceptualise this approach as realising citizens' rights to education and health services rather than paternalistic conditions to change behaviour (Britto 2005; de la Brière and Rawlings 2006).

A distinctive feature of these cash transfers is that they are clearly targeted at groups that are identified as being in need and then at individuals within these groups. There are several approaches to targeting, including income-based means testing using varying levels of sophistication to locate 'the most deserving' citizens, geographical targeting to identify concentrated areas of poverty, community-based targeting where the most needy are identified by their peers, or categorical targeting which focuses measures of need and vulnerability as defined by the state (for example, households with children under 5 years old) (DFID 2005). The most popular targeting approach is a combination of geographic targeting to identify priority districts, and then some form of means testing of citizens within the selected geographical area to identify those to be included in the programme. An example of this in practice is *Progresa* (later renamed *Oportunidades*) in Mexico in which, first, a marginality index for each locality is created based on seven variables for which data are available from censuses (Niño-Zarazúa 2010). Localities with a high marginality rating are selected for the programme, and then a second stage selection process uses a household census in the locality to determine household income.

In general, once a group of households has been selected, the individual targets within households are primarily women and school-age children, with other dependants in the household seen as secondary beneficiaries. The rationale behind targeting women and children is that cash transfers are intended to achieve holistic long-term development aims of poverty reduction and social inclusion. The transfer is paid directly to women as it is recognised that women are more likely to invest in their children. Female schoolchildren may also be targeted to redress the gender balance in education – and the assumption is that educational achievement will be linked to improved future opportunities. In such cases keeping the cash transfer depends on meeting a range of conditions related to school attendance (de la Brière and Rawlings 2006).

Mexico's *Progresa (Oportunidades)* programme has been seen as 'revolutionary' in its multidimensional approach to poverty reduction focusing on human capital development through conditions related to school attendance and health checks, and whose design sought to challenge the previous emphasis on food aid (Niño-Zarazúa 2010). This linking of cash transfers to the education and health sectors runs throughout all Latin American programmes to a lesser or greater extent and has been replicated in transfer programmes in Asia and Africa, such as those in Bangladesh and Pakistan (Köhler *et al.* 2009), and Malawi, Kenya and Ghana (Nino-Zarazúa *et al.* 2010).

One of the recognised tensions in these programmes is between large-scale transfer programmes managed by national government ministries, and programmes which have a more decentralised approach. In general the trend is towards centralisation. For example, in Brazil, the *Bolsa Família* programme's management responsibilities have been modified from its predecessor *Bolsa Escola* in which a greater number of responsibilities were decentralised. The centralised model is exemplified by Mexico's *Oportunidades* programme which is administrated by a federal coordinating agency within the Ministry of Social Development. This centralised structure was designed in response to clientelistic behaviour by local authorities and civil society organisations involved in previous programmes (Levy and Rodríguez 2004). However, it does not entirely avoid the problems of clientelism which may also occur in nationally led programmes. In both Chile and Mexico, it appears that concerns about clientelistic behaviour within public authorities and civil society have resulted in programme designs that avoid citizen participation (Teichman 2008). However, despite this and other difficulties, there has been a consistent interest in citizen involvement in both middle- and low-income countries.

The expansion of transfer programmes into urban areas has taken place in recent years (Johanssen *et al.* 2009). Many of these programmes were developed for rural areas. One factor in Brazil and Mexico in expanding programmes to urban areas is the hope that increasing levels of urban violence can be curbed through tackling poverty and inequality (Britto 2005). In China the introduction of the Minimum Living Standards transfer programme in urban areas reflected political support to address a lack of social cohesion in urban areas (DFID 2005). Box 3.2 elaborates on the target procedures followed by two of these programmes. An Inter-American Development Bank study examining rural transfer programmes that were expanded into urban areas suggested that achievements are lower in urban areas (Johanssen *et al.* 2009), highlighting the need for redesign. Key issues in the transfer of such programmes include:

- Geographic targeting and income-based means testing using household data is difficult due to migration, intra-city mobility and lack of information about the residents of informal settlements.
- Communicating details of the programme may be difficult: for example, in Mexico only 40 per cent of eligible households applied for the programme in the two-month window of registration, with one-third of eligible households unaware of the programme (Behrman *et al.* 2011).
- The cost of living is higher in urban areas; therefore the transfer amounts delivered to rural beneficiaries are not sufficient (Johanssen *et al.* 2009). The opportunity costs of children being in school rather than working are also higher and therefore greater compensation may be needed.
- There are also administrative challenges. In Mexico, 8 per cent of urban beneficiaries drop out of the *Oportunidades* programme, and an estimated 25 per cent of these are due to administrative as opposed to behavioural reasons (Heracleous *et al.* 2010).
- Urban beneficiaries appear to be less likely to comply with conditionalities than rural beneficiaries. For example, in urban Mexico, each year between 2002 and 2007 50 per cent of beneficiary households did not meet conditionalities (Heracleous *et al.* 2010).

Box 3.2 Urban targeting in Brazil and Mozambique

Brazil uses the same system for targeting beneficiaries in rural and urban areas, the *Cadastro Único* which is a database of all potential beneficiaries. Despite increasing centralisation, the *Bolsa Família* still gives municipalities the responsibility of identifying potential beneficiaries with some autonomy over the process. The Ministry of Social Development prescribes what constitutes a household unit, the level of identification households must produce and the composition of the *Cadastro Único* questionnaire which potential beneficiaries complete, with all other elements of the approach decided upon by the municipality. In general, municipalities with largely urban populations target by using poverty maps to locate low-income neighbourhoods, employing vulnerability measures or multidimensional indices of living conditions. Other innovations include replacing household visits with on-demand registering in a public place (Lindert *et al.* 2007). This system has led to successful urban targeting; for example, in 2004 19 per cent of households in São Paulo received *Bolsa Família* transfers, and of that group 95 per cent of the beneficiaries came from the lowest-income category (Figueiredo *et al.* 2006).

Mozambique's Food Subsidy Programme (*Programa Subsídio de Alimentos*) delivers a cash transfer exclusively to urban households in absolute poverty and is run by the National Institute for Social Action (INAS). The programme targets urban households in which the household head is unable to work, is disabled or chronically ill, has an income below MT70,000, is a woman over 55 years of age or a man over 60 years of age, or the household contains a malnourished pregnant woman. Despite this broad categorical targeting and the programme operating in all provinces, coverage is only 1 per cent of the population. This is due to a complex system that includes checking household composition and the health status of residents, and an income assessment and measurement through home visits and reviews by the provincial department. This is a time-consuming process and difficulties are exacerbated by the programme's low budget and limited administrative capacity. Another issue experienced in this process is the requirement that beneficiaries must have an ID card or a birth certificate, despite an estimated 70 per cent of urban dwellers in Mozambique not possessing these identification documents (Devereux *et al.* 2005).

Source: Wain (2011)

Contribution to poverty reduction

What have these programmes achieved? Mexico's *Oportunidades* programme has had a particularly high level of impact evaluations and one consistent conclusion is that transfers have a more significant impact on the severity than the incidence of poverty; i.e. cash transfers decrease the gap between a household's income and the poverty threshold but do not then go on to raise households above the threshold (Niño-Zarazúa 2010). While there has been an increase in the proportion of the population in Mexico living below the food-based poverty line between 2006 and 2008, from 13.8 per cent to 18.2 per cent, this is considered to be due to the global economic crisis and is not seen to be a failing of cash transfer programmes. More recent studies confirm this conclusion (see Escobar Latapí and González de la Rocha 2008). However, despite *Oportunidades*' limited impact in terms of moving households out of poverty, there is evidence that the programme prevents households from falling into extreme poverty. In 2008, the Mexican government increased *Oportunidades* transfer amounts by 20 per cent in reaction to the crisis, and evidence suggests that this prevented 2.6 million people from entering extreme poverty (Niño-Zarazúa 2010, p. 18). This pattern of a relatively low impact on reducing the incidence of poverty and a higher impact on reducing the severity of poverty is replicated in studies of other Latin American conditional cash transfer programmes. In addition, qualitative studies of *Oportunidades* found that beneficiaries accumulate assets more rapidly than non-beneficiaries, with cash transfers spent on improving housing, utilities, land regularisation and other assets (Escobar Latapí and González de la Rocha 2008).

Cash transfers have been important in addressing the needs of low-income households and they illustrate the critical role of government in helping citizens address their needs. A further example not explored here due to lack of space but also important are pensions which provide essential support for older people (see Chapter 2).

However, a careful analysis raises a number of concerns for the realities of urban poverty and informal settlements as understood here. As shown by the analysis in our earlier volume (Mitlin and Satterthwaite 2013), there are nine critical dimensions of urban poverty, and lack of a safety net and lack of income are two of these. The other dimensions, namely inadequate assets, insecure tenure, lack of basic services, lack of infrastructure, lack of rights and access to law, lack of voice and lack of recognition, are unlikely to be addressed by cash transfers or many other social protection measures (although conditions on health checks and educational participation may draw attention to a lack of provision). Britto (2005) analyses the programmes in Mexico and Brazil, and suggests that a weakness is that they only target those households close to education and sometimes health services. Auyero (2010) describes the ways in which the terms and conditions associated with welfare support, including cash transfers and housing subsidies, may maintain and even exacerbate other dimensions of urban poverty. He interviews those waiting in the welfare office in Buenos Aires (Argentina) and elaborates:

> More than half (59 per cent) of our 69 interviewees do not know if and/or when they will receive the benefit they came to ask for. In other words, in the indeterminate waiting that defines the interactions between poor people and the welfare bureaucracy, we witness the daily reproduction of a mode of domination founded on the creation of a generalized and permanent state of insecurity.
>
> (Auyero 2010, p. 18)

In addition to individualised experiences, concerns have been expressed that conditional cash transfers individualise relations with the state with negative implications for collective political strategies, including efforts to secure improvements in basic infrastructure and services which necessarily demand action by groups of citizens to secure the investments that are needed. This replicates similar concerns expressed in the context of public housing and/or relocation (see Posner 2012). Analysis of the political implications of cash transfer programmes and national electoral platforms suggests that there is evidence that they may lead to short-term vote-winning agendas that take attention and resources away from the need for long-term development programmes (Hall 2006). Hall (2006) goes on to argue that conditional cash transfers may reduce pressure on the state to commit to programmes of broader social spending and investment in long-term social infrastructure. They may also relieve pressure on the state to address structural transformation of the economy such that there is greater productive investment, more employment opportunities for low-income households and a reduction in income inequalities (Britto 2005, p. 26).

The consequences of individualisation are the concern of the women's nutrition organisations in Peru that were introduced in Chapter 2 (see Box 2.6). These women-led groups emerged prior to the current phase of cash transfer programmes and have long been identifying beneficiaries for food support: their work to provide food supplements is partly financed by the government. These women's groups build strong collective organisations at the neighbourhood level. In recent years the Ministry of the Economy and Finances has been demanding that these women's groups use the Ministry's identification system for targeting rather than their own system. The government favours the approach to poverty reduction currently promoted by international agencies in which selected families are identified as needy and worthy, and are 'led out of poverty' by the professionally designed interventions described above. The organisations argue that they should decide the targeting method and hence the groups of beneficiaries included in the programme. They believe that the collective interest is best realised when a wide variety of people benefit, including the women who assist in preparing the food (who may not be state 'targets') and who build strong, supportive organisations at the neighbourhood level. The leaders recognise that they are contesting values (selective, individual and centralised versus collective, neighbourhood and decentralised), and are challenging the state to recognise their capability to manage the programme and its distribution of resources (Bebbington *et al.* 2011).

Further criticisms are made in respect of the nature of conditionalities as families are encouraged to make particular choices about education and health care. Critics include Freeland (2007), who argues that conditionalities are an ethically unsound paternalistic tool as they condition basic rights of protection. He suggests that the conditionalities are unnecessary, stating that there is no clear evidence that the low-income households act in prescribed ways due to the conditionalities rather than simply due to having the extra finance which permits choices that were previously unaffordable (Samson *et al.* 2006, quoted in Freeland 2007). Moreover, implicit in the design of the programmes is that education will improve employment options but if jobs are limited by poor national economic performance it is not clear that welfare as a whole will be improved (hence the need for structural improvements in the economy, as argued by Britto (2005)). This is not to argue against the importance of investment in education, but to emphasise that these programmes need to take place alongside measures to increase the demand for labour from households in need of employment opportunities.

Local authorities

Discussions about development have long found it problematic to include the role of local authorities – this is especially true for international agencies, as their engagement in defining and determining funding flows is with national governments. Yet, as discussed earlier in this chapter, much of what development assistance agencies seek to provide or improve falls within the jurisdictions of local governments. This is especially the case for most urban contexts, where many aspects of poverty would be reduced if city or municipal governments were willing and able to fulfil their responsibilities – especially for infrastructure and service provision. In addition, when aid agencies and development banks began to recognise the importance of 'good government' and 'good governance', the focus was usually on national government rather than on local government.

Perhaps more than anything else, it was the return to democracy (or its strengthening) in many Latin American nations that provided a reminder to development specialists of the important role of local authorities in development. This role was enhanced in many nations by national reforms, and new or amended legislation that strengthened the role of local authorities and also required elected mayors and city governments. In some nations, the revenue base of local authorities was also much strengthened. All this took time to attract notice outside the region; a book documenting this, published in 2003, was entitled *The Quiet Revolution* (Campbell 2003). An important part of this was a new generation of mayors who were from outside the conventional political parties and who brought strong commitments to addressing backlogs in infrastructure and services and making city government more accountable (see e.g. Almansi 2009; Dávila 2009; Gilbert and Dávila 2002; Hordijk 2005; López Follegatti 1999).

In some nations, these changes have been underpinned by changes in the national constitution (see e.g. Melo *et al.* (2001) and Fernandes (2007) for Brazil). In Brazil, this was also backed by the setting up of a new Ministry of Cities following sustained pressure from social movements (Fernandes 2007). Some city governments increased their capacity to act by raising funds independently and making investments of their own; this was the case in Rosario in Argentina. This took place, in large part, because from 1990 its mayors came from a different political party to the one in power at the national government level (and for many years also the one in power at the provincial level), which meant its policies received little support from higher government levels (Almansi 2009).

The significance of local government for the reduction of urban poverty means that we have broken this discussion down into three parts – comprehensive upgrading of informal settlements, the provision of basic services, and participatory budgeting.

Comprehensive upgrading

Local government support for the upgrading of informal settlements implies an acceptance by local government politicians and civil servants of the residents' right to live there. This may extend to recognising the need to make legal the land occupation and ownership and over time address other aspects of illegality – for instance, of housing structures and street layouts. It represents a fundamental change from local governments seeking only to bulldoze informal settlements or simply ignoring them.

In much of Latin America, it seems that upgrading has become far more widely accepted by local governments as a conventional policy response to informal settlements. Perhaps as importantly, upgrading is not just low-cost minimalist provision of, for instance, some communal water taps and street lights but aims to incorporate informal settlements into (for instance) trunk infrastructure for water, sanitation, electricity and roads. Thus upgrading becomes more comprehensive in terms of what is provided or upgraded and in terms of being served by conventional utilities. It also implies protection against eviction which minimalist upgrading does not and often includes measures to legalise tenure for the residents.

'Comprehensive' upgrading implies a far stronger relationship between government bodies and residents, and much more possibility of residents using social accountability mechanisms as they become registered property owners with legal addresses and official (and conventional) connections to piped water supplies, sewers, electricity, health care and schools. This stands in very strong contrast to what was evident during the 1970s and 1980s when illegal settlements were seen as contravening the law and bulldozed or at best ignored (Hardoy and Satterthwaite 1989). Although evictions of residents from informal settlements do still occur in Latin America (see e.g. COHRE 2006), they seem to be less common and the scale and scope of upgrading within the region seems to have increased dramatically. Upgrading of informal settlements is now more widely seen as a conventional part of what city or municipal governments do, although city-wide support for it tends to fluctuate, depending on the political party in power – as in São Paulo (see Budds and Teixeira 2005). Rio de Janeiro was well known for the scale of informal settlement bulldozing in the 1970s (see Portes 1979) and then for the scale of its upgrading programmes during the 1990s (see Fiori *et al.* 2000) – and now it seems to have returned to informal settlement bulldozing, justified by preparations for the Olympics and the World Cup for soccer.[8]

This wider acceptance of upgrading by municipal governments would help explain why the proportion of Latin America's urban population with good-quality provision for water (water piped into people's homes) and connection to sewers and drains increased from the 1970s or 1980s to the present. It is now common for Latin American cities to have universal or close to universal provision for these services, even when higher standard definitions are used for such provision than those favoured by the UN such as water piped into each dwelling and toilets in each dwelling connected to sewers.[9]

The change in official attitudes to more support for upgrading was also served by the wave of innovation in city governments in this region with more participatory and accountable governance – including participatory budgeting and provision to second representatives from urban poor groups on to government committees; although also important here has been reflection among social movements and greater strategic intent (Abers 1998). In some countries there has been a profound change in relations between the residents of these settlements and governments, as the residents of these 'illegal' settlements are seen as having the right to government-funded infrastructure and services. The importance of city authorities having the autonomy to act is emphasised by Heller and Evans (2010, pp. 440–441), who conclude that citizens may be less able to realise their rights and secure development options if power is not decentralised downward to municipalities (see also Almansi (2009) for a discussion of this in relation to Rosario).

Of course, strong citizen pressure and the influence of grassroots organisations and their networks and federations have been important. Here too, the changes may be ascribed both to greater emphasis on democratic elections and to measures to hold the state to account.

Informal upgrading

It is not possible to assess the scale of upgrading programmes in Latin America that has occurred independent of national programmes. All we have are particular case studies that are usually the result of some researcher choosing to document these programmes or sometimes included in local authority documentation. But there are also local processes that over time contribute to upgrading (and even comprehensive upgrading). Residents living without infrastructure and services in an informal settlement can come together and do what is possible to improve this situation, and to negotiate support from their local government (and perhaps other local civil society organisations) (see Box 3.3). This may prepare the way for more formal comprehensive upgrading programmes or may continue without this. Local residents engage with local councillors and politicians eager to lobby for votes in elections, and there are possibilities to negotiate for assistance. What emerges is a wide range of different arrangements in which the state passes resources to low-income communities to enable improvements in local infrastructure and services. In some cases the local authority is responsible for installation but in other cases it may lend machinery (such as diggers and graders) and/or provide materials (sand, cement, timber) for collective construction efforts. The authority may also provide technical assistance, for example, in the design of drainage and other infrastructure. In some cases, there may be existing programmes to help finance such initiatives, such as councillor funds that can be used to support works in his or her constituency. Alternatively, there are monies at the level of the local authority that individual community groups can apply for in order to finance local level improvements. In such cases these modalities of support would not be seen as co-production as such which is a term used in much more formalised programmes of collaboration with more specific and consistent outputs (see Chapter 2). This informal co-production is a by-product of resource scarcity, personalised relations and the impossibility that self-help solutions (even when collective) can address major infrastructure needs among the low-income population. The majority of such work is neither formalised nor even recorded. However, it has been essential in developing neighbourhoods that provide their residents with the necessary goods and services.

Box 3.3 illustrates the evolution of improved approaches to informal settlements through describing one experience in Guayaquil. The upgrading processes were slow and erratic with multiple unconnected interventions. Nevertheless, over time, urban neighbourhoods with a full range of basic services and infrastructure have emerged. However, this example also points to some limitations, including the growing lack of security described by Moser (2009) as drug-related gang activity has increased in recent years.

Box 3.3 The upgrading of Indio Guayas (Guayaquil, Ecuador)

In the late 1970s, families began the informal occupation of land in Indio Guayas which was then a waterlogged mangrove swamp. They bought plots 10 by 30 metres and erected shacks made from bamboo and wood, and these were linked by catwalks above the water. They were without piped water, sewers for sanitation, electricity, health or education services.

A barrio committee was rapidly established – and in the early years the majority of leaders were men. The women then became frustrated at the slow progress and began to take over the leadership. The committee had to deal with many issues within the community and a major task was to address the need for infrastructure; this required the committee members to build relations with political parties and negotiate votes for political support (ibid., p. 79). The committee began with an alliance, together with another 20 communities, with a new political party, the *Izquierda Democrática*; this was a 'marriage of mutual convenience' rather than any deeply shared political ideology and, for the residents of Indio Guayas, improved infrastructure was a priority (ibid., p. 80).

In the following years, success was achieved, elections were won, and four infilling programmes took place in 1976, 1977, 1978 and 1980. In 1983, plot and street infilling was followed by paved roads. The exercise was not free of problems. Municipal work was often substandard, and the community had to lobby the authorities and monitor the truck drivers themselves to ensure that the construction materials and work were of sufficient quality. Moreover, there were internal conflicts and difficult disputes between households about the sequencing of infilling within the settlement.

Although there was no piped water supply when the families first moved on to the site in the mid-1970s, a tanker company provided erratic services. The company prioritised their industrial customers and came to the households living in informal areas with the water that was left. Piped services began in 1983 – when an election meant there was political interest in improving provision in lower-income areas. By 1992, only 35 per cent of households were dependent on tanker supplies, and by 2004 this figure had fallen to 6 per cent. In 1992 the laying of sewage pipes began (although 15 years later this had not been completed).

Electricity was extended to the settlement in 1976 but legal connections only began in 1979 following three further years of political lobbying. By 1992 all the households had electricity (although only three-quarters of connections were legal). In 2000, public street lighting was provided.

By 1980 there were three government health clinics and a local state primary school in the barrio but the quality was very poor. During the following two decades government provision did not improve, and additional support was provided by international development assistance, first by UNICEF (for health and pre-school education) and then by Plan International (a Northern NGO).

Source: Moser (2009)

Contribution to poverty reduction

Clearly, upgrading contributes to reducing one or more of the nine aspects of urban poverty. But it may involve only minor improvements in one aspect (for instance, some communal water taps) that may bring little or no increase in tenure security; many informal settlements that have been bulldozed in Delhi have had some public provision (Bhan 2009). At the other end of the spectrum, it involves good-quality provision for infrastructure and services, support for housing improvement and tenure. But even here, as a resident from El Mezquital, an informal settlement in Guatemala City that had relatively comprehensive upgrading, said, you are 'putting a roof over my poverty' (Díaz *et al.* 2001). In this upgrading programme, as in many others, there is a recognition of the need to increase incomes and employment opportunities but it is far more difficult to do so. Other relatively comprehensive upgrading programmes have considerably improved conditions but have not addressed crime. More recently there has been a greater interest in security issues and some local authorities have begun to include these aspects within their approach to informal settlements (see Cerdá *et al.* (2011) for Medellin, Colombia).

In both partial and comprehensive and formal and informal upgrading as well as in their support for new settlements, local government builds on the practices of self-help described in Chapter 2. Greater understanding of the ways in which low-income communities have turned undeveloped land into urban settlements has helped to introduce a number of both partial and integrated policies and programmes. Experiences have been consolidated to help improve government programmes in support of low-income households and replication demonstrates the importance of these programmes. The example of the Community Organization Development Institute (discussed in Chapter 4) providing national government support for local community-managed improvements in which local governments are a contributing agency is a further illustration of the continuity of this model.

Access to basic services

Many local government interventions for upgrading are not close to being either integrated or comprehensive. What is striking is the partiality of many initiatives, i.e. the lack of consideration of the broad urban management framework. This trend is exemplified by the emphasis on private sector involvement in water and sanitation utilities; while investment is needed in this sector, in practice informal settlements need water and sanitation to be provided alongside other services such as drainage, roads and pathways, electricity connections and improved access to health and education provision.

The trend towards greater private sector involvement in water and sanitation was led by a package of donor assistance that was seeking to address deficiencies in services (Mehta and La Cour Madsen 2003, pp. 7–8). Despite a belief that greater private sector involvement would improve the scale and quality of delivery, over 80 per cent of water supplies remain in the ownership and management of the state (Estache and Kouassi 2002, p. 4; OECD 2003, p. 58). In more recent years the trend towards privatisation for large utilities has not continued, as private sector operators found it difficult to secure a sufficient return on their investment. Fluctuating currencies combined with low economic growth have reduced incomes and profits. However, there has been an increasing trend towards corporatisation and a greater orientation to market approaches and commercial interests within company management. These approaches make the integrated provision of services through comprehensive upgrading less likely. In some nations, they also rule out the expansion of sewer and comprehensive storm drainage networks to informal settlements, as these are considered too expensive for cost recovery (see e.g. the discussion in Nilsson (2006) in relation to Kampala).

For the urban poor, despite the limitations of partial sector-focused approaches, access to water and sanitation is important. The hope was that more efficient business-oriented (private or public) companies would extend access, perhaps to areas denied supplies under the state because of their informal status. Some such investment has taken place; one example of an initiative taken by a public utility to extend service provision to informal settlements is the work of the Social Development Unit set up within the Bangalore Water Supply and Sewage Board (Connors 2005). About 10,000 households in Bangalore's 'slums' are reported to have water piped to their homes and they have benefited from reduced connection fees (that can also be paid in instalments), acceptance of alternative proof of residence for getting connection to land tenure documents (which most do not have) such as ration cards, voter ID, and ID issued by the Karnataka Slum Clearance Board and a cheaper tariff for low-income consumers.[10] However, other experiences are less positive, reflecting the significantly different opportunities for investment. In Harare, the basic monthly charges for low-income households in high-density areas to cover the property, council administration, water, sewerage and refuse add up to US$28 with additional charges for metered water consumption on top.[11] This is in a city in which the official minimum wage for a live-out domestic servant is US$80. Many households cannot afford to pay and end up sharing connections. However, an equally serious problem is the absolute lack of services in some areas with water often failing to flow through the pipes. Communities such as the Zimbabwe Homeless People's Federation have begun to provide themselves with alternatives but these are costly and only available to those with development assistance funds. In Walvis Bay (Namibia), residents explained that while the monthly costs charged by a mutual fund giving them a housing loan are N$200 a month for a two-room house, they are paying N$400 a month for water and N$800 a month for electricity. The consequences of the corporatisation of water services in Zambia are analysed by Dagdeviren (2008) and she concludes that quality has fallen, water is unaffordable for the majority of urban dwellers and cost recovery is not being achieved. Dagdeviren (2008, p. 106) also documents the considerable real price increases that took place between 1990 and 2006 which varied between 1.5 to sevenfold for low-income households: the trend of increasing real prices is also noted by Moser (2009, p. 84) in the case of the informal settlement she studied in Guayaquil.

As formalisation increases, service charges become significant for larger numbers of urban dwellers who previously lived in informal settlements. The ways in which city authorities are seeking to address housing needs in Asia in a context of high economic growth and increasing land pressures are illustrated by a study from Mumbai which analyses the consequences associated with resettlement from an area within the Sanjay Gandhi National Park (see Box 3.4). Vaquier (2010) describes the acute needs of the urban poor in Mumbai, a city with very high residential densities and around half the population living in informal settlements. Large-scale evictions have been taking place since the 1960s but there are also instances of upgrading programmes and site and service programmes, and of relocation for those evicted. To understand how needs can be better addressed and to access the consequences of relocation for families, Vaquier (2010) surveyed 200 households, including those remaining in the informal settlements (yet to move) and those who had already moved. The report provides us with insights into the monetary and non-monetary costs of formalisation.

Box 3.4 Resettlement to Chandivali and the costs of formalisation

The resettlement project at Chandivali accommodates 12,000 households being relocated from the Sanjay Ghandi National Park. By 2008, just over 4,000 households had already moved. The low incomes of households that remain in the informal settlements awaiting relocation are evident from their assets: although 65 per cent of households have a mobile phone and 50 per cent have a television, only 6 per cent have a fridge and 5 per cent have a bicycle.

Eighty-four per cent of households who have moved into the new housing (in small apartments of 21 square metres in eight-storey blocks) say that they are satisfied – most explain that this is mainly because they now have tenure security. Households in the informal settlements primarily use public taps (93 per cent) (ibid., p. 85). In the resettlement apartments, there is piped water in all but one block and 91 per cent are satisfied. Electricity provision is 100 per cent formal for those relocated, and mainly illegal in the informal settlement (although 85 per cent have access). All of the apartments in Chandivali have toilets but in the informal settlements only 2 per cent of households have private toilets and 73 per cent of households can only access public toilets (for which they pay an average of Rp.73 per family per month).

Key problems in the resettlement area relate to the lack of access to education (58 per cent of interviewees in the resettlement area are dissatisfied), health services (69 per cent dissatisfied), and lack of public transport with planned extensions of bus and metro networks not having taken place. In terms of incomes, 67 per cent of resettled households said that their income was constant, 27 per cent believe that it has fallen and 7 per cent that it has increased. Ninety-five per cent said that expenditure on basic necessities has increased following their relocation. Monthly costs for water, sanitation and waste disposal in informal settlements are Rp.119 and this increases to Rp.215 with formalisation. Only electricity costs fall from Rp.267 a month for an illegal supply (for those who access it) to Rp.230 for a legal connection. For individuals in work, transport costs have increased significantly due to the lack of good access to public networks.

Source: Vaquier (2010)

In some cases, access to basic services has been provided through serviced site schemes developed by local authorities. In the town of Ilo in Peru, for example, the mayor acquired land and subdivided it to allow low-income households to get legal land plots on which to build – with a plot with basic infrastructure costing the equivalent of US$60 (López Follegatti 1999). The women's savings group in Harare that negotiated a land site with plots and basic infrastructure on which 230 units were built got the land for the equivalent of US$18,000 – which means around US$78 per household; although this was also due to historic pricing of the plots by the local authority at a time of high inflation.[12] In Windhoek (Namibia), the city authority has designed a number of 'levels' of service quality and housing development to facilitate affordable access for all newcomers to the city. The city has sought to provide affordable land plots with tenure in response to significant migration from rural areas and smaller towns following independence in 1991. This policy sought both to provide for the upgrading of existing informal areas and to pre-empt further informal settlement through more affordable semi-formal provision. After recognising that existing policies were not working, in 2001 the City of Windhoek introduced seven service standards for land sales in an effort to respond to the lack of affordability (Muller and Mitlin 2004, 2007). Lower levels of services (communal toilets and water taps) enable residents to purchase land that can be upgraded over time; families are allowed to stay in shacks for seven years until they can afford housing construction. Community groups are allowed to buy blocks of land, thereby circumventing plot sizes and enabling them to settle with higher densities and hence lower costs. The households settling and sometimes purchasing this land are made up of those who have recently come to the city and who otherwise rent accommodation, or who squat on the periphery of the city. They move from being tenants to greenfield developments with basic services, or more commonly they are allowed to settle with very minimal services, and provision is subsequently upgraded. This policy has increased the affordability of access to tenure and enabled low-income families to have tenure security, and upgrading as finance becomes available. For those households that are slightly better off, there is additional support available from the Build Together Programme, a loan programme with subsidised interest rates for those building a four-roomed house that is financed by central government and managed by local authorities.

Contribution to poverty reduction

The scale of the deficiencies in provision for water, sanitation, health care and other basic services is described in detail in the companion volume to this (Mitlin and Satterthwaite 2013) and summarised in Chapter 1. In many nations, low-income urban dwellers are actually worse served than they were 20 years ago The lack of provision of basic services, and attempts to address this lack, highlights some considerable contradictions for utilities and local government providers. First, these services are represented as a right, particularly in the case of water, and as an essential good for health and well-being. Second, utilities or companies are under pressure to secure cost recovery and essentially to manage these services as a business broadly exemplifying a market approach to development. The shift towards private sector involvement and corporatisation discussed in the subsection above reflects this objective and the market approach to development. However, a third logic is also evident, as services are essential for a prosperous economy and need to be planned at the city scale; hence urban management considerations come into play. Finally, the political dimension is always evident, both through social movement activities related to service acquisition, and through clientelist exchanges with informal settlement residents desperate to secure access.

The discussion above highlights the commercial benefits to the state from formalisation with rising costs for households and may be indicative of the dominance of this prevailing logic. In the case of Mumbai discussed in Box 3.4, households pay less for electricity but this money is now secured by the company rather than through illicit payments to third parties. Here the state also benefits through regaining control of the land on which the households were originally settled. The research also highlights the difficulties faced by families being resettled (rather than upgraded in situ), as it is rare for services to be managed consistently and key facilities (in this case transport, education and health) lag behind. The impacts on household income are also important, although as Vaquier's

(2010) study demonstrates, the net effects are complex and in this case are linked to proximity of full-time stable jobs in the original and receiving areas, the ease and costs of commuting, the sensitivity with which those with home-based enterprises are relocated (particularly if there is a move to tenements or apartments), and the size of the units and hence the ease with which extended family members can be incorporated into the household unit.

Participatory budgeting

Organised communities frustrated with the poor-quality policies and programmes that are exemplified by the partial provision of essential services and with self-interested interventions by local authorities have pressed for greater inclusion in decision-making. As noted in Chapter 2, participatory budgeting is one substantive example of participatory governance modalities. First developed in Brazil, it has been applied in over 250 urban centres around the world (Cabannes 2004, 2013; Menegat 2002; Souza 2001). Many replications are in Brazil, but participatory budgeting initiatives are also flourishing in urban centres in other Latin American nations as well as in some European nations.

Participatory budgeting gives more scope for citizen groups and community-based representatives in setting priorities for local government expenditures; it also implies a local government budgeting system that is more transparent and available to public scrutiny (Cabannes 2004). At its core are citizen assemblies in each district of a city that can influence priorities for the use of a portion of the city's revenues and a city government that makes information widely available about its budget. In effect, community residents have more influence at the expense of bureaucrats, the local executive and local councillors, and this, combined with a more open process, helps make city government investments more linked to local priorities and helps limit clientelism and corruption (Souza 2001). There are many differences between cities where it is implemented in, for instance, the form of participation (for example, everyone entitled to participate and vote in assemblies or mainly for delegates and leaders from social movements, neighbourhood associations and trade unions). There are also differences in which organisation is in charge; in Brazil, this is usually a council of the participatory budget, and in non-Brazilian cases this may be within existing political frameworks with neighbourhood priorities being taken by established bodies. The extent of control over how public funding is spent also varies considerably – from overall influence on the investment budget to a small proportion of it (Cabannes 2004). Some cities made special provision within participatory budgeting for groups that have difficulties getting their priorities heard (for instance, committees for women or children and youth). Some have delegates elected for particular groups (for instance, the elderly, adolescents, indigenous groups and the disabled). Participatory budgeting generally means more funding going to the lower-income neighbourhoods and an increase in expenditure in social provision (for instance, education and health care). The process also provides possibilities for low-income households to see themselves as citizens and is thus an important step in building democratic institutions (Souza 2001).

Contribution to poverty reduction

There is agreement that the processes of participatory budgeting result in a transfer of resources to previously excluded areas and neighbourhoods that is often significant. Cabannes (2004, pp. 39–40) exemplifies this with a report on Montevideo where low-income areas contribute to 21 per cent of the city budget, and receive over 88 per cent of the budget for highways and 79 per cent for sanitation. In Porto Alegre, the participatory budget is also widely considered to be a success, delivering significant benefits in terms of investment in infrastructure and services within low-income areas. Hence, there is evidence to suggest that they result in a significant shift in local government priorities in favour of basic services for those who have a stable residence and who have sufficient tenure security to receive state investments in infrastructure and services.

However, other views are less positive about the potential of participatory budgeting to transform clientelist practice. In terms of the inclusion of the lowest-income citizens, in Porto Alegre there is an underrepresentation of those with low education in decision-making; this group makes up 55 per cent of the city population and 60 per cent of general participants but only 35 per cent of elected delegates and less than 20 per cent of elected councillors (Baiocchi 2003).[13] Avritzer also argues that the lowest-income citizens are less likely to participate in this city and that when they do participate they are less likely to speak and speak frequently (2006, pp. 627 and 630). In addition to the analysis of participation in leadership, Souza (2001, p. 169) compares Belo Horizonte and Porto Alegre, and argues that 'the idea that PB [participatory budgeting] has produced a generalised empowerment of the unorganised and of the poor has been challenged'. Studies have considered the impact on both material conditions and political inclusion. Despite evidence of the redistributional impact of participatory budgeting where there is not a strong associative movement, the gains in democratisation do not take place (Avritzer 2006, p. 631; Melo *et al.* 2001). Context is important, and the effectiveness of participatory budgeting in Porto Alegre and some other larger cities may be related to the pre-existing political culture, while in other cities such as Belo Horizonte this is lacking and clientelist relations prevail even within the participatory budgeting process (Baiocchi 2003, p. 66; Souza 2001, p. 181). Problems include resistance from technocrats, clientelism and the continued dominance of political elites (Hordijk 2005; Rodgers 2010; Souza 2001).

A comment by Martin Pumar, a former mayor of Villa El Salvador (one of the municipalities within Lima), summarises some of the difficulties faced when he introduced participatory budgeting.

> The municipal structure and bureaucracy were not yet capable of dealing with the changes. First of all the participatory budgeting of course implies relinquishing power, also the everyday power of councillors, municipal

> workers. Personal favours, clientelistic relations are part and parcel of our municipal culture. So there was quite some resistance in the municipal apparatus. Yet even for those who understand and support the change it was not easy. All of a sudden urban development received tens of project proposals to be implemented, where the municipality had to develop all the technical plans to prepare the construction.
>
> (Hordijk 2005, p. 226)

Some general conclusions

Most urban governments have responsibility for addressing many aspects of poverty. Fulfilling these responsibilities can make a critical contribution to poverty reduction. Their proximity to the urban poor and their responsibility for critical governmental processes such as zoning, planning, building regulations and standards, environmental health and in many cases the provision of basic services and the management of utilities means that the level of their capacity and competence makes a major difference to the quality of life of the urban poor, particularly those in informal settlements. These are the kinds of goods and services that assist people to find accommodation of sufficient quality and may help improve their access to livelihoods that secure adequate incomes. However, in a context marked by very low incomes and lack of assets, alternative modes of delivery to conventional systems are needed. The discussion above highlights the importance of incremental development. At the same time, subsidy finance and the strengthening of public accountability systems are required if outcomes are to benefit the lowest income groups. Otherwise higher-income households use their social networks and political connections to legitimise their own claims on resources and their own vision of the city and what it should provide.

Local government is often lacking in vision, resources and organisational development. This means that goods and services are rationed and if there is a context of weak accountability then clientelist politics prevails. The kinds of personal relations clientelism supports, as discussed in Chapter 2, are vertical relations of dependency in which protest, if present, is hidden in subterfuge and public acquiescence. Such relations support a culture of subservience, division and stratification – and 'agreements' may be reinforced through coercion and violence. Clientelist politics have resulted in some support for self-help initiatives and the slow progress of upgrading at the local level but the benefits at neighbourhood levels are limited and partial, and only some benefit. What is notable about the examples above is that they all represent ways to make this negotiated engagement more rewarding for low-income groups, and to occur at a greater scale.

Despite limitations, and as evidenced above and elaborated in the chapters that follow, local government is notably more responsive than more distant central government agencies and often open to negotiations with the organised urban poor. Chapter 4 explores the ways in which organised communities have been able to negotiate both informal and formal arrangements with a range of local government agencies, while Chapter 5 discusses strategies by which the organised urban poor have sought to advance their needs and interests.

Official development assistance

A review of the role of official development assistance agencies in urban poverty reduction can point to many innovative projects or interventions going back to the 1970s. But this is within a context of a low priority to urban initiatives and usually a lack of any coherent urban policy. It is also generally within a context in which a low priority is given to some of the deprivations associated with urban poverty, including improving housing conditions, provision for water, sanitation, schools, health care, safety nets and the rule of law.[14] There are also many new urban initiatives launched by different international agencies – for instance, for healthy cities or sustainable cities or child-friendly cities, or for highlighting the importance of urban management or the role of mayors – that receive little support and/or which soon disappear. There are a few examples of international agencies with a coherent urban policy with clear links to poverty reduction that then get dropped – as in, for instance, the urban policies of the Swedish International Development Cooperation Agency between 1990 and 2008 (see Tannerfeldt and Ljung 2006), the urban basic services programme of UNICEF and the World Bank's support for serviced sites. US AID had a substantial engagement in urban housing but this was shut down in 1993, and there has been little interest in urban issues since then (Shea 2008).

It is also difficult to track the nature and extent of support for externally funded initiatives that have relevance to urban poverty reduction because of the limits in how agencies report on their work. IIED's Human Settlements Group monitored the priority given to initiatives that addressed urban poverty for a range of international agencies from the late 1970s to the mid-1990s by reviewing every project commitment each agency made each year. From this, it was possible to assess the proportion that went to poverty reduction in urban areas – that included housing initiatives for low-income groups (including 'slum' or squatter upgrading, serviced site schemes and core housing), housing finance targeted to low-income households, improving or extending provision for water, sanitation, drainage and household waste collection, provision for health care services and measures to control or prevent diseases, support for primary and basic education programmes including literacy programmes, conditional cash transfers, socially oriented public works programmes and community development (Satterthwaite 1997, 2001). Overall, these data showed the very low priority given to these programmes for almost all of the agencies reviewed. This was especially the case for initiatives to improve housing conditions in informal settlements – which received no support at all from many agencies. However, such an approach to monitoring development assistance flows became increasingly difficult as more development assistance was allocated to sector reforms, basket funding and/or cross-sectoral initiatives with no details of what this funded on the ground (thus it was not possible to assess the proportion of funding that contributed to reducing urban poverty). The OECD Development Assistance Committee statistics show the low priority given by total bilateral commitments and by most bilateral agencies to (for instance) water, sanitation, basic health care and basic education but with no details as to its distribution between rural and urban areas. The limited amount of relevant data on this from other sources (Milbert 1992; Milbert and Peat 1999; Shea 2008; Stren 2012) suggest that urban poverty reduction still receives very little attention or funding.

The World Bank is important in urban poverty reduction both for pioneering support to what were to become standard interventions that sought to improve housing conditions for low-income groups – 'slum'/squatter upgrading and serviced site schemes – and for the increasing proportion of its funding allocated to initiatives to reduce urban poverty. With regard to upgrading and serviced sites, the Bank's support helped move these from being controversial to being standard responses that came to be implemented in many urban centres with domestic resources and little or no external funding. With regard to the full range of interventions for urban poverty reduction mentioned above, the proportion of World Bank funding commitments going to these doubled – from 7 per cent for 1981 to 1983 to 15.9 per cent for 1993 to 1995 and 15.4 per cent for 1996 to 1998 (Satterthwaite 2001). This is substantially higher than other agencies assessed, including the Asian Development Bank and the OECF (then the largest bilateral agency and which later became part of the Japan Bank for International Cooperation). From the early 1970s to 2000, the World Bank support shifted from specific initiatives (upgrading and serviced site schemes) to housing finance schemes (that had potential to reach more low-income groups) and then to improving the management of cities – the expectation being that city governments with more capacity would address urban poverty (Buckley and Kalarickal 2006; Cohen 1983, 2001). The Bank also supported many multi-sectoral initiatives that sought to address different aspects of urban deprivation, although these proved complicated to implement and – with high staff costs – were unpopular with workers (Cohen 2001). The World Bank also came to champion the role of cities and urban systems in building stronger, more competitive national economies (World Bank 2008). But it is perhaps notable that in the reporting by sector of its funding in 2012, with 70 different sectoral categories, no mention is made of upgrading, and the Bank no longer reports on its funding to urban development as a category.[15] In addition, a review of the projects categorised as 'housing construction' which is a subcategory under 'industry and trade' shows very few that support upgrading.

The Inter-American Development Bank is also notable for its long-established support for urban poverty reduction and, within this, for supporting slum/squatter upgrading. During the 1980s and early 1990s, among the multilateral development banks, it had by far the highest proportion of its concessional lending going to urban poverty reduction (Satterthwaite 1997). It is difficult to draw on this Bank's official statistics to get a sense of whether this priority continued; what is reported as funding for urban development has received a relatively low proportion of funding in recent years, yet many initiatives it helps finance such as social protection and water and sanitation will be contributing to urban poverty reduction. The Bank's project database[16] includes details of 221 projects labelled as urban development and housing neighbourhood upgrading, and these include many projects in recent years.

From 2000, the explicit focus given by most international agencies to meeting the Millennium Development Goals certainly increased a focus on interventions that matter for low-income (urban and rural) dwellers. But in urban areas, this would need a focus on strengthening and supporting the role of local governments in addressing key targets – for instance, around improving and extending provision for water, sanitation and health care (with a strong focus on maternal and child health), reducing hunger and extreme income poverty and 'significantly improving the lives of slum dwellers'. There is little evidence that the MDGs have generated more support for this in urban areas. Reports on MDG achievements within nations tend to focus on rural areas or report on urban achievements in which international donors played only a minor role, and where the accuracy of the data on the achievements is in doubt – for instance, the claim that tens of millions of slum dwellers moved out of poverty in India between 2000 and 2010.[17] Stren (2012) notes how development assistance has shifted to more specific targets that imply a high moral imperative because they are issues involving life or death for millions of extremely vulnerable people – for instance, programmes on alleviating climate change, on HIV aids, famines, child labour, violence against women, etc. Perhaps some of these programmes have brought benefits to urban poor groups but they do not address the structural causes of their poverty – including their capacity to act and to organise. The OECD DAC statistics show some small increases in the proportion of funding going to basic education and health since 2000, although these still receive a small proportion of total funding. It is likely that most of this funding (and increase in funding) is in rural areas.

More generally, official development assistance agencies have been giving little attention to urban poverty with needs in rural areas being awarded a significantly higher priority. For example, the sustainable livelihoods framework used by DFID has an orientation towards rural livelihoods and does not apply easily in urban contexts (Satterthwaite 2002; Satterthwaite and Tacoli 2002). When bilateral development assistance agencies have engaged with urban development, they have tended to adopt a partial approach. For example, there may have been investment in infrastructure (particularly water and roads) or in services (such as health and education). More recently attention has also been given to governance and public accountability. In a significant number of cases this focus has not been primarily concerned with poverty but has concentrated on issues related to growth and economic prosperity with an emphasis on the opportunities cities offer as economic centres. Some international agencies have supported measures to increase the social accountability of city governments although with mixed success, and often with no benefits to large sections of the low-income population. You have to be served by a piped water connection by a water utility or have access to an official health care service to be able to report on their inadequacies and demand better provision.

One particular experience notable for the depth and complexity of its design and ambition was the integrated approach to informal settlement upgrading established in five countries in Central America (Costa Rica, El Salvador, Guatemala, Honduras and Nicaragua), with the financial and organisational support of the Swedish International Development Cooperation Agency (Sida). The programmes have sought to address the lack of adequate shelter and have included measures related to tenure security and access to improved infrastructure and services (Stein 2001; Stein with Castillo 2005; Stein and Vance 2007). Each national programme has been led by a specialist agency; in some cases these have been created through negotiation between Sida and the government, while in other countries an existing agency has been used. All these programmes deliberately bring together a number of different institutions: central and local government, NGOs, micro-finance agencies and community-based organisations. Development-assistance funds have been used for three components: technical assistance; loans for housing improvements, housing construction and technical assistance; and institutional capacity building and development for those organisations involved in delivery.

The contribution of these programmes to addressing shelter needs has been small but significant (Mitlin 2013). By the end of 2010 they had benefited approximately 153,000 low-income families, representing some 3.3 per cent of the total urban population of the five countries and about 9 per cent of the total urban poor population. The programmes have reached an estimated 18 per cent of the urban poor in Costa Rica, 14 per cent in Honduras and 10 per cent in Nicaragua. Perhaps the most notable successes have been in the lower-income countries of Honduras and Nicaragua where there has been a considerable depth to the support provided. Across the five countries, the average investment cost per person is about US$100 and, with an emphasis on basic services, substantive improvements are possible. Families with incomes too low for loan finance have been helped mainly through financing for water, sanitation, drainage and other such improvements. The intention is to reach very low-income households with basic services alone, and to provide mortgage or home-improvement loans to higher-income families. The intervention has three integrated core components: financial assistance to secure land tenure, infrastructure improvements (perhaps with a community contribution), and micro-finance loans for housing improvements to those households that can afford to take out loans. Subsidy finance can be easily blended into this model at a number of stages, as has been the case in Central America. The model has been applied in a number of other Latin American countries (IDB 2006, p. 1).

Contribution to poverty reduction

The scale and depth of urban poverty for so much of Africa and Asia and parts of Latin America that we describe in detail in the companion volume to this reflect in part both the very low priority given to this issue by official development assistance and the difficulties that bilateral agencies and multilateral development banks face in being effective in urban poverty reduction. Not enough attention is given to these organisational difficulties. Five issues need highlighting.

First, these agencies have little or no possibilities of reducing many aspects of urban poverty – for instance, they cannot change labour markets to make sure these provide higher wages for unskilled workers. They have limited capacity to ensure the rule of law in the localities with high concentrations of low-income groups. They also have limited capacity to support the voice and agency of low-income groups – or perhaps more accurately here they have chosen not to invest.

Second, they cannot meet needs directly; they have to rely on intermediaries to implement poverty reduction. No staff from official development assistance agencies actually dig ditches to (for instance) install water pipes, sewers and drains, or set up and run health care facilities and schools in low-income urban areas. They can only fund others to do so. Thus they are only as effective as the intermediary institution that receives their funds. Their funding usually goes to national government ministries or agencies whose staff also rarely if ever engage in actual implementation on the ground. Thus they too rely for their effectiveness on whatever local entity they fund. Sometimes this is a local government agency or a local office of a national agency. But these in turn often rely on other organisations for their implementation – for instance, through contracts to private enterprises. Throughout this process, the low-income groups whose poverty is meant to be the reason for the international funds have very little power or voice (see Satterthwaite 2001). In addition, the enterprises or government agencies that carry out the implementation are more worried about keeping the national government and/or funders happy (accountability upwards) than delivering for the urban poor groups.

Third, what they fund has to be agreed with national governments who will usually have different priorities and may indeed have very little interest in addressing urban poverty. For 60 years it has been difficult to shake the inaccurate assumptions of senior government staff and politicians about the urban poor ('they are recent rural migrants that should return to their villages') and prevailing attitudes to informal settlements ('these are illegal, they damage the image of the city and its capacity to attract investment and they should be bulldozed').

An important part of this institutional difficulty for international agencies is how much urban poverty reduction depends on working with the residents of informal settlements and people working in the informal economy – both of which are seen by governments as contravening laws and regulations. Thus national governments are often unwilling to support work on urban poverty reduction. So too are many local governments. Alternatively, local governments may be forbidden to provide infrastructure to informal settlements and services to their residents (for instance, if they are on land owned by a national agency which forbids them to do so).

Fourth, to be successful, international agencies need to be able to promote and support the roles, capacities and finances of urban governments – in effect supporting successful decentralisation. Yet this is often contrary to what higher levels of government want or prioritise. It is politically contentious, especially if it includes increased resource flows and revenue-raising powers to local governments (and especially local governments that are from different political parties from those in national or provincial/state government). External agencies also face difficulties in working with local governments. Here, too, inaccurate assumptions about the urban poor often predominate. To address the needs of those in informal settlements and to provide legal alternatives to such settlements that are affordable, they need local governments with the capacity and commitment to manage land-use changes in the public interest (especially in the interest of low-income groups). But there is so much money to be made by not doing this both by real estate interests and by government staff.

Fifth, aid agencies and development banks have difficulties engaging with the very people whose needs they are meant to be addressing, namely the urban poor and their organisations. National and local governments certainly do not want international agencies talking to or funding urban poor groups direct. They will insist that they are the elected representatives of their citizens and as such have the right and the mandate to make decisions about how funding is used.

Then there are also the inaccurate assumptions embedded in the prevailing approaches by the international agencies themselves – for instance, the much over-stated hope that economic growth would have widely distributed benefits (see Mitlin and Satterthwaite 2012, ch. 5). At least for water and sanitation, the faith placed in privatisation as a way of getting more capital investment into the sector and of giving more accountability to low-income groups (who were now 'clients') proved unfounded in many instances (see e.g. Budds and McGranahan 2003).

Non-governmental organisations and associated civil society organisations

There are a vast number of civil society organisations working on urban development issues including homelessness, informal settlements, water and sanitation, informal waste management, urban livelihoods (such as urban agriculture and street vending), citizenship and the right to the city, the needs of disadvantaged groups including street children and old people, urban governance and political participation, and city-wide urban development including transport and urban well-being. Much of this falls within what is often termed the brown agenda (which focuses on environmental health). There are also many civil society organisations working on the green environmental agendas (for example, greener cities, recycling and waste management, and the 'green' economy). Only some of these organisations aspire to catalyse significant social change in their work. Most come with their own agendas and these may not coincide with urban poor agendas or priorities.

Other agencies within the sector may focus on service delivery providing the necessary goods and services to the urban poor, filling a gap created by a lack of state institutions, and/or the urban poor's own limited capacities and activities. Some are micro-finance agencies which restrict their contribution to improving the effectiveness of the financial markets and/or particular goods markets, hence seeking to help achieve pro-poor economic growth and social mobility for individual households without substantive social transformation. Others have a cultural orientation and are concerned with the arts, history and other aspects that help to form identities and traditions, and which may provide protection for particular disadvantaged groups.

In terms of social transformation and progressive politics, the importance of civil society has long been recognised. Its significance as a counter-hegemonic power has been highlighted by Gramsci (Bebbington *et al.* 2008), who argued that civil society makes its (potential) contribution through being the institution that challenges prevailing influences on the structure of social relations, and specifically the ways in which exploitation and oppression takes place. In the present context in which low-income and otherwise disadvantaged groups are usually exploited and often dispossessed by capitalist forms of development, one of the most important contributions of civil society is to provide an institutional 'home' for the nurturing of an alternative vision of social justice, and associated systems and structures. The importance of ideas and the ways in which they emerge and are legitimised is manifest in the modalities and relations that influence the quality of life in urban areas, and particularly the quality of life for the most disadvantaged citizens. Less explicit but none the less important is the influence of civil society agencies on the moral values and the principles by which society organises itself. For an example, see Box 2.3 (Chapter 2) and the activities of sex workers in Bangladesh to challenge the discriminatory practices of both the neighbourhood and the police (Drinkwater 2009, p. 149).

There has been extensive discussion and analysis of the ways in which the shift towards neoliberal economic policies and the reduction in the role of the state and towards a market-based approach to development has affected civil society and particularly the NGO sector (see Banks and Hulme (2012) for a recent review). As governments reduced their expenditure in the financial and economic crisis of the 1980s and 1990s, NGOs found the needs within their target populations deepening and the numbers in these target populations rising. While development agencies may have provided additional monies for NGOs to undertake activities previously provided by the public sector (Bebbington *et al.* 2008; Lewis 2007), such funds did not directly substitute for state services and very few NGOs had the capacity to address infrastructure deficits. It should also be remembered that for most informal settlements, state provision for infrastructure and services had been partial or non-existent.

More positively, NGOs in some countries in the Global South have sought new opportunities in the context of democratisation which included the extended provision of welfare services and greater recognition of citizenship and citizens' rights. Civil society (and particularly NGOs) had sometimes provided a home for the political opposition and this helped to support new roles and enhanced relations between the NGO sector and the state. In Chile and the Philippines, NGOs became significant contributors to state programmes. NGOs have been encouraged to extend their activities as politicians and professionals who had been located in civil society sought to protect democracy with the creation of new quasi-civil society agencies such as the Human Rights Commission in South

Africa. From the mid-1960s, northern NGOs have been increasingly significant within international development assistance, and these trends have replicated this in the context of southern governments.

For northern NGOs, as their role increased there was a shift in emphasis towards investment in southern-based institutions. More recently, some other challenges associated with increased official development agency funding for NGOs have become evident. With increasing pressure on state agencies in North and South to be accountable for expenditure, more precisely defined quantitative indicators have been introduced with more stringent reporting for the northern NGO sector (Bebbington 2005).

In terms of the specific contribution to urban development, it is helpful to note the distinctive contributions of NGOs from the Global North and the Global South. General trends in the northern NGO sector over recent decades have been an increased share of official development assistance and a greater involvement in contractual activities for official development assistance, in addition to dedicated funding streams for civil society and NGO activities. There has been greater recognition of the potential role of groups such as trade unions and workers' associations as well as a growing role for transnational network civil society agencies (Edwards and Gaventa 2001), and this has led to greater recognition and support for social movements. There has also been an increasing interest in urban activities. At least up until the mid-1990s, many northern NGOs did relatively little in urban areas of the Global South because of a much more sustained commitment to rural development (Hall *et al.* 1996). The growing interest in micro-finance did something to counter this trend (as the urban context is often important here) but this area of work was not seen as relevant for many agencies. The 1980s saw the growth of a number of sectoral specialist agencies, particularly in the UK, in part reflecting a long-standing interest among professional groups in supporting development, and in part because of the structure of funding and the incentives for co-financing. These agencies have developed around particular sectors, such as water, environmental health, housing and engineering, and have somewhat favoured technical approaches to development within their own sphere of interest.[18] Some of these had largely urban agendas (for instance, Homeless International) while others explicitly developed urban agendas (for instance, WaterAid and Comic Relief). The growing interest in the rights-based approach to development encouraged northern agencies to build new alliances with a wider range of citizen groups, including social movement organisations active in both livelihoods and settlement level issues (Hickey and Mitlin 2009; Molyneux and Lazar 2003).

The remainder of this section focuses on the contribution of southern NGOs which, embedded within local political relationships, have a greater possibility to be considered legitimate political agencies.[19] We focus our discussion on organisations that have a particular interest in supporting pro-poor political change, i.e. that are not concerned solely with small-scale charitable welfare and/or market approaches. We elaborate three groupings of activities and then discuss the contribution of these NGOs to urban poverty reduction.

Campaigning for rights

In the Global South during the 1970s and 1980s, there was a significant constituency of local NGOs that focused on urban development and in particular the issues facing informal settlements and their residents, including evictions, basic services and related struggles for tenure and upgrading. The approach used is broadly consistent with the rights-based approach but significantly pre-dated its adoption by northern-based development agencies (Moser *et al.* 2001). For instance, in many Latin American nations during the 1970s, new local NGOs emerged to work specifically with urban poor groups, although often with considerable difficulty and danger as they worked in nations under dictatorships (see, for instance, CEUR and IIED-America Latina in Argentina, Sur and Taller Norte in Chile, DESCO and others in Peru, COPEVI and CENVI in Mexico, Ciudad in Ecuador, etc.). Indeed, many of their staff had been ejected from universities by governments.

One key change evident from the late 1980s was a shift from making demands for housing and basic services on the basis of need, to making demands on the basis of rights (see Hickey and Mitlin 2009; Leckie 1989). It was hoped that the use of the law and more scrutiny of governments' accord with international conventions and treaties would attract more attention to urban poverty reduction. Some NGOs favoured radical strategies, and in part this was influenced by the prevailing anti-democratic and anti-poor practices prevailing in some countries. In other countries such as India the issue was the continuing failure of democracy to deliver secure tenure and adequate housing to the urban poor (Mageli 2004). NGOs sought to challenge state practices by adopting confrontational strategies, including support for protest rallies and marches, critical reports and challenges to state practices. Confrontation was often encouraged by the uncompromising stance of the government. In Kenya, the offices of a Kenyan NGO Kitua Cha Sheria were attacked when they were contesting government policy in the 1990s and the NGO still continues to work in contentious areas such as urban landownership (Klopp 2008). In terms of success, as noted above, there was a significant wave of democratisation from the 1980s, and while specific attribution is hard to apportion, the contribution of civil society has been recognised. However, in other cases success has been harder to achieve; Mageli (2004) discusses the limitations of the strategies of Unnayan and the National Campaign for Housing Rights in India, while Thorn and Oldfield (2011) discuss the way in which a legal challenge disempowered and closed down alternative options for a group fighting eviction in Cape Town. Irrespective of immediate prospects for success, some NGOs continue to challenge repression and injustice in shelter issues such as, for example, in anti-eviction struggles (COHRE 2006).

Other agencies have concluded that a confrontational and often defensive stance may be of limited value, particularly if there is a democratic government with a strong claim to legitimacy. Moreover, NGO staff came to recognise that protests against injustice often failed to secure a change in policy. Even if they secured policy changes these were not then taken into changes in resource distribution and state practice. As Racelis (2008, p. 200) describes in the Philippines, NGOs also learned how the state and elites defend their interests and maintain political control despite promises of change. Other experiences also suggested the need for more complex political strategies. Alejandro Florien (then director of FedeVivienda) explained how, in the early 1990s, the social movements in Colombia had pressed for the right to housing to be included in the constitution. After massive mobilisation they were successful. However, the government interpreted the right as one of enabling access, and introduced a subsidy programme for 30,000 households a year; the movements, in Florien's analysis, were exhausted by their previous efforts and unable to challenge the outcome (personal communication, August 2003). Drawing on these experiences and perspectives, some NGOs followed other directions, and began to explore and diversify their tactics.

The contribution of professionalism

A second strategy taken by southern NGOs working on urban development issues has been 'problem-solving' on the ground working in informal settlements – with an underlying assumption that the state has a potential willingness and capacity to respond to the interests of the urban poor once solutions emerge. Such NGOs work in particular on sectors such as housing, land tenure and basic services; areas in which local government has particular significance. NGOs focused on particular technical capabilities, particularly the 'urban' professions such as architecture, planning and engineering. While this work appears to have a close fit with the urban management approach to urban poverty reduction, it focuses on the needs of the lowest-income groups and advocates direct intervention to improve opportunities rather than support for urban growth in general. Connolly (2004) discusses the work of the Mexican NGO sector in the 1980s, showing how agencies used their operational experiences to influence government housing policies and establish innovative participatory state-financed programmes to support community-led developments. The work of IIED-AL (*Instituto Internacional de Medio Ambiente y Desarrollo – América Latina*) in Argentina and the urban sector network members in South Africa are also illustrative of types of professional interventions (see Almansi *et al.* (2010) for the former).

For example, the IIED-*América Latina* began an engagement with the residents of an informal settlement (San Jorge) in the municipality of San Fernando on the periphery of Buenos Aires in 1987 when staff helped design and supervise the building of a mother-and-child centre there. This then developed into a long-term permanent engagement that included upgrading programmes (for housing and infrastructure), land tenure and the acquisition of a seven-hectare plot next door where a proportion of San Jorge residents could be rehoused – both to reduce density in the original site and to allow roads to be built. Other programmes have been developed, but with the external support agency always seeking to keep decision-making within the residents' organisations. These included a building materials yard to support residents in building, improving or extending their homes and a credit programme. The success of these interventions led to requests from neighbouring informal settlements for similar kinds of support. IIED-America Latina also worked with residents' organisations in informal settlements in mapping their settlements, when a new programme by the (then) private water utility sought to connect these neighbourhoods to the water distribution network. Here, the intention was always to engage local government and to show local government the capacities of residents and their organisations. This later led to work with other local governments – notably Moreno, where the scope of the interventions was municipality-wide (Almansi *et al.* 2011).

The Orangi Pilot Project (OPP) (see Chapter 4), operating in the very different context of Karachi (Pakistan), provides further insight into how this work added value through the development of new technical and organisational designs (Hasan 2008). The work of the OPP also shows how these technical and professional agencies began to align with citizen groups in formal and informal alliances for change. A further illustration is evident in Nairobi where research of the African Population and Health Research Centre highlighted how urban poor groups in Nairobi face very serious health problems. In the past decade, the work may have been taken up by other civil society agencies such as citizen groups. For example, resident associations in Nairobi that began as protests against the inadequacies of government by resident associations in specific areas (mostly middle-class areas) have developed into a Nairobi-wide and Kenya-wide association of residents' associations.[20]

For these professionally orientated agencies, experiences over the past two decades have led to a deepening and refinement of the strategies. There is now a greater recognition of the limitations of the approach and the necessary conditions for success. Key lessons appear to be the need to build up sustained relations with the state and (sometimes private) agencies responsible for basic services. There is also recognition that this part of the NGO sector may have most to offer when there is an evident commitment of the state to improve its work in the sector (e.g. health, education) but, on their own, are unlikely to have the political influence required to secure political commitment and a large-scale redistribution of financial resources.

NGO alliance-building

A third strategy, NGO alliances with people's organisations, has strengthened over the past ten years. This may be in part due to the adoption of a rights-based approach to development (Hickey and Mitlin 2009) but this direction also has a much longer-standing presence, and over 20 years ago Korten (1990) identified this as a 'fourth-generation' NGO strategy. Such alliances seek to influence state policies to be pro-poor and ensure that gains are cemented, with policy change being carried into programmatic and practice reform. Two further factors encouraging this strategy have been the need to reconcile contradictions as NGOs have been invited into partnerships by governments following broadly neoliberal policies and have failed to reduce poverty and inequality (Dagnino 2008, p. 59); and the constraints faced by those practising the 'professional' approach who found that they could win the argument but have little impact on policy (see Hasan (2008) and Racelis (2008) for Pakistan and the Philippines, respectively). And in Latin America, the depth of organised social movements led to many alliances and collaborations (Foweraker 1995). Box 3.5 summarises an example from Pakistan where the Urban Resource Centre in Karachi has networked together many of the community groups working with the Orangi Pilot Project as well as reaching out to other residents' associations in the city. As discussed more generally in Chapter 2, supporting social movements has emerged as being a distinct approach within urban poverty reduction. The programme interventions introduced in Chapter 4 and analysed in Chapter 5 all involve intense collaborative relations between social movement organisations and professional support agencies in alliances to shift political outcomes in favour of the urban poor.

Box 3.5 Urban Resource Centres in Pakistan

The first Urban Resource Centre was set up in 1989 by urban planning professionals and teachers, NGOs and community organisations to serve as a centre of research, information and discussion for all civil society groups within the city. The Centre provides an opportunity for regular meetings of a wide range of citizen groups operating in the city, including community groups.

As a part of its work, the Centre reviews all proposed major urban development projects from the perspective of low-income communities and interest groups, and makes these reviews widely available – for instance, through quarterly reports, monographs and a monthly publication, *Facts and Figures*. It organises forums that allow different interest groups to discuss key issues relevant to Karachi – and by doing so, has been able to develop much more interaction between low-income, informal communities, NGOs, private (formal and informal) sector interest groups, academic institutions and government agencies. For instance, research and forums have examined in detail the problems faced by flat owners, scavengers, theatre groups, commuters, residents of historic districts, working women, wholesale markets and transport companies. Related to this, there are discussions and negotiations between civil society groups and political parties and different tiers of government.

This Urban Resource Centre and the network of NGOs of which it is a part helped to get the Lyari Expressway stopped (it would have uprooted 100,000 people and caused immense environmental damage to the city) and replaced with a less damaging alternative. Its proposal for the extension of the Karachi circular railway into Orangi and other areas of Karachi has been accepted. It has also supported many other initiatives that changed government policies or the way government agencies work. Comparable Urban Resource Centres have been set up in other cities.

Source: Hasan (2007)

Other kinds of alliances have also been made between different groups within civil society. A wide diversity of civil society organisations and especially citizen groups have been recognised as relevant to urban poverty reduction, including worker cooperatives, insurance associations, community user groups such as water associations, burial societies, trade organisations, ethnic groups, faith-based associations, mothers' clubs and youth groups (Chen *et al.* 2007). As noted in Chapter 2, trade unions make a major contribution to the labour movement, although this has limited relevance for the lowest-paid and most vulnerable workers in the Global South as they are predominantly informal, even if they are casually employed by formal sector companies. Some civil society organisations have made considerable efforts to build collective organisations among informal workers who may also self-organise to reduce their disadvantage: however, in general, this has been limited (Pearson 2004, pp. 137–139 and 146). As noted in Chapter 2, civil society activities (generally trade associations for informal workers working with NGOs) have also been important in contesting regulations that have made livelihoods difficult for informal workers. Many of these activities have been concentrated in the sector of street vending. Other initiatives have included prostitution (Thorp *et al.* 2005, p. 913) and recyclers (Fahmi 2005, p. 158; Fergutz *et al.* 2011).

The Self-Employed Women's Association (SEWA) whose work is described in Box 3.6 is unusual in that it is registered as a trade union. As the name suggests, it is a union of women workers and it includes those who work from home (for instance, weavers, potters, handicraft makers) or who are hawkers, vendors or manual workers. It was among the first organisations to recognise how most low-income groups worked in the informal economy. SEWA's initial focus was on strengthening members' economic activities and much of its work today is on getting better conditions and incomes for its members. But it has also developed to support other neighbourhood-level improvements in services and other aspects of well-being (see Box 3.6), Two international groups, HomeNet and WIEGO, collaborate with SEWA and also support informal settlement workers. Their activities include assisting local groups, and more appropriate international processes such as greater recognition for informal workers within the International Labour Office (ILO) (Batliwala 2002). However, such international activities may have limited relevance for exploited workers within their locality as they are very often isolated and powerless to challenge adverse working conditions and low pay.

Box 3.6 The Self-Employed Women's Association (SEWA) in India

The Self-Employed Women's Association (SEWA) is the largest independent trade union in India and any self-employed woman can become a member. It was established in 1972 by a group of self-employed women textile workers as a union and organisation for self-employed women in the city of Ahmedabad. Its main objective was to strengthen its members' bargaining power to improve income, employment and access to social security. It campaigns for increased wages and better working conditions for self-employed women working in many different trades or sectors. SEWA sees itself not merely as a workers' organisation but as a movement. It might be considered to have emerged from three movements – the labour, cooperative and women's movements.

By 2009, it had 1.26 million members. It had expanded beyond its initial focus on urban (63 per cent of members were in rural areas) and on Ahmedabad and the state of Gujarat (half of its members are in other states). Among the financial services it offers are micro-finance (through SEWA Bank), micro-insurance and pensions. It has a range of programmes that support livelihoods (for instance, with market linkages to help sell products), health care and schools.

Improved housing and infrastructure is a pressing need for SEWA members, and SEWA provides shelter finance, legal advice and technical assistance. It has supported upgrading in various 'slums' in partnership with urban authorities and has also supported the training of construction workers. This has included working with the city of Ahmedabad and community-based organisations in the Parivartan (slum networking) programme and with other city or municipal authorities in improving infrastructure and services. For most self-employed women, there are obvious links between working conditions and the living environment as they work from home. In addition, those who work outside the home, such as vendors and rag-pickers, use their homes to store, sort and process their products.

SEWA participated in the Parivartan Programme to upgrade slums in and around Ahmedabad through the joint participation of government entities, non-governmental organisations, the private sector and low-income residents themselves. The Programme provides a water supply to every house, an underground sewerage connection, toilets in the home and an efficient storm water drainage system. Further benefits are street lighting, paved roads and pathways and basic landscaping, together with solid waste management. Costs are divided between the residents and the municipality (respectively Rp2000 (US$42) and Rp8000 (US$170)), and SEWA helps low-income residents meet their share of the costs by providing loans.

Source: Biswas (2003) and Cities Alliance (2002)

Contribution to poverty reduction

Over 20 years ago, Edwards and Hulme (1992, p. 15) identified four strategies for NGOs seeking to 'scale up' and expand the scope of their activities: working with government, linking to the grassroots with lobbying and advocacy, advocacy in the North, and increasing impact through organisational growth. NGO responses have included building and strengthening the capacity of local community organisations to engage with political agencies and processes to achieve both participatory governance (and inclusion in decision-making) as well as the improved delivery of affordable services. Arguably it is working with government that has led to a shift from the professional to the alliance approach, as some NGOs have realised that no matter how convincing their analysis and recommendations, significant redistribution of resources is unlikely without political momentum. NGOs, for example, have pushed for the reform of basic service delivery (e.g. water, sanitation) and won the arguments, but have still failed to secure the required resources. In other contexts, including Chile and South Africa, there has been political momentum and a strong engagement between the state and NGOs with the movement of NGO staff into government, but an underlying political commitment to local development and empowerment has not emerged. For these and other reasons, NGOs have broadened their links to grassroots organisations beyond lobbying and advocacy, and arguably many NGOs are now investing in stronger relations with grassroot organisations, even if they choose not to operate within very close levels of interdependency.

At the same time as explicit support for social movements, NGOs from both North and South have come to combine service provision with advocacy in recognition that there is no simplistic alignment of service provision with a charitable (non-political) response. While some radical agencies see service provision as supporting a neoliberal state, others have identified several advantages. On the one hand, innovations in services can lead to interesting engagements between the state and citizens such as co-production (Evans 1996), and help develop new solutions to address poverty and its consequences. At the same time, strong grassroots organisations demand practical assistance for their members in addition to participating in strategies for longer term political transformation – and provide evidence that the material matters (Chapman *et al.* 2009). In Chapter 4 we discuss some of the most interesting evolutions that have taken place in towns and cities in the Global South as community organisations have been encouraged to innovate and strengthen their interactions to build strong city-wide processes able to develop a new urban politics.

A number of lessons emerge from civil society organisations' work in urban poverty reduction. One first conclusion is that time matters. International development fashions are fickle and although serious political transformation is a long-term project, there are relatively few circumstances under which support is offered for a sufficient time. All the programme interventions discussed in the following chapter have evolved over decades. This is critical in a context in which relationships have to be developed, and solutions tested and revised. However, these are the exceptions in the broader field of NGO programming, and three to ten years of support is much more common.

A second conclusion is that a major problem for many initiatives is that the scale and scope are limited. Hence in larger cities it is not uncommon for several agencies to divide up the city with differential conditions for access in different neighbourhoods. A study of NGO activity in Faridpur (Bangladesh) (Saha and Rahman 2006) illustrates this through its discussion of the variety of services, terms and conditions, and opportunities in neighbourhoods across the city. A related problem also evident in Faridpur is that some areas are left out entirely. Moser's (2009) discussion of Plan International's work in Guayaquil illustrates the ways in which partial NGO programmes work within clientelist political systems to select some neighbourhoods (and leave out others).

Third, and as already discussed above, the involvement of the lowest-income and most marginal households and individuals remains difficult. In a context in which NGOs have limited success in securing state redistribution, many activities have limited relevance for the lowest-income and most vulnerable households. Many NGO activities, although not fully orientated to the market approach to development, still require local residents to make financial contributions to goods and services, and there are evident difficulties for universal inclusion (Hanchett *et al.* 2003). It is these exclusions that are part of the motivation for a more inclusive form of programming that requires political will and commitment. However, the earlier experiences remain important, and experiences show that NGO innovations in terms of providing lower cost incremental designs for new urban development options are a significant contribution.

Fourth, despite the commitment of many NGOs to a participatory process of development, the difficulties involved have been widely recognised (Cooke and Kothari 2001; Hickey and Mohan 2004; Illich *et al.* 1977; Wilson 2006). Hence while the contribution of professional agencies is widely acknowledged to be necessary for grassroots organisations to advance their needs and interests (for a recent discussion about the importance of this role, see Piper and von Lieres 2011), such relations are fraught with difficulties (see Bolnick 2008; Gazzoli 1996). Professional perspectives on urban development can exacerbate the challenges faced by grassroots organisations, especially if NGOs are working closely with local government and state agencies that favour conventional 'modern' models. While Laurie and colleagues (2005) discuss how indigenous movement organisations in Ecuador have been able to develop alternative modalities to interact with academic institutions and professionals, this has been little discussed in an urban context, despite the high degree of professionalisation within urban development. Chapter 4 returns to this theme with new models of accountability between professionals and networked community groups.

Conclusions

In this chapter we have sought to explain how the approaches to addressing issues of urban poverty described in Chapter 2 are transposed into specific programmatic interventions. As is evident from the discussion, there are a wide range of understandings, approaches and actions. However, success in poverty reduction has been limited, especially in the case of the lowest-income households and individuals (Mitlin and Satterthwaite 2013). Despite the investments of many agencies, very large numbers (and significant proportions of the total urban population) remain in situations of acute deprivation. Even when figures for income poverty fall as a result of transfers such as in South Africa or Chile, it is not evident that longer term structural disadvantage is addressed. Many other aspects of absolute poverty and of inequality such as access to basic services, enhanced voice and greater social and spatial inclusion may not be addressed. There is too little interest in ensuring universal access to the resources or services required to address basic needs, as evidenced by the partial targets within the Millennium Development Goals as well as the replication of partial programmes of service delivery at the level of national and local governments.

Some consistent factors emerge from the analysis of agencies in this chapter. This concluding section discusses these factors and how they seem to underlie the lack of success in agency efforts, and the continuing depth and scale in urban poverty and inequality. These suggest that the problem is more to do with the structural failings in policy and programme design than the poor execution of policy and programmatic interventions that are substantively robust. This chapter's key conclusions are summarised below and elaborated in the paragraphs that follow:

- However committed any national government may be, their programmes rarely survive a change in administration (in part because of the association of reform programmes with particular politicians or political parties).
- While democratisation is important in enabling redistribution, voice and accountability, many democratic governments do not appear to be pro-poor and/or to challenge inequality.
- The market dominates many transactions and market-based approaches are a significant component in many programmes. However, as suggested in Chapter 2, they have limited relevance for very low-income households who struggle to take part in these programmes.
- Clientelism is powerful and ubiquitous, despite efforts to prevent it, and it reappears in many development activities (in part because of the scarcity of resources that it does little to address).
- Most programmes do little to support urban poor groups' own priorities and capacities, and many reinforce hierarchical authorities that exacerbate these groups' lack of agency. This is particularly significant given the need for persistent grassroots pressure on the political system if more pro-poor outcomes are to be secured.
- Further to this, there is a need to recapture the collective in development activities, as it is only through collective action and collective pressure by urban poor groups that political progress can be achieved in reducing urban poverty. Too few development programmes have placed organisations of the urban poor at the centre of their process, investing in their capabilities to be representative and accountable to their members, and building the relations they need to have a substantive impact across towns and cities.

To improve the situation for many of the urban poor and particularly those living in informal settlements, there is a need for development assistance agencies to go beyond a passive static engagement with the state and other elites to more grounded negotiations with leaders that have strong and accountable relations with the residents of informal settlements. If low-income and disadvantaged citizens are to be included, they need to build their capabilities for collective organisation and action, as it is this that offers a source of power able to contest adverse processes within the state and the market. In particular, there is a need to build the institutional space and capability to negotiate and renegotiate outcomes with other political agencies. This makes for the sustained, measured and reflective interventions that open up the political system and enable representative groups to participate in the kinds of politics that redistribute resources to those in need and to change the rules around allocations and procedures (Leftwich 2008, p. 6).

National governments may be committed to programmes that address urban poverty, but the experience with respect to housing-related interventions is that such commitments do not last. Successful policies have been secured from the state and particular politicians have demonstrated a strong ideological and pro-poor commitment, but government appears inherently unreliable as programmes change in nature and content. Successful programmes are associated with politicians who lose power and their successors avoid reinvesting in the same activities, as they do not want loyalties to be confused; in these cases, others are unlikely to see the merit in financing activities that are identified with their predecessors – and often the opposition. Alternatively, programmes that show initial success are captured and changed as incumbent parties seek to build their following from their identification with such initiatives. Even if the sponsoring party remains in power, the balance may change once elections are won as clientelist forms of politics come to dominate the allocation of benefits. There appear to be particularly intense pressures on programmes that have transferred land and housing because of the scale of the assets involved and the attraction of 'capturing' such benefits for individual gain (see e.g. Rolnik's (2011) analysis of Brazil). These outcomes suggest that the urban poor depend on the commitment of governments at their peril. However, some initiatives are popular enough to survive changes in government – for national programmes and for city programmes.

In recent years, the design of programmes (both housing and cash transfers) has sought to avoid both clientelist practices and outright corruption, but the government's response has been to increase the level of technical rules and managerial direction (see, for example, the discussion above on the capital housing subsidy programme in South Africa). Strengthening technical components means that the programmes often have a poorer fit with the realities of those living in informal settlements and who work in informal jobs. While such technical components may be important in maintaining the quality of the intervention from government officials seeking to be more pro-poor, they frequently reduce the likelihood of grassroots organisations being able to control or even influence the activities and the final product, and hence they reduce the effectiveness of these expenditures for poverty reduction.

In the African and Asian nations that moved out of colonial rule in the past few decades, social transformation was anticipated following national independence and greater democratisation. Social transformation was also anticipated in the many Latin American nations that moved to or returned to democratic governments. The symbolic and material significance of the universal franchise is evident, but there have been manifest shortcomings in its influence on development options. Many of the urban poverty reduction programmes of both national governments and international development assistance agencies have given too little consideration to ways in which activities can include and strengthen local organisations of the urban poor. Civil society organisations have tried a number of different strategies to respond to this situation. Some community groups, social movement organisations and the professional agencies supporting them have placed less emphasis on formal and explicit political agencies (national government and political parties), and given more attention to capacitating and strengthening local collective action. At least in some places they have been able to contest political outcomes at neighbourhood and municipal levels and in a way that has been sustained over time both at and between elections. Some civil society groups are also placing less emphasis on engaging the state and more on building a citizen process (through residents' associations) to protect commitments and existing activities, and to encourage the state to move in particular ways. In other words, the response to a limited or fragile political commitment for social transformation has been that it is necessary to put in place institutional 'checks and balances' that keep the process on track. Chapter 4 looks at the evolution of these strategies and efforts to consolidate strong local organisations.

The belief in the importance of market-based approaches remains evident in many programmes – as does the need to find approaches that support scale but which ensure that the lowest-income groups are not excluded or seriously disadvantaged. The popularity of both direct-demand housing subsidies and shelter micro-finance in addressing housing needs suggests that innovation has been skewed towards those that are most able to address their needs through some level of integration with the market. There are many households that have been unable to benefit from the opportunities that are available because of low and unstable incomes. Further shortcomings in market approaches to service delivery are evidenced by the declining interest in private sector involvement in basic services for low-income groups, as low-income households simply cannot afford to pay enough for a piped water or sewer connection, health care or regular solid waste collection to make it profitable.

At the same time, the need for scale, particularly in addressing shelter needs with a context of relatively high-cost unit interventions, has shown many agencies and activists that there is a need for an engagement with the market. Three points need highlighting. First, many of those designing housing solutions for low-income households have recognised that the extreme dearth of options means that lower-middle-income households occupy solutions for low-income households unless they can have access to alternatives that better address their needs. A proportion of low-income households can afford to pay and there is a need for suitable provision. The challenge of design is one that allows these households opportunities to move ahead but at the same time creates a context in which lower-income households are not excluded from access to improved shelter. Second, the scale of need combined with the costs of housing makes financial contributions from households essential. Moreover, many urban households spend considerable amounts of money on housing. These resources are important to securing improvements, and will need to be incorporated. Third, there is the recognition that financial incentives are likely to be important in designing effective solutions. Or, put another way, providing free goods in a context of significant resource scarcity sets up a scale of competition that does not result in outcomes that favour those on low incomes.

The issue in regard to the market for basic services is equally difficult. Utilities and local governments appear to view formalisation of service provision as a source of revenue generation with increasing demands on the urban poor to pay significant charges.[21] This has evident implications for their well-being as it increases the pressure on their household budgets and reduces the amounts available for other essential goods and services. It may result in non-payment and the cutting off of access (with obvious negative effects on both individuals and in some cases the general condition of the neighbourhood). The consequence of these experiences is a recognition both that there is a need to blend market resources with other sources of finance (particularly state subsidies) to ensure that outcomes can also benefit low-income and disadvantaged households, and that market allocation systems need to blend with other allocation processes. Equally, public organisations (either state or non-state) need to have explicit social justice objectives and be accountable for their actions, all of which intensifies the pressure on social movement organisations and support agencies to develop effective political strategies that are then realised.

A further conclusion relates to the discussion in Chapter 2 which considers the implications of clientelist politics for the nature of authority within low-income settlements and between residents' associations and more powerful political agencies. The ways in which hierarchical vertical authorities impact on low-income and otherwise disadvantaged households is negative, reinforcing their sense of impotence and their secondary social status, and encouraging households to act as passive supplicants. Clientelist politics encourage the urban poor to be beneficiaries of an elite-dominated and self-interested political process. As soon as households engage through these relations with anti-poverty programmes, the programmes lose effectiveness because, although they may deliver essential resources, they do not empower the households to use them to best effect. Although resources are secured, and arguably at least some of the urban poor are encouraged to be pro-active in setting up 'deals', others are excluded from the benefits and relations between residents and 'leaders' may be coercive and violent. As discussed in Chapter 2, while some resources are secured, a wide range of studies have identified potentially negative impacts from such political relations. Such vertical authorities may also emerge from the developmental state in its various manifestations, including welfare delivery, managerial technocracy and as a powerful arbitrator. Top-down government programmes may be delivered on a very different basis, but in some guises they are also disempowering. Agency is not encouraged and hence benefits offered by the programme cannot be blended with the resources of individuals and households to improve their livelihoods and accumulate assets. Appadurai (2001) highlights the importance of addressing these themes when he articulates the need for a 'capacity to aspire' among disadvantaged communities. He argues for the importance of development activities that nurture this capability, and illustrates this through the work of the Indian Alliance, whose contribution to urban poverty reduction is elaborated in Chapter 4.

Thus, there is a need to recognise and rebuild the collective. Collective action remains a critical component of pro-poor political change for multiple reasons. The experiences recounted above show the importance of the collective in terms of strategic knowledge, capability to engage with processes of urban development and to negotiate with the state, changing awareness and understanding of disadvantaged individuals, challenging prejudice and discrimination among better-off citizens, and securing universal rights and needs. However, they also point to the importance of getting political strategies right, which includes combining protest with practical measures to address need, and building political reputation and relationships alongside a strong mass base organisation with a demonstrated 'social worth' and staying power.

Our argument is that it is the interaction between an unreliable (and often weak) state, too great a confidence in the capacity of low-income groups to participate in market relations, vertical authorities, and unaccountable and unrepresentative local organisations that accounts for the failure of many urban poverty programmes to date. What is notable is the relative lack of a scholarly literature on these problem areas. Despite this, they appear to have been widely discussed, deliberated on and acknowledged among urban activists working in the 1970s and beyond. These individuals have been working alongside and within these programmes and have experienced their shortcomings at first-hand. An analysis of the problems with the programmes discussed in this chapter has been combined with a knowledge of other experiences, including the failure of the state to address urban poverty with public housing construction and the need to build on manifest realities such as self-help initiatives and strategies to address clientelist politics if scale is to be achieved. It has also been combined with more recent trends, including an increased role for transnational civil society, disillusionment with the contribution of international processes to secure justice and poverty reduction and, at least outside of Latin America, an apparent trend towards greater income and asset inequalities. The experiences of these individuals and the ways in which they have been interpreted have led to experimentation with urban poverty reduction programming that began in the 1980s with the Orangi Pilot Project in Karachi and SPARC in Mumbai, and which continue to this day. Chapter 4 discusses the emergence and recent manifestation of such alternatives, and Chapter 5 analyses these interventions.

Notes

1 This may be seen in Abrams (1964), in the work of John F.C. Turner and William Mangin (see Mangin 1967; Turner 1966, 1976), and in the 132 governments that

formally endorsed the Recommendations for National Action at the UN Conference on Human Settlements in 1976; Hardoy and Satterthwaite (1981) reviewed government policies and practices in this in 17 nations; Hardoy and Satterthwaite (1989) described how such policies and practices had changed during the 1970s and 1980s; see also Angel 2000; Ward 1982.

2 Singapore is an exception to this, as housing conditions and access to infrastructure and services was greatly improved by public housing construction – but this is a nation that had among the fastest economic growth rates over long periods and that also had virtually no rural population, so the growth in the city's population was very low in relation to the growth in the economy. In addition, at independence, the state owned a high proportion of the land needed for public housing.

3 The poverty line in urban areas in Niger is defined as those households with an annual income below CFAF 144 750 (equivalent to about US$290).

4 In many Latin American nations, eligibility thresholds for programmes that are meant to be for low-income groups are often set at some multiple of 'the minimum wage' that seems generous (i.e. for those earning less than 2.5 times the minimum wage). But the figure for the minimum wage has usually not been adjusted for inflation or it may be set unrealistically low, so that it becomes far less than what should be 'the minimum wage' if poverty is to be avoided

5 For example, the Solidarity fund for low-income housing.

6 http://www.housing.gov.za/default.htm media release, 15 December 2008. Accessed 15 February 2009.

7 This section draws on a background paper prepared by Ross Wain.

8 See http://rioonwatch.org/.

9 See Chapter 6 for more details of how the very low standards set for what constitutes 'improved provision' for water and sanitation by the United Nations hides the scale of the deficiencies in provision in urban areas.

10 http://www.adb.org/water/actions/ind/bangalore-slums.asp.

11 Discussions with Zimbabwe Homeless People's Federation groups in Crowborough and Hadcliffe, and with staff of Dialogue on Shelter, January 2012.

12 Interviews with staff at Dialogue on Shelter for the Homeless in Zimbabwe Trust, Diana Mitlin, January 2010.

13 Annual meetings of residents (participants) elect delegates on a neighbourhood basis. Delegates determine priorities and then elect councillors, who sit on the Municipal Council of the Budget.

14 See, for instance, the OECD Development Assistance Committee's database showing the proportion of official bilateral commitments by sector over time. Reviewing the period from 1990 to 2010, water and sanitation receives between 3 and 5 per cent of commitments for most years while basic health receives 0.3 to 2.7 per cent from 1993 to 2004 and 3.0 to 4.1 percent for 2005 to 2010. Basic education received less than 2 per cent for the years 1993 to 1999 and less than 3 per cent between 2000 and 2010, except for 2004 when it received 3.3 per cent.

15 http://www.worldbank.org/projects/sector?lang=en.

16 http://www.iadb.org/en/projects/advanced-project-search,1301.html?sector=DU&subsector=DU-NEI&nofilter.

17 This is discussed in more detail in Chapter 6.

18 For example, WaterAid, Homeless International, Water for Kids, RedR.

19 This section draws on Mitlin (2011a).

20 See www.kara.or.ke for more details.

21 http://www.iied.org/basic-service-provision-shouldn-t-just-be-money-maker. Accessed 14 October 2012.

4 Citizen-led poverty reduction

Introduction

The economic realities of the 1970s and 1980s challenged the priorities given by governments and many international agencies to economic growth. At the same time, the experiences of professionals working on the ground challenged their understanding of social and political change. But the struggle of low-income urban dwellers for housing, services, inclusion and entitlements – or more broadly for well-being – continued, usually un-assisted and often opposed.

Urban informal settlements continued to be sites of multiple forms of resistance, particularly in nations in which the urban poor were seen as a source of political opposition by repressive and anti-poor governments. In addition, in many nations, demand for land from economic and urban growth brought pressures to evict informal settlement residents from inner city land or other well-located sites – for instance, demand from speculative construction or land for infrastructure. In some places, the use of democratic systems was seen as the best way to advance the needs and interests of the urban poor by their own organisations and other agencies concerned with social justice and poverty reduction, and alliances were created between the more politicised residents' associations and political movements to secure these systems. In other cases, democratic states had been in place for years or decades but substantive pro-poor response and redistribution was lacking.

As noted in previous chapters, in the absence of formal alternatives, informal settlements provide accommodation for large sections of the low-income population. Low-income groups need affordable shelter that is as well located as possible in regard to livelihoods and access to services – which in large urban centres generally means a need for effective transport systems (as locations within walking distance of these centres are no longer possible). Investment in bulk infrastructure is needed to enable household connections at relatively low cost – for water, waste water removal, sanitation, drainage, electricity and all-weather roads and paths. Urban development requires high-density accommodation options to reduce the scale of funds required for infrastructure (per person or household served) and to minimise the demands on the limited supply of well-located (and thus valuable) land. It also requires access to construction materials that can be used safely at high densities and that prevent the shack fires that are a devastating but regular event in so many informal settlements. A comprehensive provision of affordable shelter also requires an effective response to overcrowded inner city formal homes where one or more families or several adults rent a shared room because of their need for well-located space and lack of better-quality and still affordable alternatives. In summary, new settlement formation and neighbourhood design have to be thought about and planned at the city scale if they are to be efficient and effective (and housing prices kept down). In the absence of this, households have to find accommodation wherever they can and to accept far from adequate solutions.

There are many examples of important contributions made to addressing this problem within the history of responses to the needs and interests of low-income groups. In this chapter, we highlight five particular programme interventions, each chosen because they are indicative of particular traditions and because they help elucidate the present generation of responses to urban poverty. In this selection, we are aware that we are neglecting many others that may be considered equally worthy of inclusion. However, we believe that these five programmes help provide an understanding of how strategies have evolved, why they have been effective within their particular social, spatial and political context and temporal moment, and how their evolution led to a more substantive framework of intervention. The five programmes we look at here are considered in historical sequence. They have been active alongside the agency interventions described and discussed in Chapter 3 and some share similar features. However, the programme interventions described here are significant in their own right. None of them depended on funding from international agencies and some did not draw on such funding. We argue that it is support for these approaches that underpins the new generation of emerging poverty reduction programmes elaborated in Chapter 5.

We have termed them 'programme interventions' due to the need to find a term that goes beyond programme and organisation to encompass both formal agency components and the broader set of social agencies that engage with them. The term 'intervention' includes a broad array of structures, agencies, systems, processes and relations. We use this term to avoid implying that these interventions are simply organisations (although they all have formal organisations as a part of their work), social movements (although they include social movement organisations and less formal associations, and alliances with other organisations and associations), and are a part of civil society that is separate from political society (there are overlaps). Each of the programme interventions involves deliberate learning processes that have sought to refine their practice such that it results in social and political change that advantages the urban poor and those living in informal settlements.

These programme interventions offer insights into what it takes to shift politicians, political parties, civil servants, state agencies and political institutions (with their norms, values and ways of behaving) to be more pro-poor. They are seeking to change politics from the bottom up. This includes both action on the ground to address needs and renegotiating relations between citizen and state (and community organisations and the state). This is to secure political inclusion while taking account of the ways in which state agencies function and balance between representative and participatory democracy. It includes changing behaviours of both citizens and politicians, and redefining state politics, programmes and practices such that they nurture a more inclusive kind of politics. All five examples support the agency and collective capacity of low-income groups. Many of these interventions are also notable for their tenacity – their sustained efforts to resolve problems that arise – more than for their immediate success. Their histories include many revisions to their approaches as strategies are refined because circumstances change or because a particular approach proved less effective than was originally hoped.

The National Slum Dwellers Federation in India[1]

Background and urban poverty problem

There have been many low-income communities that have organised against the evictions that have threatened homes and livelihoods but there are fewer examples of city-wide and national organisations doing so. In this first example of programmatic interventions we give an account of the emergence of the National Slum Dwellers Federation in India, its efforts to build support for improved living conditions and the relations that have enabled this initiative to catalyse new approaches to the reduction of poverty and inequalities. Jockin Arputham was president of the National Slum Dwellers Federation during this period and subsequently became president of Shack/Slum Dwellers International SDI (see below). He also founded this Federation. In the subsection below, we draw on his personal story (see Arputham 2008, 2012) to begin the account in the late 1960s. The formation of this Federation's alliance with *Mahila Milan* and SPARC (termed the Indian Alliance) is then described as the strategic development that has enabled the process to go to scale. Later subsections describe the events that followed.

A perspective of the situation of pavement dwellers in India is captured in the analysis that Jockin offered during this first visit to South Africa in March 1991. He had been invited by the Southern African Catholic Development Association (SACDA) which called a five-day conference for South African shack dwellers to reflect on how they could realise their needs and interests as South Africa moved towards its first democratic government (Baumann *et al.* 2004). Jockin stated:

> The Indian Congress fought the British for independence and one person one vote. They promised the people that when the British were gone there would be milk and honey for all. The British has been gone for 44 years but all the poor get from their government is shit. A change in government does not mean a change for the poor and the homeless.
>
> (ibid., p. 198)

As described below, the past few decades have been a period of many challenges for the urban poor in India. However, despite multiple forms of disadvantage and exclusion, they have organised to challenge these outcomes and the 'solutions' provided by the state. There have been significant successes as the state has acknowledged that options have to be offered to the informal settlement dwellers displaced by economic development and, more recently, support for in situ slum upgrading. Remaining challenges are considerable as multiple levels of governments fail to provide for the needs of the urban poor.

Initial intervention[2]

Soon after India achieved independence, informal settlements formed within the island city of Mumbai. Those near the docks and major markets were evicted and their residents dumped at two locations which began to be known as Janata Colony Goregaon and Janata Colony Chembur. In 1967/1968, the Bhabha Atomic Research Centre gave notice to the people living in Janata Colony Chembur to vacate their land. Jockin (then aged 23) lived there, working as a carpenter (although he did not have a home and slept outside); he had already been involved in a range of community-driven activities including organising an informal school to help children there, getting a service to remove solid wastes, securing water connections and improving toilet maintenance. Jockin became involved in mobilising the 70,000 members of the community against this eviction threat. He talked to all the different community organisations in Janata Colony – cultural, religious, political – and helped bring them together. The Atomic Energy Commission were not only seeking to evict them but were also trying to avoid paying compensation. Jockin's work brought people together to oppose this; a central committee was formed with representatives from many different groups – Tamil groups (a large part of the Janata's population were from the state of Tamil Nadu), different churches/religions and different political parties. An all-party committee was formed with a mandate being no eviction, no negotiation, and a huge demonstration followed.

Jockin explains: 'The management of the Atomic Energy Centre treated us like shit; they demanded that we get out. They would not negotiate with us, they would not recognize us.' To contest the eviction, it was important for residents to prove that they were a permanent settlement and that they had the right to be there. It took six months to work out a memorandum that described how the settlement was formed so that residents could fight the eviction through the courts. The Colony had been formed in 1947/1948–1950 with 300-square-foot plots being allocated and everyone paying rent (2 rupees, 50 paise) to the municipality. Documentation proved that Janata was a permanent settlement, not a temporary one; the regular payment by residents to the municipality was proof that they were not illegal. To demonstrate to outsiders that they were a legitimate settlement, community organisations documented the range of infrastructure and enterprises there – for instance, the telephone poles, electric meters, shops and other enterprises, ration shops, and flour mills.

At this time, there were two main places in Bombay (later Mumbai) that the authorities wanted to clear – Janata and Dharavi. The clear space around Janata meant that it was possible to see the authorities coming and no one could enter Janata Colony unseen. Even the Shiv Sena could not enter Janata Colony, as it was a predominantly South Indian population. Jockin documented the history of the settlement, building up a range of contacts that would support them in their struggle against eviction, including the municipal commissioner – who had arranged the land occupation of Janata Colony by those evicted from a more central site – who was retired and living in Goa. The chief minister of Tamil Nadu also backed the residents against the eviction.

By 1970, Jockin was travelling all over Mumbai, visiting other slum communities (many of which were threatened with eviction). The Bombay Slum Dwellers Federation had been formed in 1969 to provide a unified body to defend the urban poor. When the Bangladesh war started, many refugees came from Calcutta (now called Kolkata). Jockin was invited to go to Kolkata by the Director of CARITAS and he was put in charge of housing in a refugee camp with 120,000 people. He mobilised 10,000–15,000 of the refugees to provide drainage and build houses, using bamboo, bamboo mats and gunny sacks. Jockin also arranged for people to be waiting in the jute fields just by the border to help the refugees as they streamed across the border and sought to avoid the army (see Arputham 2008).

When Jockin returned to Mumbai in 1973, the residents of Janata were facing renewed attempts to evict them. The NGO BUILD (Bombay Urban Industrial League for Development) sought to support his efforts, and it provided many international contacts in Sri Lanka, the Philippines and South Korea. BUILD established a famous left-leaning study group in 1972 that included many well-known professionals. Jockin was frequently asked to be present at their study meetings but after six months he challenged them: 'What have we achieved? After almost six months, I had had enough; this ideology flowing up and nothing getting done' (Arputham 2008, p. 329).

In 1973, with the Atomic Energy Commission pushing hard for the demolition of Janata, Jockin went with other community leaders to see the head of the Commission, Homi J. Bhabha. Bhabha refused to talk to them and there was much political action and lots of demonstrations to get the attention of the municipal government. Two days before the demolition was due to take place, the courts provided a stay order. Jockin knew that this was a temporary success. The Atomic Energy Commission went to court three days later to claim that they had followed the regulations – but the judge ruled that they had failed to serve notices to each individual household who would be forced to move. Thus, when the municipal employees came the following week to serve notice of the eviction, the activists had organised for all the residents to be out so that the eviction notices could not be served. This delayed the process (see Arputham (2008) for more details), and in this and other actions municipal employees supported the residents. Jockin argues that organising the settlement was easy because there were many young people who were unemployed.

When the eviction orders had finally all been served, the activists went to the citizens' court and won. The Atomic Energy Commission took it to a higher court and they won. After claim and counter-claim, the case made its way to the

Supreme Court. Before the Supreme Court met, Jockin managed to get in touch with two judges and the community won the case. But one of the judges told Jockin that it was too late to stop the eviction because the prime minister, Indira Gandhi, was personally interested. In 1974, Jockin and 22 other residents went to Delhi to try to talk to the prime minister and secure political support for stopping the eviction. After many postponements and delays, the group had a meeting with Mrs Gandhi. The meeting lasted for around 20 minutes and included five or six parliamentarians. Eventually, Mrs Gandhi agreed that there would be no demolition of Janata Colony without consultation with their action committee and Jockin demanded that this statement be put in writing. The police were waiting in Mumbai to arrest the leaders on their return but they avoided capture by getting off the train before it reached Mumbai.

By this time, Jockin had been put in jail many times – and usually kept there for three or four hours. He was arrested some 67 times. One day in 1974, he was arrested but, within an hour, more than 10,000 people had come to demonstrate. After he controlled the crowd, the police became friendlier and in this case (and others) they served him with an arrest warrant but did not arrest him.

On 25 June 1975, Mrs Gandhi declared a National Emergency; this was followed by the arrest of many leaders and the suspension of civil rights. During the Emergency, the Slum Dwellers Federation had to take care in distributing material, because passing out handbills was illegal. Supported by BUILD, they developed and printed handbills. During the Emergency, a slum eviction law ordinance had been passed and a huge demonstration took place.

Jockin was due to speak at the UN Conference on Human Settlements (Habitat) in Vancouver (Canada) in May 1976 but he was arrested. The British Broadcasting Corporation (BBC) announced the arrest. However, the court let him out again. Then, on the morning of 17 May 1976, the first demolition in Janata began, involving thousands of police and municipal workers. It was not possible to stop it and the eviction took place over 45 days. The families moved to an alternative site at Cheetah camp. All the organisation, mobilisation, pressure and use of influential contacts may not have stopped the eviction but it delayed it for many years – and the residents got a much better deal in provision for resettlement than would have been the case without this struggle. Jockin followed the advice of the police and left India for the Philippines and then other Asian countries. In Cheetah camp, the site to where the occupants of Janata had been moved, a home had been allocated to him. After returning to India from the Philippines, he travelled all over India – Karnataka, Tamil Nadu, Maharashtra, Andhra Pradesh – talking about his experience with the Janata Colony. He made many contacts during his journey – mostly with chairs of action committees, but also with staff of local NGOs.

In 1975 the National Slum Dwellers Federation was founded. Bombay/Mumbai had a large committee of slum leaders, and many other cities followed this example. All slums had action committees that were organised mainly to protest, demonstrate and march – making demands for land, water and cleaning of public toilets. Officials were elected or deputed by these organisations, but they had no funds. For instance, Jockin had a typewriter that went back and forth to a local pawn shop when there was no money available. It was during these years that he saw a need to change the approach. In the fight to avoid eviction in Janata Colony, he had encouraged and supported agitation, demonstration and militant opposition but he saw that there was little or no material benefit to the people. He had not built one toilet, nor had he been asked by the government if he or the Federation could build a toilet.

Strategic developments

The Society for the Promotion of Area Resource Centres (SPARC) was established in Mumbai in 1984 by staff who had previously worked for a conventional NGO. SPARC's staff embarked upon a quest on how to be a partner – and not a patron – of the poor who live in slums and on pavements. Through a process of dialogue and enquiry, the focus on land, housing and basic amenities emerged as a collective priority. As explained in this NGO's name, the initial purpose was to support area resource centres owned and managed by communities, who would see them as safe and secure places to meet, discuss their issues and invite others to meet them. An early challenge was to locate a space where this could be done. The first area resource centre was in a garage behind the Meghraj Sethi municipal dispensary in Byculla, right in the midst of the pavement dwellers' community with day and night access as the dispensary for low-income families. It was accessible by train and bus to those who lived slightly further away.

In June 1985, India's Supreme Court announced its judgment in a case about the city's right to evict pavement dwellers. It showed sympathy to the plight of pavement dwellers but allowed the city to evict them. The first Area Resource Centre in Byculla was inundated with pavement dwellers seeking advice, and the women who SPARC staff had met began to come to the centre. There was a huge contradiction in a choice of the way forward. The men connected to other NGOs in the city wanted to fight with the eviction squads, while the women's collectives wanted a negotiation that would allow them to live in the city. They had nowhere else to go and they knew that a violent demonstration would mean that the men would be jailed and money would have to be found for their bail. The women insisted that SPARC staff visit various municipal and government offices – and when they did this they realised that many myths about pavement dwellers circulated in officialdom. SPARC decided to do a census of pavement dwellers and this was titled *We the Invisible*. This showed how to do a census with slums and pavement dwellers (who rarely get recorded in official surveys or censuses); the results documented that pavement dwellers are not 'here today and gone tomorrow' but households with adults employed, living long term in the city. The women's collectives told SPARC that they wanted to explore why they would never get a house and also how to deal with the constant threat and fear of demolitions and evictions. This exploration to understand all the issues related to getting a house resulted in a 'self-managed housing training programme' for SPARC and the pavement dwellers.

Initially, the male leadership of NSDF showed their irritation with pavement dwellers. Jockin recalls: 'the pavement dwellers have always been considered by us as being foolish … why waste long term energies squatting on locations where you know you are in sight of the city and will face evictions?' Jockin and the other NSDF leaders observed how SPARC and the pavement dwellers undertook the survey and worked together. The tenacity and confidence of the leadership of women living on pavements impressed them. In 1985, SPARC entered into a partnership with NSDF. Then, in 1986, NSDF and SPARC together initiated the first pavement women's organisation, *Mahila Milan* (Women Together), concerned with finding ways to ease their daily struggles. *Mahila Milan* was initially a federation of six clusters of pavement dwellers who began to save, first for a crisis credit scheme and then for housing solutions (Patel and d'Cruz 1993). SPARC and NSDF believed that the slum and pavement dwellers and their organisations (not the NGOs or politicians) needed to be at the centre of the process. Urban poor communities themselves needed to develop the capacity to identify credible alternatives and to negotiate for them. The women pavement dwellers working with SPARC and NSDF began to strengthen their collectives and create a new learning and knowledge that would build such capacity.

There were no mass evictions of pavement dwellers in November 1985. The networks of pavement dwellers began to work with SPARC to examine how they could get a home of their own which was secure. The presence of NSDF in this alliance made the possibility of exploring alternatives for housing a real possibility through the experienced male leadership who had a much better grasp of these issues.

SPARC, *Mahila Milan* and the National Slum Dwellers Federation (NSDF) continued to collaborate to explore how professionals and community leaders could jointly produce strategies by which communities can seek 'the right to the city'. The 'right to the city' encompasses a spectrum of rights with the most critical focus for these organisations being to explore ways to build the capacity of the urban poor themselves to ensure secure land, housing and basic amenities. The alliance of SPARC, *Mahila Milan* and NSDF is foundational to the very basis of the narrative of how the poor seek to obtain their rights to the city. The processes of evolving these strategies are deeply embedded in the ways in which each of the three organisations were created, how they sought to align with others and what they all negotiated in this relationship with each other, with the wider body of informal settlement dwellers, the city and the city and state government.

The first major demolition that took place after the formation of *Mahila Milan* was a critical milestone. Unlike in the past, when each woman, terrified and fearful of the demolition squad, begged and pleaded separately, a collective strategy was discussed. The women sought a dialogue with the men sent to demolish their houses. They also reinterpreted the role of the police to maintain law and order which meant the police only interfered if there was violence on either side. The women encircled the demolition squad and engaged them in conversation while a few of the younger boys went to neighbouring settlements to spread the word so that within a few minutes there was a much larger number of women gathered together. The women volunteered to break down their own houses, telling the squad that this way it would not destroy their utensils and other household goods. With the help of the other women, the houses were dismantled and all vital materials, like wood and tarpaulin, were put aside and waste materials were dumped into the vans. The street looked as if the demolitions had taken place but, unlike past situations, their valuables and goods were safe, there were no arrests and the morale of all was high.

In the following months, Jockin and NSDF began working with SPARC and *Mahila Milan* to help them understand the politics of housing in the city. It was recognised that knowledge and understanding alone could not give land security to the urban poor, and that all land allocations were a matter of political choices, which communities of the poor had to decide they were going to influence. During 1986, NSDF met with *Mahila Milan* and SPARC and, without deep discussion or formal agreements, *Mahila Milan* became a sister organisation of NSDF: the partnership became known as the Indian Alliance. For SPARC, it was as if the search for a vibrant social movement to partner with had ended. NSDF, by agreeing to support the cause of pavement dwellers, had embraced their cause which, up until then, they had chosen to ignore.

Now, by engaging with SPARC and *Mahila Milan*, NSDF had not only included the pavement dwellers in their organisational focus, they had also initiated a very important process of dealing with women's collective leadership, which they gradually incorporated into their federations. NSDF was beginning to accept that their strategies, while very effective in the face of evictions, were hard to sustain, and one of the most significant reasons for this was that women were never part of their leadership. The respect and honour with which the leaders of *Mahila Milan* were treated by NSDF encouraged other community federations to make space for women and assist in the development of their capacities. The public recognition of the role of women entailed negotiations both with families and communities and set a benchmark for others. As a result, the partnership (or what came to be known as the Alliance) of SPARC, *Mahila Milan* and NSDF began. SPARC's role was to backstop the initiatives of the community process, open new doors (especially to government staff) and investigate new possibilities to see where they could lead. At the same time, SPARC was ready to abdicate to *Mahila Milan* and NSDF community leadership as soon as they were ready to take the innovation forward. NSDF sought to deepen and strengthen community voice and choice, and to expand its network to cover as many city federations as possible. *Mahila Milan* took over local processes like managing savings and credit activities, assisting in creating more space for women, and supporting the formation of neighbourhood organisations that could be turned into cooperatives or legal entities once they managed to gain access to the necessary resources. All activities of planning and delivery are interlinked and the three organisations negotiate more with each other than with the outside world. While each helps the other, the boundaries are managed very carefully. The legal face of the alliance is SPARC but all projects are activated and delivered by NSDF and *Mahila Milan*.

Achievements

Since the late 1980s, considerable innovations have been nurtured within and by the Indian Alliance in many cities in India. From the early 1990s, close relations were built with an embryonic process in South Africa to organise the residents of informal settlements and this resulted in Shack/Slum Dwellers International (SDI). The processes initiated by the Indian Alliance were refined and augmented through these relations and they are summarised in the section below on SDI. This section discusses the material and political gains that have been achieved in India.

The Indian Alliance has demonstrated to city governments the value of working with organisations and federations of 'slum dwellers' in which women play key roles. They have helped change the way pavement dwellers and slum dwellers are viewed. Indeed, Jockin, who was regarded by the government as one of the most dangerous agitators during the 1970s, was awarded one of India's highest civil awards (Padma Shri award) in 2011; this was also awarded to Sheela Patel, the director of SPARC.

In 1997, the Slum Rehabilitation Act came into being which states that all slum dwellers in Mumbai, including pavement dwellers who had documentary evidence of having voted in the 1995 elections, would be protected. The policy allows all of those living in informal settlements to receive a 25-square-metre apartment with water and sanitation – with this being financed by builders receiving transferable development rights. Since 1998, NSDF and *Mahila Milan* have supported the construction of over 5,000 units of such housing to demonstrate ways in which slum and pavement dwellers could take up such projects. Much of this has been possible due to a facility called the Community Led Infrastructure Finance Facility.

What began as six savings groups formed by the pavement dwellers in Byculla (Mumbai) in 1987 became a large federation of savings groups. By 2011, the Indian Alliance had extended *Mahila Milan's* savings networks to 65 cities with an estimated 750,000 savers. In terms of collective housing developments outside of relocation projects, about 3,000 units had been completed across the cities of Mumbai, Pune, Bangalore and Sholapur, and another 4,700 were underway with a further 1,500 under negotiation. An additional 1,000 individual loans had been given primarily in Hyderabad, Bangalore and Pondicherry, and a further 300 individual houses were being completed. In terms of relocation, which has taken place only in Mumbai, 83,500 households have been relocated to new housing with a further 5,000 currently in the process of relocation. The role of the Alliance varies; in some cases they may be involved in construction and in others their role is to facilitate the community process. The experience of designing relocation for pavement dwellers was useful when the government of Maharashtra and Indian Railways were negotiating a large infrastructure project with the World Bank. This required 35,000 households to be relocated from roadsides and along railway tracks to upgrade public transport. The Alliance participated in designing a policy which facilitated communities along the railway track to organise, undertake surveys and help design and manage their own relocation (Patel *et al.* 2002). This produced a scalable precedent which has meant that over 75,000 households are to be relocated instead of being evicted.

Similarly, women pavement dwellers designed and developed neighbourhood toilet blocks to address the issue of poor municipal infrastructure. Thus, when the authorities in Mumbai were negotiating a loan for sewage treatment, the reality that half of the city lacks connection to sewers and many defecate in the open produced a demand for slum sanitation. After much discussion this project stalled, but Pune's municipal council took that possibility and *Mahila Milan* constructed many neighbourhood toilets (Burra *et al.* 2003). With this example and continuing need, similar toilets were built in Mumbai and in many other cities. By 2012, over 1,000 community-designed and managed toilets with 20,000 toilet seats had been provided. Toilet provision has been spread across several states with about 90 per cent of the investment taking place in Mumbai and Pune. These toilets have been contracted through various improvement initiatives, most recently the second phase of the Mumbai Sewerage Disposal Project. The toilet blocks designed by *Mahila Milan* typically include pour-flush latrines for men and women, specially designed children's latrines (that allow for quick access and are at a scale suited to young children), and a community hall and caretaker's room integrated into the building. Other innovations introduced include separate toilets (and queues) for men and women, and large water tanks to ensure that water is still available when mains supplies are not. The communities in which these toilets are built are involved at every stage of the project – from choosing the site and planning and design to construction and supervision. Local federations take responsibility for maintenance and for the collection of fees that go towards maintenance and operation. Residents can get family passes which are much cheaper than paying per use.

The success of the Alliance in demonstrating the effectiveness of community managed development has resulted in its active participation with numerous housing projects of which three are currently ongoing. Under the Basic Services for the Urban Poor (BSUP) scheme of the Jawaharlal Nehru National Urban Renewal Mission (JNNURM), the Alliance is engaged in partnerships with local and municipal authorities for in situ upgrading of houses and infrastructure in three settlements in Puri, Orissa and Pune. In Pune, this includes the upgrading of a very high-density informal settlement. Here, as in most high-density informal settlements, residents much prefer in situ upgrading to relocation or rebuilding but there is a need to demonstrate how to do this – otherwise local governments will see it as much simpler to bulldoze the informal settlement and rebuild.

The Alliance also undertook an assessment of the Indian government's Basic Services for the Urban Poor (BSUP) programme in 11 cities. This pointed to limitations – not only in the failure to reach targets in many cities but more fundamentally the failure to involve the residents in the design and planning of what was to be done, or in decisions as to whether the slum should be upgraded (or the inhabitants relocated), or in implementation. In many cities, the approach to 'upgrading' did not involve incremental improvements to existing housing but bulldozing all buildings (with no provision for temporary accommodation for those displaced) and rebuilding, but often units were of poor quality. As Patel (2013) notes, many of the BSUP projects are simply public housing construction relabelled – and often with very inadequate provision for the 'basic services' whose improvement is meant to be at the centre of the BSUP.

The Alliance has also contributed to the demise of a redevelopment plan for Dharavi, the large, centrally located informal settlement. This would have displaced much of the population (estimated at over 300,000) and destroyed or disrupted livelihoods. Estimates suggest that Dharavi-based businesses have a turnover of over US$400 million a year (SPARC and KRVIA 2010). They are also fighting to stop the forced eviction of the 100,000 or so households who live on land adjacent to the airport (Arputham and Patel 2010).

The Alliance has also developed police stations in many informal settlements that are supported by resident committees (the police panchayats) through a partnership with the Mumbai police (and championed by its Commissioner). These bring policing to such communities for the first time, have police officers in these settlements who are accountable to the residents, and increase the effectiveness of policing and of complaint management through the role of the residents' committee of seven women and three men (Roy *et al.* 2004)

Orangi Pilot Project[3]

Background and urban poverty problem

The Orangi Pilot Project (OPP) is a local NGO that began work in Orangi's *katchi abadis* in 1980. *Katchi abadis* are informal settlements created by the unofficial subdivision of state lands, and Orangi was the location of the largest informal settlement in Karachi. The project was begun by the renowned development theorist and practitioner Dr Akhtar Hameed Khan, and the initial design was based on his concept of research and extension. After an initial period of study and discussion, sanitation was identified as the first point of intervention. The model of low-cost sanitation that evolved with sewer/drain connections for each household was rapidly adopted by the communities and had a substantive impact on the city within ten years. The 'component-sharing model', as it came to be known, places responsibility for building household and lane-level sanitation infrastructure on the residents, while the government (municipal authorities) are responsible for building and maintaining secondary infrastructure, including mains, disposal and treatment. In terms of the discussion in Chapter 2 it is a specific form of co-production. The first component is referred to as 'internal development' (or small pipes) by the OPP-Research and Training Institute (OPP-RTI) and the second component is known as 'external development' (or large pipes/drains).

By August 2012, 107,090 households had provided themselves with sanitation through 7,161 collective initiatives organised in lanes, representing 91 per cent of the entire settlement of Orangi. Collectively, communities invested P Rps. 124

million of their own money in their sewerage system, with the government investment being P Rps. 891 million.

In 1980, over 60 per cent of Karachi's population were living in *katchi abadis*. These settlements are created through a process where an informal developer occupies state land with the support and connivance of corrupt government officials who in many cases are informal partners in such development schemes. Low-income families buy and move into these unserviced plots and the developer makes arrangements for water to be supplied through tankers or through hand pumps where ground water is potable. In 2007, there were said to be 539 *katchi abadis* in Karachi (although some unofficial estimates put the figure at 702).[4] Of these, 72 per cent have been notified (that is, accepted by government and for which the process of provision of land title has been approved, so the settlement cannot be evicted). The acquisition of sanitation has played an important role in this formalisation process. The broader context has become increasingly difficult. In recent decades, Pakistan has long had a political culture of authoritarianism with a powerful military influence even when the government is democratically elected, it maintains its position through physical domination and alliances with political parties. In Karachi, political, religious and ethnic violence has been coupled with a booming real estate market where considerable commercial gains are made through construction contracts and sales.

The primary intervention

OPP staff engaged local communities through a participatory research process and identified four major problems: sanitation, employment, health and education. Sanitation was considered the most important although later programmes developed to help address the other needs. The staff were clear that foreign loans and grand state planning could not solve the infrastructure problem in *katchi abadis* – rather the solution lay in local resources and local expertise if costs could be reduced to what residents could afford. Thus, community involvement required that the costs of construction were reduced, with engineering standards modified, and residents mobilised both to finance and manage the construction of an underground sanitation system. For this to happen, OPP's work suggested that four barriers had to be removed:

1 Psychologically: communities had to be convinced that not only the house but also the street and the neighbourhood belonged to them (and that they should not wait for the state).
2 Socially: communities had to come together and organise.
3 Financially: costs needed to be affordable for households.
4 Technically: assistance needed to be available and communities had to be provided with tools, maps, estimates and technical supervision.

The internal development consists of sanitary toilets in each house connected to underground sewers in the lanes and neighbourhood collector sewers. External development consists of trunk sewers (box drains) into which the lane and neighbourhood sewers feed, and treatment plants. The funding for internal development has been generated by the community and organised at the lane level. Lanes typically include 20 to 40 houses and, as such, are small enough to be cohesive. Once a lane has applied for assistance, OPP-RTI staff survey the lane and develop a map and estimates for funding, labour and materials for the lane manager or the lane team. Lane committees are responsible for collecting and managing people's financial contribution. The lane manager or team organise the work with OPP-RTI supervision and managerial guidance. Initially, only those lanes that were near to a natural drainage channel into which they could discharge their sewage and waste water could participate. Later, lanes that were far from the drainage system began to apply. For these households to dispose into the natural drains, collector sewers were required. This led to the creation of a confederation of lanes that financed and built the collector sewers. In certain wards where the confederation of lanes was strong, the elected ward councillors funded this effort. OPP-RTI staff assisted as required.

OPP-RTI believes community self-financing for internal development is the only way to create a sense of ownership, a factor that is important in the construction phase and critically important during problem-solving and maintenance. It also ensures that the sanitation system will be used and be functional. OPP-RTI's experience from Orangi and in the many other locations where projects have been implemented shows that when subsidy is used, it often results in the collapse of the project. It creates dependence, and the community expects others to take responsibility for paying for the services, and when started in one community, this quickly spreads to other areas. It ends up with a whole population just waiting to be helped and not doing anything themselves. Moreover, community self-funding rapidly brings down project costs as activists simplify designs, negotiate prices with local skilled labour and strategise to avoid paying bribes. The cost per household for the sewage line in the lane, the house connection and the sanitary latrine pan works out at between P Rps. 1,000 and P Rps. 4,000[5] (approximately US$15–50 over the past decades). For the collector sewer, which links the lane sewers, the cost per household varies considerably depending on the length of the sewer. This cost is between P Rps. 100 and P Rps. 300, or an average of P Rps. 200 for each household. OPP-RTI staff believe that donor funding can be useful but it must be used strategically. It has also shown how if costs are kept to what households and communities can afford, there are no limits on scaling up. It also avoids the complications external funding always brings, as inappropriate conditions are often imposed by the funder.

The government's role in the 'big pipes' shows it has an important contribution to make and the principle of component- rather than cost-sharing provides a clear framework for their responsibilities and enables the effective use of their monies. It was clear to the OPP from the very beginning that the natural drains into which sewage was being disposed could eventually be converted into box trunks (converting the natural open drain into a covered, concrete drain) and treatment plants built, where they flow into natural water bodies. For such 'external' infrastructure, detailed plans and estimates are developed by OPP-RTI staff and/or its local NGO and CBO partners. NGOs and CBOs can negotiate with local government to fund external development. Local governments can support the process by building the external development and training their staff in OPP-RTI methodology and associated work with communities.

This model does not require a large number of OPP-RTI staff, as it trains members of the community and encourages them to self-monitor their work and performance. Activists do not receive a salary or compensation, but OPP-RTI offers them respect and encouragement to use their initiative and take on additional responsibilities. Activists are not closely monitored, but are supported and encouraged through various means, including information-sharing, the provision of training opportunities, and links with other government and non-governmental actors in the area of development. OPP staff seek to add value to community activities through:

- surveying and documenting what already exists – as in the careful maps prepared for each informal settlement in Karachi showing existing infrastructure on which new interventions can build (Orangi Pilot Project – Research and Training Institute 2002);
- providing a team of technicians and social organisers to support the community;
- supporting local activists who are aware of the problems, think about them, try to solve them, and are open to suggestions from others;
- offering a conceptual plan for sewerage development based on component-sharing, thereby breaking down each project into components that communities can manage, and balancing needs, resources and standards;
- ensuring that the local social organisational unit is manageable for local residents;
- respecting local communities as repositories of knowledge about existing conditions and circumstances, and with a potential to do more;
- documentation and dissemination of experiences and programmes with other professionals so as to build public support for the project and encourage contributions from academics and consultants;
- nurturing the monitoring of communication, constant feedback and transparency including public dissemination of the accounts of the organisation;
- the relating of local issues and realities to wider urban realities by ensuring that concerns are properly transmitted through dialogue to the relevant government agencies and politicians;
- avoid aiming for quick results and risk derailing the project.

Over time, OPP expanded beyond sanitation provision to cover programmes for health, credit, low-cost housing, water, education and support for communities to get secure tenure. By 1988, the OPP had evolved into four autonomous institutions to manage its expanding concerns. Since then the OPP Research and Training Institute (OPP-RTI) has been responsible for the low-cost sanitation, housing and education programmes, and more recent programme extensions including the secure housing support, water supply, women's savings and the flood rehabilitation programmes. The Orangi Charitable Trust (OCT) runs a credit programme in urban and, since 2005, mostly in rural areas (Hasan and Raza 2011), the Karachi Health and Social Development Association (KHASDA) implements the health programme, and the OPP Rural Development Trust manages some aspects of the rural credit programme. The OPP-RTI has developed an extensive network of community activists from all over Karachi and in other parts of the country where it has partner organisations implementing the low-cost sanitation programme and the credit programme. These partner organisations are part of the Community Development Network (CDN), which meets once every quarter to discuss both local issues and their links to wider national and international processes. These meetings are hosted by different partners across the network.

In the initial years it was assumed that OPP would be able to develop partnerships with government and, as such, be able to direct available government resources to supporting the sanitation programme. In practice this was not realised in the early years and this led the considerable evolution of the intervention's political strategy.

Strategic developments

The name itself, Orangi Pilot Project, captures the initial understanding which was that the government would adopt the sanitation design developed by the OPP (and what later became OPP-RTI) once its advantages had been demonstrated. But by the late 1980s it was evident that this was not going to be the case. Consultants with conventional high-cost and state-installed designs actively lobbied against the OPP proposals for city-level support and expansion, while officials and politicians were reluctant to sanction what appeared to be 'second-class' solutions. There was even a United Nations expert assessment that criticised the OPP approach as the wrong approach (Orangi Pilot Project 1995). A further problem was that the methodology disabled the ubiquitous clientelist relations between politicians and local community leaders by enabling the communities to improve their situation without small government grants; as a result there was limited political support.

OPP-RTI responded to this by continuing to support residents in their efforts to improve sanitation. Over time, this was important as the growing scale became evident to politicians. While the government seemed indifferent to OPP-RTI's professional dialogue, the NGO began to work more closely with residents' associations in the *katchi abadis*. To strengthen an engagement with and capacity to act on issues facing the city, OPP-RTI professionals worked with other interested groups and individuals (including teachers, students and activists) to establish the first Urban Resource Centre in Karachi in 1989 (Hasan 2007). Its founders were well aware that Karachi's planning process did not serve the interests of the low- and lower-middle-income groups, small businesses and informal sector operators.

This planning process was also creating adverse environmental and socio-economic impacts. The founders were aware that the city's development was being adversely affected by a process controlled by uninformed politicians, powerful real estate interests, international development agencies anxious to make loans, opportunistic national and international consultants, and profit-seeking contractors and companies. A core objective of the Urban Resource Centre was (and is) to modify planning processes through alternative research, advocacy, mobilisation of communities, and building and supporting alliances for change. It realises this through the following activities:

- collecting and disseminating information about the city and providing a resource centre;
- analysing city and federal plans for the city and debating them with all interested residents to achieve a consensus;
- identifying and promoting research and documentation on major issues facing Karachi;
- creating and nurturing professionals and activists who understand planning issues from the perspective of communities (especially low-income communities) and politically disadvantaged less powerful groups;
- providing a space for informed interaction and discussion, leading to the institutionalisation of that space.

From the late 1980s, the Karachi Urban Resource Centre nurtured alliance-building at the local level of NGOs/civil society leaving the OPP-RTI to maintain an emphasis on its professional approach and the relationship with urban poor communities and government agencies and officials. Networked together, the lane organisations and neighbourhood associations became a more effective influence on politicians. In addition to bringing together local civil society including NGOs, the Centre also provided a forum for all of those interested in urban development in Karachi including politicians, and professionals such as architects, planners and lawyers as well as the organised urban poor. This helped spread an understanding of OPP's critiques of the high-cost models of urban development continuing to be promoted by consultants and commercial interests. Meanwhile, the OPP-RTI continued to assist grassroots agencies to understand their neighbourhood development options, and the related city-wide development and planning options to help all residents to be aware of the different visions for their city (Hasan 2008).

Alongside the work of organised communities to engage politicians in investing in the infrastructure needed to make their local investments effective, OPP-RTI continued to reach out to further low-income neighbourhoods in the city. From 1989, OPP-RTI started to work outside of Orangi by documenting and mapping settlements and infrastructures and drainage systems across Karachi (Hasan 2006), and increasing levels of engagement with concerned government departments and agencies such as the Karachi Metropolitan Corporation and Sindh Katchi Abadi Authority, as well as Karachi community-based organisations.

Between 1989 and 1999, OPP-RTI began to build on the momentum established by the continued growth of sanitation by local communities in Karachi. At that time, there was a proposal for the Korangi Waste Water Management Project (KWWMP) being discussed for Karachi. OPP-RTI argued that it was flawed as well as expensive and would not solve the problems it was designed to address, and staff presented an alternative low-cost plan. They argued their case with the Karachi Water and Sewage Board (KWSB), government of Sindh departments, the Planning Commission in Islamabad, the President of Pakistan, the Governor of Sindh and the Asian Development Bank (ADB) whose loan was to part-fund this programme.

The experiences of OPP–RTI working as consultants to the Sindh Katchi Abadi Authority had shown staff that the proper mapping of the existing infrastructure of the *katchi abadis* was essential because it would be much cheaper to build onto existing systems than to build a new system. Mapping documents already exists on the ground (in terms of sanitation and drainage infrastructure); and it influences the government to align its investments with what already exists rather than to ignore it, as was previously the case. A comprehensive process of survey and mapping for hundreds of *katchi abadis* was undertaken to document the natural drains, their catchments and the work to date. The extensive documentation of sanitation infrastructure throughout Karachi, reinforced by statistics and maps, has had positive repercussions for planning efforts in Karachi and beyond, as these came to be used by government agencies, and increased OPP-RTI's standing and credibility.

Achievements

In terms of the scale of investments, in Karachi's *katchi abadis* (now known as low-income settlements because of the land tenure approval) people have invested P Rps. 161.48 million and the government has invested P Rps. 805 million in sewerage through ad hoc projects. Similarly, people have invested P Rps. 113.35 million in water lines and government has invested P Rps. 255.88 million. These households have built their neighbourhood sanitation systems, and their total investment is around one-sixth of what it would have cost if local government had undertaken the same work. Although this work has great significance in improving access to water and sanitation, this has not been documented and was previously ignored completely by governments and foreign consultants.

Outside of Orangi, by August 2012 the work of OPP-RTI expanded to 452 settlements in Karachi and 44 cities/towns as well as in 107 villages (spread over mostly the Sindh and Punjab Provinces) covering a population of more than 2 million. The replication projects have mobilized over P Rps. 779 million from local government funds to build 'external' sewage disposal systems, not only for their settlements but also for large areas of the towns and/or cities in question; the people's investment is P Rps.190 million. In two replication projects, water supply systems have also been installed on an 'internal–external' basis. In three small towns, staff working for the replication project have become a consulting team to the government for water supply, sanitation and road-paving projects that are all being built on the 'internal–external' concept.

Thus the Orangi Pilot Project works in far more places than Orangi, is not a pilot (although its approach has influenced government practice) and is far more than a project. In terms of policy impacts, OPP-RTI has both challenged conventional approaches to sanitation and drainage and put forward alternative plans. After a protracted struggle by OPP-RTI and various partners over the best solution to sanitation, the government decided to cancel a loan from the Asian Development Bank (ADB) for the sewerage and waste treatment investment (KWWMP) of US$70 million in April 1999. An alternative plan was prepared by OPP-RTI but was not taken up at the time (although it is now being implemented). This whole process unleashed a wider debate on city-wide sewerage, drainage and waste water treatment infrastructure, and this debate eventually led to the adoption, in government policies, of principles and practices espoused by OPP-RTI. Moreover, the learning from the experience, and contacts made during it, opened up new and related avenues for OPP-RTI advocacy work in the areas of housing rights, water management and governance.

Since 1991–1992 OPP-RTI has been working with government and communities for improvements in large-scale project implementation, continuing its efforts to inform government policy and practice through demonstration and research work. In 1991 the first ADB-financed large-scale project in Orangi began and this was followed in 1997 by a large-scale drainage project in one part of town serving a population of one million. In 2006 these efforts led to the government adopting OPP-RTI's plans as part of its Sanitation III project. From 2004 onward, the government implemented the city-wide adoption of the OPP-RTI designs for box drains, since now more than 70 per cent of all Karachi's sewage/drainage disposal system has been developed according to this plan. The work of OPP-RTI and its partners influenced the National Sanitation Policy which was approved in November 2006 and which has adopted the component-sharing model. OPP-RTI's influence on sanitation policies in Pakistan is illustrated by the formulation of this national policy. To secure this National Sanitation Policy with the inclusion of the component-sharing model the *most significant* factor has been the consistent support and promotion of this model by a number of government department officials and politicians (who occupy strategic posts within federal and provincial agencies) as well as by the members of WaterAid and UNICEF. As a result, government-appointed consultants at the provincial and national levels were appointed who understood and were supportive of this model; and this has been possible due to this support and understanding which has been nurtured over years of working together. The government appointed Arif Hasan (OPP-RTI's principal adviser) as the national consultant to draft the document. The policy relies heavily on the OPP-RTI model for implementing low-cost sanitation on a 'component-sharing' basis. This includes mapping as a fundamental step before any intervention, and the sharing of internal and external infrastructure development between citizens and the government. There have been numerous further examples of the acceptance of this model by the state. For example, in June 2004, engineers from the Works and Services Department of the Karachi city government requested OPP-RTI support for the conversion of *nalas* (streams) into box trunk sewers and also for the upgrading of existing drains in Karachi. The *nalas* that are to be upgraded are identified through local consultation and the OPP-RTI support consists of providing survey maps, designs and estimations. The city government's Taimeer-e-Karachi programme has allocated P Rps. 2.02 billion for the development of *nalas*. This development has been made possible by the fact that the OPP-RTI has documented 106 natural *nalas* with a total running length of 1,006,260 feet in the city.

There are many other aspects to the work of Orangi Pilot Project-Research and Training Institute that are not covered here.[6] This includes their work with the inhabitants of hundreds of *goths* (villages) that have become incorporated into Karachi and its periphery as the metropolitan area has expanded. To protect them from eviction, there was a need for them to be recognised as urban settlements and get land tenure. OPP-RTI has also been mapping available infrastructure in these settlements and advising on its upgrading while encouraging local government to provide bulk infrastructure. It has also been helping schools upgrade while OPP-KHASDA has been providing health care services. OPP-RTI has also long been active in documenting how water supplies could be managed better in Karachi in ways that not only increased available supplies and revenues to the government water utility but also improved provision and lowered costs to large sections of the low-income populations (see Rahman 2008).[7]

The programme presented a challenge to dominant development paradigms, which tend to take a prescriptive approach to development, are usually too technical, too reliant on government and donor support, and generally treat low-income communities as objects rather than drivers of development. When asked, OPP does not consider that it has accomplished anything remarkable in terms of implementation, intervention or invention. Staff considered that they evolved a low-cost and contextually appropriate system of management and implementation of local-level development. This low-cost system is built on the articulation and strengthening of what the people have been doing (in terms of addressing their development needs) through documentation and technical assistance. Many were building sewers before the organisation began its work. OPP-RTI staff focused their professional skills into working out how they could enhance these activities and then providing this support.

UCDO, CODI and Baan Mankong[8]

Background and urban poverty problem

In Thailand, the failure of economic growth to reduce inequality was evident by the early 1990s. The government responded with a willingness to introduce a particular programme orientated to addressing the needs of the urban poor living in informal settlements and with the goal to secure a more inclusive form of urban development. The programme was taken on by a leadership that recognised the significance of citizen-led development and which drew on a number of earlier innovations managed by both civil society and official development assistance agencies. Innovations included land-sharing schemes agreed between informal settlements residents and landowners (Angel and Boonyabancha 1988). Drawing on the earlier experiences across Asia, the Urban Community Development Office was set up in 1992 to provide an integrated response to the needs of the urban poor with funding windows for community strengthening, housing investment and livelihood activities. In 2000, the Urban Community Development Office was merged with the Rural Development Fund to form the Community Organizations Development Institute (CODI).

Initial intervention

In 1992, the Thai government provided the Urban Community Development Office (UCDO) with capital of 1,250 million Thai Baht (just over US$41million) to allow it to make loans available to organised communities so that they could undertake a range of activities related to housing, land acquisition and income generation. This funding also provided for small grants and technical support to community organisations.

Initially, loans were available to community-based savings and loan groups for income generation, revolving funds, housing and land acquisition (for instance, to allow communities threatened with eviction to purchase the site they occupied or land elsewhere, and to develop housing there), and housing improvement. The National Housing Authority provided some infrastructure subsidies, including to communities with UCDO housing projects. Any community could receive any of these loans, provided it could show that it had the capacity to manage savings and loans and that the loans could be used to respond to the particular needs of each group. As a result, UCDO developed links with a wide range of community organisations, savings groups, NGOs and government organisations. The loans had much lower interest rates than the other loan sources that urban poor households could turn to, although these rates were high enough to allow the initial fund to be sustained and to cover administrative costs.

As the savings groups that worked with UCDO became more numerous and larger, UCDO found it more difficult to provide support to individual groups and effectively address their needs. This difficulty in scaling up its work brought UCDO to a new stage, where it linked individual savings groups so that they formed networks at various levels. City and provincial savings group networks were assisted to join together and negotiate with city or provincial authorities, or to influence development planning, or simply to work together on shared problems of housing, livelihoods or access to basic services. UCDO loans were now also provided to networks of community groups, who then on-lent to their member organisations. The emergence of large-scale community networking brought immense change to UCDO. These networks linked communities so that they could share their experiences, learn from each other, work together and pool their resources. These networks became increasingly the means through which

UCDO funds (and later CODI funds) were made available to low-income groups. Networks might be based around occupations (for instance, a taxi cooperative), pooled savings and cooperative housing. There are also community networks based on shared land-tenure problems (for instance, networks of communities living alongside railway tracks or under bridges). When networks manage loans, decision-making processes are decentralised so that they are closer to individual communities and better able to respond rapidly and flexibly to opportunities identified by network members. Networks helped communities manage debt, and allowed UCDO to remain effective despite the economic crisis that started in 1997. Most community networks also developed their own community welfare programmes. What became evident from UCDO's work is, first, how far funding can go if organised and managed by community organisations or networks; and second, how many community-managed activities can achieve cost recovery.

The networks enabled UCDO to expand beyond loans. Over time, UCDO added other activities, including a small grants programme for community-managed environmental improvement projects (which supported 196 projects benefiting 41,000 families; see Boonyabancha 1999); help for savings groups facing financial difficulties maintain their loan repayments after the financial crisis of 1997; and community welfare funds, made available to communities for use as grants, loans or partial loans for education, income generation and other welfare (for instance, school fees, those who are HIV positive, the sick or the elderly).

Despite building a strong base among savings groups and related networks, communities faced many challenges. Many community groups seeking to improve their housing situation struggled with affordability. In Bangkok, the only low-cost land available was far from the city centre and households that purchased these plots struggled to repay, often because they had to rent a bed or room in the city centre to maintain their livelihoods. The networks recognised that a better strategy was to negotiate to remain on the informal sites they occupied close to the city centre. But infrastructure upgrading costs were often high and tenure security was difficult to achieve. Additional finance was needed for the lowest-income communities to really be able to improve their neighbourhoods. It was also recognised that the local revolving funds set up and managed at the community level were critical. Each of these is the community's own financial mechanism through which they had the flexibility to address their own needs rather than simply respond to 'windows' provided by external support agencies.

Strategic developments

The Thai government recognised the successes of UCDO and in 2002 the Community Organization Development Institute (CODI) was established to continue and extend this work. While UCDO had been located within the National Housing Authority, CODI's separate legal standing as an independent public organisation provided it with greater possibilities (for instance, being able to apply to the government budget for funds), greater flexibility, wider linkages and new possibilities for supporting collaboration between urban and rural groups. The emphasis on supporting community-managed savings and loan groups and community networks extended to 30,000 rural community organisations, and many community networks that CODI supports include both rural and urban community organisations. CODI recognised that for pro-poor development to take place, relations between low-income groups and the state had to change, and critical to that change was the establishment of representative accountable local organisations. From the outset, CODI sought to bring together different interest groups – with senior government staff, academics and community representatives sitting on its board.

Despite progress, significant challenges remained. In 2003, some 5,500 low-income communities, with 8.25 million inhabitants, were living in poor-quality and often insecure housing in 300 cities. In 3,700 of these communities, land tenure was insecure and 445 communities were under threat of eviction. Between 70 and 80 per cent of this population could not afford conventional housing, either through the market or through the government housing programmes. The government responded by introducing *Baan Mankong* ('secure housing'), a national programme for upgrading and secure tenure, in January 2003. The programme initially set a target of improving housing, tenure security and access to basic services for 300,000 households in 2,000 poor communities in 200 Thai cities within five years. Recognising the work of CODI in strengthening local organisations, reducing poverty and addressing inequality, *Baan Mankong* was passed to CODI for implementation.

The *Baan Mankong* programme channels government funds in the form of infrastructure subsidies and housing loans direct to low-income communities, which plan and carry out improvements to their housing environment and to basic services. Infrastructure subsidies of 25,000 baht (US$625) per family are available for communities upgrading in situ, 45,000 baht (US$1,125) for re-blocking and 65,000 baht (US$1,625) for relocating. Families can draw on low-interest loans from either CODI or banks for housing, and there is a grant equal to 5 per cent of the total infrastructure subsidy to help fund the management costs for the local organisation or network. A further programme, *Baan Ua Arthorn* ('we care'), was also introduced for households able to afford low-cost formally constructed housing with monthly repayments of between US$25–37 and this was managed by the National Housing Authority.

Baan Mankong was set up to support processes designed and managed by low-income households and their community organisations and networks. These communities and networks work with local governments, professionals, universities and NGOs in their city to survey all low-income communities, and then plan an upgrading programme to improve conditions for all these networks within three to four years. Once the plans have been finalised, CODI channels the infrastructure subsidies and housing loans directly to the communities. These upgrading programmes build on the community-managed programmes that CODI and its predecessor UCDO have supported since 1992, and on people's capacity to manage their own needs collectively. They also build on what these communities have already developed, recognising the large investments that communities have

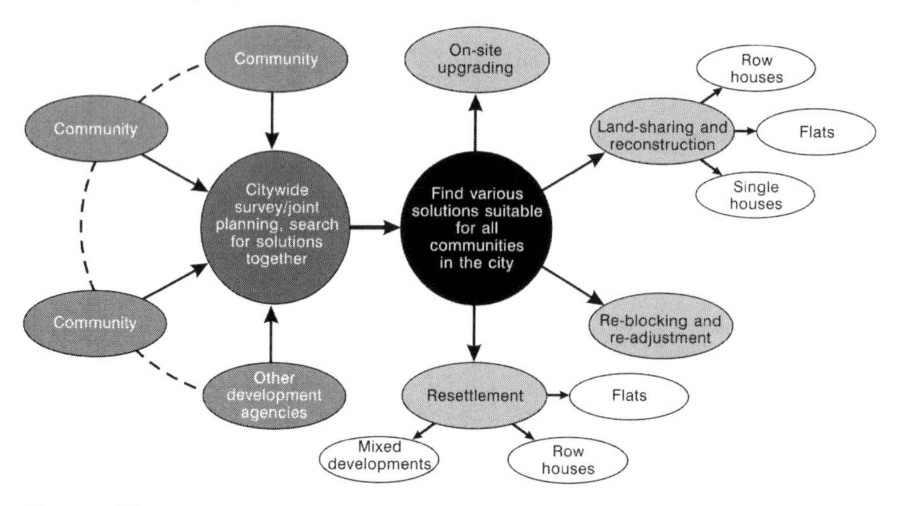

Figure 4.1 The linkages for a local housing development partnership by city-wide networks with communities and local authorities

already made in their homes. Upgrading existing settlements is supported whenever possible; if relocation is necessary, a site is sought close by to minimise the economic and social costs to households. The programme imposes as few conditions as possible in order to give communities, networks and stakeholders in each city the freedom to design their own programme. The challenge is to support upgrading in ways that allow urban poor communities to lead the process and generate local partnerships, so that the whole city contributes to the solution.

Figure 4.1 illustrates the process through which a city-wide upgrading/housing development programme is developed, bringing all key formal and informal agencies together. The design of a city-wide upgrading programme, and the city network necessary to implement it, involves certain key steps:

- identify the stakeholders and organise network meetings, which may include visits from people in other cities;
- organise meetings in each urban poor community, involving municipal staff if possible;
- establish a joint committee (including urban poor community and network leaders, the municipality, local academics and NGOs) to oversee implementation and establish new relationships of cooperation to integrate urban poor housing into the city development plan;
- conduct a city meeting where the joint committee meets with representatives from all urban poor communities to inform them about the upgrading programme and the preparation process;
- organise a survey covering all communities to collect information from all households about their tenure, housing and access to infrastructure. This

provides opportunities for people to meet, learn about each other's problems and establish links;

- from the survey, develop a city-wide community upgrading plan;
- support community collective savings;
- select pilot projects on the basis of need, communities' willingness to try them out and the learning possibilities they provide for those undertaking them, and for the rest of the city, preparing development plans for pilots, starting construction and using implementation sites as learning centres for other communities and actors;
- extend improvement processes to all other communities and individuals in need;
- integrate these upgrading initiatives into city-wide development, including providing secure tenure or alternative land for resettlement, integrating community-constructed infrastructure into larger utility grids, and incorporating upgrading with other city development processes;
- build community networks around common landownership, shared construction, cooperative enterprises, community welfare and collective maintenance of canals;
- create economic opportunities wherever possible within the upgrading process;
- support exchange visits between projects, cities and regions for all those involved, particularly community representatives and local government staff.

Source: Boonyabancha (2005)

In its implementation, CODI has to address the reality that many think that the municipality should manage the city – but city authorities do not have much power and governance systems need to be opened up so that citizens feel it is their city and that they are part of the development. Responsibility for different aspects of city management can be decentralised to communities – for instance, for public parks and markets, maintenance of drainage canals, solid waste collection and recycling, and community welfare programmes. Opening up more room for people to become involved in such tasks is the new frontier for urban management – and real decentralisation. Upgrading is a powerful way to spark off this kind of decentralisation. When low-income households and their community organisations do the upgrading, and their work is accepted by other city actors, this enhances their status within the city as key partners in solving city-wide problems. City-wide processes are now underway in many cities – and the form this takes is illustrated in Figure 4.2.

Achievements

By 2000, when UCDO's work was integrated into CODI, 950 community savings groups had been established and supported in 53 of Thailand's 75 provinces; housing loans and technical support had been provided to 47 housing projects involving 6,400 households; grants for small improvements in infrastructure and living conditions had been provided in 796 communities, benefiting 68,208

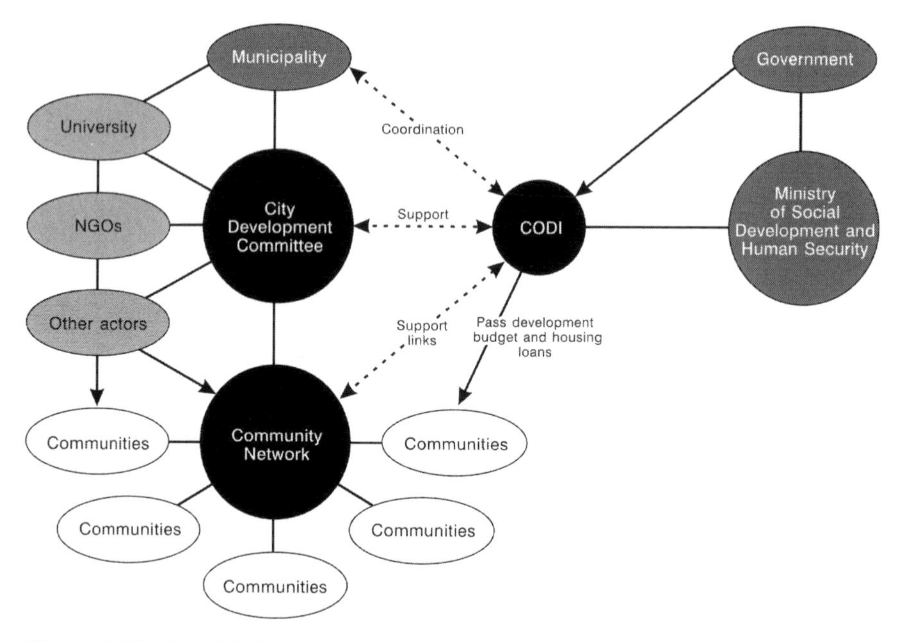

Figure 4.2 The *Baan Mankong* Programme mechanism

families; and more than 100 community networks had been set up. More than 1 billion baht (around US$25 million) had been provided in loans, and more than half of the loans had already been repaid in full. Informal estimates suggest that assets of some 2 billion baht had been generated by the projects. The special fund to assist savings groups facing financial difficulties had helped many communities and community networks to manage their debts and continue their development activities.

Between 2003 and 2010, within the *Baan Mankong* programme, CODI approved 745 projects in more than 1,300 communities (some projects cover more than one community) spread across some 249 urban centres and covering more than 80,000 households. Sixty-four per cent of beneficiaries belong to communities that were upgraded in situ with long-term secure collective tenure. Fourteen per cent of the beneficiaries relocated to new sites within two kilometres of their former homes. Those communities that moved to public land negotiated long-term collective leases to that land; the ones that moved to private land purchased the land at prices they negotiated with CODI with loans made to the community cooperative. By 2010 the total number of households reached by the programme had grown to more than 25 per cent of the numbers that *Baan Mankong* targeted, but they still represented only about 13 per cent of the 600,000 families in need within towns and cities in Thailand. During the same period, grants for infrastructure upgrading exceeded US$46 million; and loans for land and housing exceeded US$52 million. More than 82 per cent of households supported by CODI are now living in settlements that have also achieved tenure security, via long-term leases or collective landownership. By April 2012, the *Baan Mankong* programme had led to the upgrading of over 91,000 houses across 270 towns.

CODI has provided an example of how governments can support an integrated approach to poverty reduction with the simultaneous building of community organisations, informal settlement upgrading, housing and income generation. The model has been important in illustrating a partnership approach with community organisations playing a major role in both implementation and decision-making. The emphasis on city-wide approaches that seek alliances between middle-class and lower-income residents demonstrate how to pre-empt some of the more exclusionary urban politics that have been seen in other cities. They have demonstrated the contribution of multiple community networks collaborating together to promote pro-poor urban change. While the government remains of primary importance in terms of loan capital and subsidy finance, from 2010 the networks established their own savings-based loan funds following delays in recapitalisation by the government and anxieties about the reliability of continued political support. This recognition of the importance of an autonomous organising approach replicates earlier conclusions made by Castells (1983).

CODI staff have shared key lessons about their work and achievements:

- The city-wide scale that *Baan Mankong* supports is critical for the new kind of slum upgrading by people. Working on a city-wide scale makes evident the differences between informal settlements within the same constituency, and between neighbourhoods. People begin to understand these differences – for instance, differences in landownership and legal status, differences in the availability of infrastructure and in housing and environmental conditions, differences in people, and differences in degrees of vulnerability. People living in low-income communities have the opportunity to compare different experiences and realities and this sparks off the question 'why are there these differences?'
- Horizontal linkages between the communities that are their peers prevent individual communities from being isolated and draw communities together into a process of making structural changes. Almost all systems related to power and wealth in present societies are vertical hierarchical systems. Once this is recognised, and low-income groups are within a more horizontal process whereby they have a chance to think and understand – and choose together – they gain a new decision-making power. When people from the different urban poor communities look together at the city-wide scale, this new view offers a larger, clearer picture of the city in which they live. The problems of land tenure, infrastructure, housing and services are not isolated, separate or unusual but intimately linked to the larger systems of governance and the allocation of resources. As peers come together in a city platform they see how their problems (and the possibilities of resolving them) relate to the structure of the city and its governance.
- The process of choosing the pilot upgrading projects in a city needs to belong to urban poor organisations and networks. These projects show everyone in

the city something tangible. Different communities frequently use different criteria for making this choice but what is important is that the group understands the reasons for choosing the projects. This makes them a part of the pilot, even if the project is not their community. As communities watch the development of the pilot projects they chose, they begin to look at their own situation in a new way and prepare themselves and make changes. The pilots are powerful examples because they are being undertaken by peer communities themselves. Problems arise if the projects are selected by outsiders, either professionals or municipalities, not least because the communities that are not chosen feel left out.

- The *Baan Mankong* programme helps city authorities and other groups to see the problem of informal settlements as something normal, and also as something that can be improved. It changes the usual perceptions of the problem, which positions city authorities against the urban poor communities. Officials, politicians and other groups begin to look at all these people in urban poor communities as normal urban citizens who are located where they are because of the way the city has developed. With this recognition it is possible to move forward and address these problems. Once city authorities begin to improve the situation, they automatically become a part of the city development agenda. Very often, the view held by large sections of the city's society that informal settlements are not or should not be part of the city is so pervasive that the slum dwellers themselves have begun to believe it.
- The physical form of the upgrading is neither the issue nor the problem. Upgrading offers the possibility to create a change going beyond the physical aspects to transform relationships and allow the urban poor greater space and freedom. Finding technical solutions for all these communities (whether in situ improvement, re-blocking, land-sharing or nearby relocation) is the easy part. The inhabitants will also make important contributions to the physical aspects of upgrading, with lots of variety and creativity. Thus the big question is how physical upgrading can include these relational aspects.
- City-wide and nation-wide upgrading may be difficult at first. But once there is the space for urban poor communities to be involved, the process will go in the right direction almost by itself. Having many initiatives moving in the right direction is more important than having some attractive but isolated 'model' projects. There are concerns that there are not enough architects to assist in the upgrading planning processes. CODI responds by using the resources that are at hand. In many cities, the community networks have persuaded the local universities to provide technical support. Some have hired professional architects from design firms. As the upgrading programme progresses, the more sophisticated the communities become about physical design. Good design ideas are getting seen and noted, and are spread around by the people themselves.
- Doing things collectively – for and with the urban poor – is critical. Reviving this culture of collectivity in low-income communities is far more important than any physical upgrading or any housing project. These people need to

build on the collective sharing that was born of necessity. This can be nurtured if the collective process includes everyone with collective housing, collective tenure and work being shared. The goal of the *Baan Mankong* upgrading is to get everybody in that community to have tenure security, to become recognised as legitimate citizens and this includes those who cannot buy a unit of any sort, who perhaps don't have any income. This is a challenge for all residents and everybody has to pool their resources, their ideas and their creativity.

- Upgrading is particularly important in addressing illegality, lack of security and lack of rights. With improvements to rights and security, people's status in the city changes and with legality comes key aspects of citizenship. In earlier upgrading models, the main concern was physical improvements to the infrastructure and housing. These physical improvements are important, and are the most visible and easy to comprehend (and measure). But it is important that the people in that community feel they are 'upgraded'. There is recognition that if the residents of informal settlements aren't changing, then things aren't changing. Upgrading is a process in which the residents of informal settlements begin to believe in their own power and see that they are no different from all the other citizens in the city. Once they believe in their power, they start looking at things differently, and can adjust their relationships with other actors in the city. Upgrading provides a space in a city in which the local authority, the network and the community can interact and work together.
- Savings and credit are important in linking people together, collecting people to work together and think collectively. But the most important side of savings and credit activities is that they teach communities to manage finance collectively – both their own savings and outside finance. This helps ensure that the people themselves become key actors in development. When a community manages its own finances, it learns financial systems that are transparent, equitable and effective. Financial management skills support new development possibilities and enable people to work together and develop the ability to deal with the development of all the members of the group. This capability to manage finance as a group is something that has to be learned, practised, strengthened and matured. Finance is crucial because once people are able to manage finance collectively, they gain security from the market. It offers collectives a degree of maturity, which means that communities no longer have to be looked after by anyone other than themselves.

Shack/Slum Dwellers International (SDI) (and affiliated Homeless and Landless People's Federations)[9]

The formation and development of the National Slum Dwellers Federation in India was described above. This section is about the other national federations of slum, shack and homeless people that have developed and the international umbrella organisation they have formed with the local NGOs that support them – Shack/Slum Dwellers International.

Background and urban poverty problem

As Jane Weru, ex-director of Pamoja Trust and now director of the Akiba Mashinani Trust (a Kenyan fund for the Kenyan Homeless People's Federation to use for capital investments), has explained:

> The people in Shack Dwellers International, in the leadership of the Federations and in the support organizations, are mainly people who are discontent. They are discontent with the current status quo. They are discontent or are very unhappy about evictions. They are people who feel very strongly that it is wrong for communities, whole families to live on the streets of Bombay or to live on the garbage dumps of Manila. They feel strong enough to do something about these things. But their discontent runs even deeper. They have looked around them, at the poverty eradication strategies of state institutions, private sector institutions, multi-laterals and other donors. They have looked at the NGOs and the social movements from which they have come and they are unhappy with most of what they see.
>
> (SDI 2006)

It is this discontent that has become a catalyst for change, driving the formation and expansion of this alliance of people's organisations and NGOs that together are seeking new and different ways to end homelessness, landlessness and poverty. The network emerged to bring together and empower homeless and landless people's federations and their support NGOs with the understanding that existing strategies will not reach scale as they do not link to the people's efforts to improve their homes and neighbourhoods, and nurture pro-poor political relationships. But like many such initiatives it did not begin with a grand plan for a transnational network; rather it emerged from activities in a number of places. As the work of the National Slum Dwellers Federation and the Indian Alliance has already been described, we begin the story of these federations in South Africa.

In 1991, South Africa was on the path to its first democratic government with negotiations between the African National Congress and the ruling party underway. However, in low-income urban settlements, informal residents were often not included within the many civic organisations that had emerged to provide local government services to a population neglected by the state. To address this, the Southern African Catholic Development Association (SACDA) financed a meeting to bring together 95 community leaders from these informal areas – also inviting some of the support professionals working with those living in informal settlements. After this meeting in March 1991, the community leaders from informal settlements agreed that they should come together and network to ensure that their needs and interests were represented in the soon-to-be democratic country. A team of two professionals began to visit the groups that had attended the workshop and strengthen their contacts with one another. This informal network provided initial support to informal settlement dwellers who wanted to be equipped to make their own informed decisions about their urban development options and the path they wanted to follow.

After an exploration of different organising initiatives, particularly those in Asia, the organising methodologies of the Indian Alliance (described earlier in this chapter) were identified as a good place to begin community-to-community learning, and a programme of capacity-building in savings and federating was begun.

The first exchange programme took place in February 1992 when Indian pavement dwellers came to South Africa to visit the informal settlements and share organising experiences; this was followed by a return visit of South Africans to India in May 1992. As a result of their exposure to the work of *Mahila Milan* and the National Slum Dwellers Federation in India, the women living in South African informal settlements began to organise their own savings schemes. By 1993, 100 such schemes had been established and the activists agreed to come together and set up their own collective, the South African Homeless People's Federation. A small NGO, People's Dialogue on Land and Shelter, was established to support this initiative. Community exchanges between pavement dwellers from India and informal settlement dwellers from South Africa provided a rich learning platform. At the same time, these groups had an opportunity to engage with policy-makers. In a context in which the government of South Africa was anxious to address the housing deficit, there were opportunities to link to the Minister of Housing (Joe Slovo) and officials then designing a housing programme that provided a capital subsidy (this was described in Chapter 3).

Back in 1991, the emerging savings schemes in South Africa had linked up with a group in Namibia. Here, savings schemes had been established in 1987 when the UN International Year of Shelter for the Homeless had provided the opportunity to establish a revolving fund for housing linked to an existing credit union. The fund had financed housing construction by Saamstaan (Standing Together), a group made up of self-managed communities of the urban poor living in Katutura (the low-income area of Windhoek). In the mid-1990s, the South African groups began exchanges with Zimbabwe which catalysed the establishment of savings schemes and then, in 1998, the formation of the Zimbabwe Homeless People's Federation. In the same period, links between the Indian Alliance and the Asian Coalition for Housing Rights (see below) stimulated an exploration of savings-based organising involving savings groups from the Philippines, Nepal, Cambodia and Thailand. Federations of savings schemes in informal settlements emerged in the first three of these countries.

By 1996, a decision was taken by the Indian and South African community leaders to form Shack/Slum Dwellers International (SDI) as a network of federations of the homeless and landless in towns and cities of Asia, Latin America and Africa. These federations bring together residents in informal settlements and those living informally in formal areas, enabling them to identify and realise a range of strategies to address their needs and interests. These national federations link to small professional agencies (NGOs) that provide them with support services.

Initial intervention

The core form of organisation within the slum/shack/homeless people's federations that formed SDI remains savings schemes, local groups that draw together residents (mainly women) in low-income neighbourhoods to save, share their resources and strategise to address their collective needs. These local groups and the larger federations to which they belong are engaged in many community-driven initiatives to upgrade informal and squatter settlements, improving tenure security and offering residents new development opportunities. They are also engaged in developing new housing that low-income households can afford, and installing infrastructure and services (including water, sanitation and drainage).

With most savers and savings-group managers being women, these savings groups help address the multiple forms of disadvantage, oppression and exploitation that they face (see Chapter 5). The immediate focus and localised orientation to collective savings provides them with a different role and one which is supported by their peers. This challenges and helps overturn discrimination and limited social expectations as women engage with each other as activists (rather than remaining subservient to male and/or older household members), public agents (rather than enclosed in the household) and strategic thinkers (rather than passive). The collective nature of savings helps to ensure that women are nurtured as they develop a new understanding of themselves and their capabilities. As women take up new leadership roles in providing essential goods and services centred on the home and neighbourhood, an engagement with the state begins. Relations with the state, including local councillors, officials, and sometimes traditional authorities, are essential if urban deprivation is to be addressed and development takes place – even if these are often clientelist.

These federations and their local organisations learn from and support each other at the neighbourhood, city, national and international scale. The initiatives or precedents undertaken by savings schemes demonstrate how shelter can be improved for low-income groups, and how city redevelopment can avoid evictions and minimise relocations. The strategies (shared across the networks of savings groups within and across nations) build on existing defensive efforts by grassroots organisations to secure tenure. But they also add to these existing efforts new measures designed to strengthen local organisational capacity and improve relations between the urban poor and government agencies. Savings schemes defend themselves against eviction threats and negotiate for secure tenure as well as exploring strategies to improve their members' livelihoods using a variety of methods and approaches. These have several key components that are summarised below.

The emphasis on local savings emerges from a commitment to strengthen social relations and social capital between some of the most disadvantaged urban dwellers living in informal settlements without legal tenure. Women, more than men, see the multiple benefits that arise from coming together in small groups and collecting available finance (pennies, cents, rupees). Savings scheme members form active local organisations able to consider how best to address their own needs and those of their families. The savings groups are immediately useful, providing members with crisis loans quickly and easily. The accumulation of each member's savings provides the group with a fund for housing improvements or income-generation investments. Particularly significantly, the collective management of money and the trust it builds within each group increases the capacity of members to work together on development initiatives. Finance, rather than being a means of exclusion, becomes a trigger for the formation of strong local organisations, as women combine to find ways to aggregate, protect and enhance their small change.

The individual members, working within the savings collective, develop the confidence and skills to identify and realise their ambitions (Appadurai 2004). As a result of organising among some of the lowest-income women living in informal settlements, a strong emphasis on shelter-related activities has emerged. Women take on most domestic and child-rearing responsibilities, often completing the associated tasks alongside home-based income-generation activities. Many of the savings scheme members do not have secure land tenure and are at risk of eviction. They do not have access to basic services, such as regular good-quality water supplies and toilets. In this context, improved shelter is a priority.

Each federation works with a support NGO, staffed by professionals who assist in a range of tasks related to grant management, technical development services and documentation for a professional audience.

Each federation engages in exchanges as representatives from different savings groups visit each other. These exchanges may be experiential (for example, seeing how a savings group has negotiated for land), related to the development of specific skills, and/or with a political purpose by bringing politicians, officials and community members together for a visit to development activities in another community, city or country. Most of these exchanges and dialogues take place within cities or between cities within nations. But international exchanges have also been important. Exchanges and the sharing of experiences between informal settlement dwellers in India and South Africa in the early 1990s resulted in growing bilateral links between the Indian and South African federations and an awareness of the value of international networking. Exchanges help to ensure that ideas come from the urban poor and are not imposed on them by well-meaning professionals (Patel and Mitlin 2002). As a consequence, emerging strategies are embedded in the proven practices of the urban poor. Savings scheme members and others learn what is effective through their own experience, supported by that of other communities around them. Learning, rooted at this level, consolidates individual and collective confidence among informal settlement residents in their own capacities. Moreover, the consistent horizontal exchanges build strong relationships between peers, adding to the effectiveness of local negotiations.

Community-managed enumerations, surveys and maps create the information base needed for mobilisation, action and negotiation (*Environment and Urbanization* 2012; Karanja 2010; Patel *et al.* 2002; Weru 2004). Enumerations are in effect censuses – as each household is interviewed and data are collected on them and their needs, along with maps prepared to show all buildings and infrastructure. But the process of enumeration is much more than data collection. These enumerations are part of a mobilising strategy, drawing in residents who want to participate in a locally managed identification and verification of their shacks and plot boundaries. Managing these processes strengthens existing savings groups and encourages new savings groups to form. Equally important is that once the findings are assessed, local residents have the opportunity to set collective priorities through neighbourhood and settlement meetings. Neighbours come together to look again at their settlement through the enumeration data and assess what needs to be done. These are not easy discussions but they are essential in developing an awareness of potential priorities, the practice of intra-community negotiating and agreeing on a way forward. As household (and other) data are provided to local authorities, stronger relations are built; this information helps to change the attitudes and approaches of governments and international agencies. These are settlements that are ignored in official documentation with little information being available. By providing verified data on these areas, federations both challenge this exclusion and shift the terms of the debate. As a result they provide local communities and city federations with a negotiating advantage as, in many contexts, politicians and officials recognise the federations' capacity to provide a fair and accurate information base widely accepted by residents; and this is required for upgrading and housing development – and for the inclusion of informal settlements in the maps and plans of local governments.

Projects that federation members take on to improve shelter options, including investment in tenure security and physical improvements, provide precedent-setting investments that can be scaled up. Through a set of specific activities related to planning of land (often with some re-blocking to improve road access) and installation of services, and sometimes construction of dwellings, members of savings schemes illustrate how they can improve their neighbourhoods. This also demonstrates their understanding of the costs associated with this process, and supports their learning that allows them to develop more ambitious proposals. City governments and some national governments have become interested in supporting these community-driven approaches, recognising their potential contribution to poverty reduction and urban development. As elaborated below, Urban Poor Funds are set up to provide and manage finance once these kinds of activities grow to a significant level.

The federations that are SDI affiliates seek a development partnership with government, especially local government. Affiliates recognise that large-scale programmes to secure tenure and provide services are not possible without government support. As most of the homes and settlements in which federation members live are illegal, such relationships are essential if security is to be achieved. Often the groups are squatting on land belonging to a state agency, and require the government to acknowledge their right to stay and to provide them with tenure. In other cases, they are on private land and need state support either to negotiate tenure or to find an alternative location. In addition to legal tenure, various local state institutions control aspects of shelter development. Local government agencies implement land-use controls (including zoning regulations) and building regulations, and these are often responsible for putting affordable housing beyond the reach of most citizens. For land for housing and shelter improvements to be affordable, such regulations need to be renegotiated. The purpose of precedent-setting investments is to demonstrate the kinds of regulatory amendments that are required for an inclusive city as well as to explore the scale of finance required and the kinds of cost-sharing arrangements that may be necessary. When local government officials see the quality of a new housing development undertaken by the federation that has plot sizes smaller than the official minimum-plot size regulation or cheaper forms of infrastructure, the possibility of allowing other settlements to use these is greatly increased.

The federations are aware that governments face the problem of managing the city, including dealing with squatter settlements, some of which are located on land needed for infrastructure (such as road or railway reserves or along natural drains), coping with the additional pressure on existing services that accompanies in-migration, and handling the fact that poor-quality settlements are often judged to compromise the image of the city. In some cases, government agencies can be persuaded to be partners in precedent-setting investments: federations and city governments collaborate to identify improvements in which both groups have a stake. The challenges that governments face draw them into an engagement with federations; often they are open to working with federations if they are persuaded that federations can help them address such challenges. Exchanges of community residents, politicians and government staff provide a platform to explore these issues within some kind of neutral space. Government officials responsible for, for instance, zoning and land-use management or water and sanitation become more open to innovations suggested (or implemented) by the federations if they find that their peers in the city they visit have accepted or even supported these measures.

Strategic developments

The federations recognised that for the urban poor to have an impact on the urban development agenda, consolidation and growth was required. In 1996, the federations and their support NGOs were ready to explore the formalisation of what was already a close working relationship between federations in several countries. Slum/Shack Dwellers International (SDI) was established to promote and support international exchanges between member federations so as to strengthen their activities, and to support emerging federations in other nations. From the beginning, the network recognised that the core issue was to build local processes able to engage effectively with local government to secure resources and amend legislation that compounded the difficulties faced by those living and/or working informally. At the same time, SDI's leaders sought to influence the policies and practices of international agencies such that they are more supportive of the local agendas of the urban poor.

Since SDI's inception in 1996, the network of federations that make up SDI has grown from the six founding members (South Africa, India, Namibia, Cambodia, Nepal and Thailand) with the addition of Bolivia, Ghana, Kenya, Malawi, Tanzania, Uganda, Zambia, Zimbabwe, Sri Lanka, Philippines and Brazil. Table 4.1 provides a summary of the growth of the network. Exchanges have taken place with groups in many other countries.

Fully fledged federations exist in the following countries:

- *Asia*: India, Nepal, Philippines, Sri Lanka
- *Africa*: Ghana, Kenya, Malawi, Namibia, South Africa, Tanzania, Zambia, Zimbabwe, Uganda
- *Latin America*: Bolivia, Brazil

Table 4.1 The growth of the SDI network

	Date began	*Federation formed*	*Urban poor fund established*
India	1976	1976	
South Africa	1991	1993	1995
Cambodia	1992	1996	1998
Thailand	1984		1992
Namibia	1987	1998	1999
Philippines	1995	1996	2000
Kenya	1996	1998	2003
Swaziland	1997	2001	
Zimbabwe	1997	1998	1998
Lesotho	1998	2009	
Nepal	1998	2003	2004
Sri Lanka	1998	2000	2004
Colombia	1999		
Tanzania	2000	2006	2008
Zambia	2001	2006	2006
Uganda	2002	2002	2004
Malawi	2002	2003	2005
Indonesia	2003		
Ghana	2003	2003	2007
Angola	2004	2012	
Brazil	2004		
East Timor	2004		
Sierra Leone	2008	2011	
Bolivia	2009	2012	
Botswana	2011		
Burkina Faso	2011		
Peru	2012		
Haiti	2012		

Informal settlement communities have formed savings groups in the following countries, although fully-fledged federations have yet to emerge:

- *Africa:* Angola, Botswana, Lesotho, Liberia, Mozambique, Nigeria, Sierra Leone, Swaziland
- *Latin America*: Peru

Countries exploring options to engage the SDI network as an affiliate include:

- *Africa*: Burkina Faso, Democratic Republic of the Congo
- *Asia*: Pakistan, Thailand, Indonesia
- *Latin America*: Argentina

Thailand is unusual in that it has long had networks of grassroots organisations and a substantial and highly supportive national government agency (UCDO and then CODI) whose work is described earlier in this chapter and also long-established links with SDI affiliates.

As the international network developed its work, other opportunities emerged. As professionally designed urban development programmes have been exposed for their failures (for example, a lack of local ownership meant that investments are not maintained), so the model developed by SDI has received increased attention. International agencies such as Cities Alliance have sought to collaborate with the Uganda federation. The UK government's Department for International Development (DFID) financed the Community Led Infrastructure Financing Facility (CLIFF) which is managed by Homeless International and piloted by the SDI affiliate in India (2002), spreading to Kenya in 2005 and then to the Philippines in 2007. Table 4.2 summarises the growth of members, and savings activities both in respect of daily savings and the national urban poor funds established by SDI affiliates.

Over time, and working with particular donors, the network of federations and support NGOs that form SDI secured access to donor funds which it is allowed to allocate itself (rather than being directed by the donors). This has had a profound impact on the network. A rapid growth in members and deepening of strategy were key results of the SDI board,[10]coordinators and secretariat accessing the resources to enable the urban poor to make choices and learn from the results. These funds are now referred to as the Urban Poor Fund International (UPFI) and are located in SDI's secretariat. The progenitor of the fund was initiated in 2001 as the International Urban Poor Fund and located within the International Institute for Environment and Development (IIED) and at the SDI Secretariat (see Mitlin and Satterthwaite (2007) for a discussion of the operation of the fund based at IIED). These monies supported new activities and functions within SDI,

Table 4.2 SDI affiliates: cities, savers and savings (2011)

	Cities	*Savers*	*Total of daily saving*[1] *(US$)*	*UPFI savings (US$)*
India	65	750,000	850,000	
South Africa	25	23,800	268,000	229,000
Namibia	84	19,000	1,230,000	47,000
Philippines	17	24,600	604,000	898,000
Kenya	11	63,000	700,000	297,000
Swaziland	10	7,400	25,000	
Zimbabwe	53	42,300	247,000	235,000
Nepal				
Sri Lanka	24	53,100	-	32,000
Tanzania	6	7,500	86,000	18,000
Zambia	33	45,000	40,000	125,000
Uganda	6	24,200	168,000	15,000
Malawi	28	10,000	148,000	34,000
Ghana	7	12800	108,000	39,000
Angola	10	7,400	25,000	
Brazil	3	348	22,000	
Sierra Leone	4	2,400	15,000	1,000
Bolivia	2	219	4,000	

[1] Not that these are the present totals in savings groups; some funds are out on loan and are not included in these figures.

including discussions about the most effective strategies for Urban Poor Funds. More significantly, the grants have helped support the growth of the network with an increasing number of affiliates and a deepening awareness of the contribution of network activities in adding value to local development. These funds have helped support the emergence of a number of local funds.

The first two years of the SDI/IIED Fund (2002–2003) demonstrated the efficacy of small project funds upon which the network of federations could draw, and showed how external funding goes much further when the monies go direct to grassroots savings groups, which usually leverage additional local resources. Initial resources came from annual contributions from the Sigrid Rausing Trust. Between 2003 and 2007, other funders, including the UK Big Lottery Fund, became interested in how they could add value to their funding through supporting this work. The scale of available funding increased, and SDI federations accessed funds for a range of activities, including:

- tenure security (through land purchase and negotiation) in Cambodia, Colombia, India, Kenya, Malawi, Nepal, Philippines, South Africa and Zimbabwe;
- 'slum'/squatter upgrading with tenure security in Cambodia, India and Brazil;
- bridge financing for shelter initiatives in India, Philippines and South Africa (where government support is promised but is slow to be made available);
- improved provision for water and sanitation in Cambodia, Sri Lanka, Uganda and Zimbabwe;
- enumerations and maps of informal settlements in Brazil, Ghana, Namibia, Sri Lanka, South Africa and Zambia that provide the information needed for upgrading and negotiating land tenure;
- exchange visits by established federations to urban poor groups in Angola, East Timor, Mongolia, Tanzania and Zambia (in Tanzania and Zambia, these visits helped set up national federations);
- community-managed shelter reconstruction following the 2004 Indian Ocean tsunami in India and Sri Lanka;
- federation partnerships with local governments in shelter initiatives in India, Malawi, South Africa and Zimbabwe;
- the emergence of a number of local funds, including those launched in Africa after 2001.

In 2007, as the network responded to the increased scale and depth of activities, SDI established a council of federations which is the ultimate governing body, as it is responsible for electing the board and making key decisions (for example, the allocation of capital funds between affiliates). The council is made up of three representatives from each member federation. It meets about every six months around a particular event in an affiliate organisation. The increasing engagement of the council in SDI strategising means that national affiliates are thinking through their projects with greater care, testing out new approaches and learning about how to deepen their own process and take it to scale.

In 2008, SDI built upon the International Urban Poor Fund and created the Urban Poor Fund International (UPFI) as a platform to access finances from international sources. Major donors to the UPFI now include the Bill and Melinda Gates Foundation, the governments of Norway and Sweden, and the Rockefeller Foundation (UPFI 2012). The SDI secretariat built up the capacity to manage project funds and these are variously used for housing projects, technical assistance and 'federation strengthening' (for example, with the launch of savings schemes and the conduct of enumerations). Between 2008 and 2010, the Fund supported investments in land development, housing and basic services in more than 22 towns and cities. In India, Kenya, the Philippines and South Africa, the Fund has provided finance for developments involving thousands of people who were renting or squatting in shacks without secure tenure. This Fund has produced a new way of financing community-led development, and encouraging and leveraging support from local and national governments.

Achievements

SDI affiliates have achieved access to financial services, secure tenure and improved shelter for some members and policy and regulatory reforms that offer an improved political context for others in need. They have also begun to inculcate within both the state and other civil agencies an attitude that favours new relationships with the urban poor (especially low-income and disadvantaged women) based on recognition of their equality in regard to dimensions such as capability, development contribution and status. The network has added to the contribution of affiliates, raising resources to assist them to expand the scale of their work, building links between federations through exchanges to augment the experiential learning and skill sharing that is already taking place. The network also supports the emergence of new federations, and provides an international voice for the urban poor so that they can present an alternative perspective to the professional discourses that dominate such debates. This means challenging discriminatory practices and presenting new opportunities to donor agencies and governments searching for new and more effective strategies to realise pro-poor and equitable urban development.

The figures reported below are from June 2011 unless otherwise specified (UPFI 2012). Both the national federations and the international network seek to support the work of local groups and the figures below reflect this, as they aggregate achievements at all levels of the network.

Financial services: The core SDI network involves over 16,000 savings groups with an average membership of 70 per group. Consistent saving is, for SDI, the most important indicator of membership. This is because their interest in engaging the lowest-income urban citizens means that an individual's total savings can be a misleading indicator. Activities include information gathering and exchanges to strengthen local savings schemes.

A key mechanism for investing in improved tenure and services has been Urban Poor Funds which are established by SDI affiliates as they begin to undertake precedent-setting investments. Just under US$10 million is currently in the savings accounts of the federations from daily savings collections (see Table 4.2). Much of this finance is locally circulated within savings schemes, as loans are given to members (for consumption, emergencies and small enterprise loans) and then repaid. These figures are very much an underestimate, as the scale of money kept and circulated at a local level is captured only intermittently. An additional US$2 million is community savings in national Urban Poor Funds.

Secure tenure and improved shelter: Just over 200,100 households have secured formal tenure (either individual or collective) as a result of this work. However, these figures understate the number of people reached as some have greater security but not formal ownership of the land. One example of this is the upgrading of 160 shacks within an informal settlement on a road reserve in Sheffield Road (Cape Town) that has resulted in greater safe internal space (for children to play) and better access to sanitation. Another example is current negotiations underway over the number of residents entitled to remain in Hadcliffe Extension in Harare (where residents are currently installing eight communal sanitation blocks, as there are hundreds of families and no existing facilities).

Policy and regulatory reforms: The challenge is to transform the improvements that individual settlements and/or city federations secure into regularised repeated interventions through changes in policy and/or practice including the operation of state programmes (see Box 4.1). A first step is often the establishment of a formal partnership with the state and eight SDI affiliates have such agreements with the national government. Across the network, there are 102 agreements with provincial or city authorities which establish a dialogue with a potential for a more equal relationship between the authorities and the communities. In some cases, these partnerships build on or lead to deeper engagements. For instance, in Ghana, Federation members have been brought into the technical committee currently reviewing national housing policy where issues of the urban poor's housing needs are being prioritised. Another example of this is Federation members in Harare participating with the city in a programme of activities to upgrade Mbare, a low-income settlement around the central market. There are many other examples of explicit changes in policy driven by federation initiatives but these have not all been collated.

Box 4.1 Examples of policy change catalysed by national federations

One early example of policy reform was the creation of the People's Housing Process (PHP) by the South African government in 1998 as an adjunct to its housing programme. As described in Chapter 3, the South African government had set up a programme to provide subsidies to housing for low-income groups but it was usually contractors who got the subsidy and who often built housing of poor quality on land far from income-earning opportunities. The PHP offers greater scope for communities to make decisions for themselves in the use of the subsidy, allowing them to provide voluntary labour and to undertake project management activities. As explained by the then-Minister of Housing, 'This policy and programme encourages and supports individuals and communities in their efforts to fulfil their own housing needs and who wish to enhance the subsidies they receive from government by assisting them in accessing land, services and technical assistance in a way that leads to the empowerment of communities and the transfer of skills' (Mthembi-Mahanyele 2001, p. 4). Attributing changes in policy is always difficult but the People's Housing Process option emerged in part because the local communities linked to the South African Homeless People's Federation demanded a more community-driven collectivised process

Source: Baumann (2003); Khan and Pieterse (2006)

Many other changes in policy have been catalysed or influenced by the federations. These include the adoption of a new approach to informal settlements in Windhoek influenced by Federation-supported exchanges that reduce housing costs by allowing smaller minimum plot sizes and lower infrastructure standards (Muller and Mitlin 2007). They include the acceptance of eco-sanitation within the regulatory standards in Zimbabwe and Federation-style eco-sanitation also being accepted within the National Sanitation Policy in Malawi. They also include changes in tender processes to facilitate community construction in Bhubaneshwar and other cities in India, municipal land rules and regulations relaxed in the case of housing in Moratuwa (Sri Lanka), and the agreement of the Kenyan railways company to resettle 11,000 households squatting in the rail reserve in Nairobi. In Kenya, Malawi, Namibia, Tanzania and Zimbabwe, federation groups have been able to reduce plots sizes to below the present standard both through securing agreement for double occupancy and/or by using a collective land title to negotiate exemption from the regulations (Mitlin 2011b).

There are also cases where the federations negotiate for the application of existing policies rather than the introduction of new policies. In the Philippines, for example, the federation has helped to negotiate a Shelter Code in Iligan City (with a financial commitment of PhP50,000,000 or just over US$1 million) and a Shelter Plan in Kidapawan. This Shelter Code is provided for within existing legislation but, without the active engagement of the Philippine Homeless People's Federation, it is unlikely to have been realised. The Plan recognises the federation communities (and other low-income groups) and provides land for the housing needs of those due to be relocated due to living in danger zones and land needed for major infrastructure projects; it seeks to encourage low-income residents to play an active role in improving their neighbourhoods. There is also a commitment to making finance affordable and available, and to recognise alternative building materials such as bamboo.

Policy influence and change are seen as involving dynamic experimental processes. SDI affiliates are aware that they need to shift formal policies and subsequent programmes and practices. Many policy initiatives take place at the city level as local groups try out new approaches, drawing in politicians and officials in co-learning with subsequent planning and implementation. As city-wide approaches are found to be effective, then national leaders and SDI respond by negotiating changes in national and international policies to support and consolidate inclusive pro-poor urban development.

ACHR and ACCA[11]

Background and urban poverty problem

The Asian Coalition for Housing Rights was formed in the years immediately following the establishment of the Orangi Pilot Project and SPARC (together with *Mahila Milan* and its alliance with the National Slum Dwellers Federation). In 1987 there was a decision by a global network of housing professionals and activists, the Habitat International Coalition, to set up regional groups. In Asia, there was already a network of people and organisations working in this area that knew about each other because of SELAVIP (the Latin American and Asian Low-Income Housing Service). This produced a six-monthly newsletter in which communities shared their experiences and built up a knowledge of organisations and individuals active in Asia. SELAVIP itself emerged from a decision in the mid-1970s by Asian Jesuits to prioritise work on urban poverty in the region. Father Jorge Anzorena, an Argentine Jesuit trained as an architect based in Tokyo, was spending several months each year travelling around Asia to meet and then write about the activities of different community and local initiatives. Stencilled typewritten pages were sent around Asia to all these groups and others, sharing the information. Those receiving the newsletter began to have an understanding about what was taking place elsewhere, and along with this, their regional consciousness and sense of identity were strengthened.[12]

The first meeting of the Asian Coalition for Housing Rights took place in 1988. In a review of ongoing activities, participants were particularly moved by Fr. John Daly's description of evictions taking place in Seoul (South Korea) as the city prepared for the Olympics. The Coalition's first project was designed to expose the problem and bring it to the attention of the government (see ACHR 1989). The project had an impact in South Korea, provoking greater awareness of the shelter problems of the urban poor and resulting in the launch of a new housing programme. The following year a further regional meeting took place, named the Asian People's Dialogue: agreed activities included exchanges between NGOs, a campaign against evictions in Asia and an analysis of the impacts of aid projects in the region. A committee was established and a Secretary General appointed. There was a joint agreement that the Coalition should be decentralised and bottom-up. The deliberate emphasis was on networking to support a diversity of city-based agencies, not on the creation of a further organisation with its own programme and direction. The members of the Coalition included professional NGOs established and managed by architects or planners (i.e. urban professionals), faith-based agencies, a range of organisations within the Alinksy tradition of community organising, academics from university departments specialising in urban development and the built environment, and members from other traditions including credit unions. A small office was set up in Bangkok to house the secretariat for the regional network: this currently has three full-time and four part-time staff.

Initial intervention

ACHR developed several work areas as its programme consolidated. In the early years, emphasis was given to housing rights and problems of evictions in Asian cities. International fact-finding missions that put pressure on governments to stop evictions were organised to South Korea (twice), Hong Kong and the Philippines with positive outcomes. Between 1991 and 1993, the Coalition focused on developing solutions to eviction problems. A third phase (1994–2000) centred on the Training and Advisory Program (TAP) supported by DFID (UK). TAP was a package of support activities for ACHR partners with cross-country learning, exchange visits, regional workshops, exposure to key regional projects, new country action programmes and research. This added significantly to the knowledge and capacity of the regional intervention process. This regional platform allowed for sharing and mutual learning between groups in different countries. Most conventional development processes by international agencies or government are characterised by vertical (or hierarchical) free-standing interventions; evidence suggests that this does not build a substantial and sustained process. The ACHR platform helps support contextual understanding rooted in the region and in each city while mutual involvement in activities strengthens community, city and country capacities to secure change. TAP gave particular attention to sharing ideas about community-based savings development, equitable city development and housing rights with fact-finding missions to expose the abuse of housing rights, as well as city studies and other research to elucidate Asia's political realities, its feudalistic social and behavioural patterns, and its institutional traditions.

ACHR's fourth stage, from 2000, represents a more mature process and a broader scale of intervention. This included the introduction of community savings and credit activities and the development of many Community Development Funds which have been able to influence new forms of development change in Lao PDR, Cambodia, Vietnam, Nepal, Mongolia, Sri Lanka, Thailand, Philippines and India. Regional responses to the 2004 Indian Ocean Tsunami devastation developed with the promotion of community-driven relief and rehabilitation. In terms of policy advocacy, the contribution of ACHR key activists has increasingly been recognised with requests to meet and share knowledge with policy-makers, including government ministers. This reflects increasing interest in possible collaboration and support for urban community development and/or community upgrading programmes. City-wide and country-wide slum upgrading in Thailand and India has been used to boost and support learning for structural change in other countries. Further areas include community-led disaster relief and mitigation, including post-tsunami work, support for Cambodia to develop community-led options for shelter improvement and urban development, and an exposure programme for young Asian professionals. Some ACHR interventions and support have been integrated into government policies, while others have been successfully upscaled without government support. The upgrading of 100 slums per year in Cambodia and the community savings and credit groups initiatives interventions in Lao PDR and Mongolia have developed into large-scale programmes. Continuing activities include the production and widespread dissemination of international publications, including the *Housing By People* series, *Eviction Watch* and *Understanding Asian Cities*.

In 2008, ACHR launched the Asian Coalition for Community Action (ACCA) to catalyse change in Asian cities with regard to slum upgrading. This programme builds on the tradition of work within the Coalition and the particular contribution of ACHR's secretary general (Somsook Boonyabancha) where she has been involved in the senior management serving as deputy director of the Thai government's Urban Community Development Office and then deputy-director and director of the Community Organization Development Institute (see earlier sections on these organisations). ACCA seeks to support hundreds of community initiatives that then catalyse city-wide upgrading and partnerships between community organisations and local governments. By September 2012, it had helped fund initiatives in close to a thousand settlements in 165 cities in 19 different Asian nations. In each city, small grants support several community-led initiatives. These encourage city-wide networks to form, as communities visit each other and as members share skills with each other and learn to negotiate with their local governments. Further support was available as local governments engaged in and then came to support this process, including the formation of jointly managed community development funds.

Strategic development

ACCA enables community groups to be the primary doers in planning and implementing projects within which they tackle problems of land, infrastructure and housing. In each country, the support for community initiatives is channelled through ACHR members that are already working on issues of urban poverty and housing. These groups share a common belief in a large-scale change process that is led by the people. Many of these groups already support federations and networks of low-income community people, and most have already cultivated some kinds of collaborative links with local government agencies. ACCA offer new tools to these groups to enhance, strengthen and scale up the work they are already doing to create a collaborative, city-wide mechanism for bringing about change in their cities.

The core activities of the programme, which account for 60 per cent of the budget, are the small upgrading projects and larger housing projects that are being implemented in low-income communities by their residents. The plans for these projects, as well as the city-wide surveying, saving and partnership-building processes they are part of, are developed and implemented by the local groups. The budget ceilings for the upgrading projects are very small (a maximum of US$3,000) but offer flexibility in how community organisations use those small resources to address what they choose. These budgets give people something to begin negotiations, while forcing them to economise and think of what other resources can be mobilised. The expectation is that if communities plan well and use these funds strategically to link with other resources, these modest amounts can help 'unlock' people's power to negotiate with other actors for more resources, more land, more support. The funds for each city involved in ACCA are as follows:

- $15,000 for at least five small upgrading projects, in five different communities in each city;
- $40,000 for one big housing project in each city, with a maximum of about seven or eight big projects per country;
- $3,000 per city for city process support, to cover a variety of joint development processes within the city, like surveying, network-building, support for savings activities, local exchanges and meetings;
- $10,000 per country per year for national coordination, meetings and exchanges.

The programme supports the setting up and strengthening of collaborative mechanisms to build linking, learning and mutual support structures. The regional committee set up at the start of the programme helps coordinate this process. The 15-member committee meets every two to three months and is the key regional mechanism for learning, sharing, assessing, supporting the cities, organising exchange visits, establishing forums of communities and community architects and linking with international organisations. Some subregional groupings have also emerged, in which groups in neighbouring countries assist each other more regularly and more intensely (especially in Indochina and South Asia). National joint committees have been set up in several countries, which link community groups, government officials and NGOs to work together to make decisions, learn, assess, advocate, build joint capacity and make policy changes. In other countries, this step has yet to be taken as the processes are evolving at a slower rate. In most of the cities, some kind of joint working group has been established at the city level to provide a platform for community networks, city governments, civic groups, NGOs and academics to plan, to manage the upgrading and city development fund process, to look at land issues and to support change in the city. These city committees are seeking a new kind of partnership and participatory governance process, which is distinctive, since it emerges from the development activities being undertaken in the informal settlements. Underpinning these committees are community networks that link low-income communities in the city, helping them work together, support each other, pool their strengths, learn from each other's initiatives, survey and map their settlements, strengthen their community finance systems, formulate their upgrading plans, negotiate collectively for land and for various other resources and changes, and plan joint activities in collaboration with other groups.

One further innovation is the way the programme is assessed. When the programme was being designed, ACHR recognised the dysfunctionality of conventional, *supply-driven* development projects and their assessment. Here it is common for the funding agency to hire outside professionals to assess the project, with or without the participation of the communities and implementing groups. The assessment is done according to some pre-prepared list of objectives and outputs that were agreed upon in the original project document. The external assessor has no knowledge or experience of living on a very low income or in an informal settlement, or fighting or actually experiencing eviction. ACCA uses an alternative assessment process, where the assessment of community initiatives is done by their peers (Boonyabancha *et al.* 2012). This involves a new, horizontal system for comparing, assessing, learning from and refining the ACCA projects in different countries, through a series of intense visits to ACCA projects within certain countries and discussions with the people who are implementing them. These 'joint' assessment trips include a mix of community people, support professionals and sometimes even a few supportive local government officials that can usefully be exposed to the ACCA process in another country. All participants are actively and jointly involved in implementing ACCA projects in their place of origin, and hence they are able to raise many questions, doubts, problems and ideas.

ACCA's new option (introduced in 2011) is to lend money to community development funds able to demonstrate loan management capacity. It is predicted that this further step will both encourage new capabilities and enable networking community organisations to demonstrate what they can do to persuade formal financial institutions (both state and commercial) that have previously shown little interest in the urban poor. At one level, these loans simply increase the financial resources available to community development funds, but they also demonstrate to the networks the potential of loan capital and build up the practices of repayment to external investors. In the future it is anticipated that community development funds will access both private sector and state capital – working out ways in which the flows of external capital can be blended with local agencies which nurture reciprocity such that the needs of the urban poor are addressed while vulnerabilities are not increased. The ACCA regional fund has offered five loans to city processes in four countries with a financial commitment of US $178,500.

The rigour in ACCA's processes may be seen in their adherence to 10 operating practices that provide a core around which experiences and experimentation take place:

- All action should be collective action to address mutual needs.
- There should be an immediate focus on activities as capacity development is best realised through practical actions with demonstrated results.
- Surveying and mapping informal settlements city-wide is the best way to begin a process, breaking down the isolation of individual settlements and enabling the community activists to see the city as a unit.
- The selection and prioritisation of projects should be done by those communities involved in the programme, as a collective decision.
- Budgets should be flexible and easily accessible, but small.

- Funds should be controlled by the people themselves with horizontal public systems of accountability.
- There should be multiple small projects to create many opportunities.
- Low-income communities should be networked at the city and national scale.
- A platform for negotiation and then partnership with the city authorities should be established.
- Each city should include at least one housing project to demonstrate the link between small projects and larger scale process of shelter improvement and to explore how such development can be secured.

Achievements

ACCA activities have rapidly been taken up across Asia. By October 2012, the programme was supporting activities in close to 1,000 settlements in 165 cities in 19 Asian countries. By August 2012, a total of 111 big projects had been approved with a budget of US$3.9 million. Funding approved for small projects totalled over US$2 million. Community development funds have been established in 107 cities; 70 of these are cities in which the process of fund establishment has been directly linked to ACCA investments. There are an additional 19 projects in eight countries addressing the situation following disasters and helping communities respond effectively; US$481,000 has been budgeted to address these needs.

The nature of the small upgrading projects that communities have identified is summarised in Box 4.2. As is evident, basic services is a key area of activity with improvements in water, sanitation, drainage, solid waste management and electrical services all taking place. The retaining walls are linked to drainage and flood protection. Initially surprising was the investment in roads and bridges; however, community leaders explained how important it was to link their settlements into the transport network of the city (although in many cases the means of transportation is walking or cycling). The community centres and playgrounds all offer places for communities to meet. The projects for the rice banks enhance livelihoods and stretch incomes further while the health centres help reduce expenditure. Libraries and trees help to create a positive environment.

Box 4.2 What have people built with ACCA small grants?

227 road-building projects
174 water supply projects
141 sewers and drainage projects
136 toilets and bathrooms
89 community centres
52 electricity and street lights
37 playgrounds
29 house repairs

20 solid waste management
17 bridge-building projects
17 schools
12 library projects
11 rice bank projects
14 retaining wall projects
Others include markets, fire protection, tree planting, retaining walls to reduce flooding, clines, livelihoods.

Source: Preparation for third-year report, November 2012

Table 4.3 demonstrates the increasing contribution of the state between the small project and big project stage of ACCA. The small projects both build community capability and increase the visibility of their collective potential. Community capability is supported both at the neighbourhood level and also, as small projects take place more or less simultaneously in five or more locations, at the city level. The city networks share understanding and develop a strategy to engage the local authority. The small projects, placed in the public eye by the network, attract state interest and enable a negotiation between the network and state authorities to take place that results in a higher level of government contribution. The countries in which this redistribution is particularly significant are those in which ACCA was able to build on existing activities demonstrating the potential influence of this model. In Cambodia, Nepal, Philippines, Vietnam, Sri Lanka, Thailand and Lao PDR there have been broadly consistent initiatives to support people's led development.[13] In Fiji the government has been exposed to the work in other countries and has, on this basis, responded positively; the challenge for communities is to consolidate this interest with future commitments based on local activities. In India, the group was able to benefit through established government programmes.

In 57 per cent of the big projects thus far, the land has been provided by the government under a variety of tenure arrangements. These communities have been successful in negotiating further resources from the state. Analysing the budget shares, 4 per cent of funds have been provided by ACCA, 13 per cent from the communities themselves, 3 per cent from other sources (e.g. northern NGOs), and the remaining 80 per cent has been provided by government agencies including both local and national government (see Table 4.4).

One of the most important objectives of the ACCA Programme is to develop new financial systems for low-income households that work well within the realities of their lives and that they can manage themselves. The most basic building block of a people's financial system is the community savings group, in which they build, use and manage their own resources. Community savings and credit is being practised in 101 of the ACCA cities so far. Many city-level community development funds are emerging now, most seeded with capital from the

Table 4.3 Summary of big and small ACCA projects at the end of year 3.5, including financial contributions

	Number of projects completed or underway	*Households directly benefiting*	*Budget contributions to projects (US$)*				
			Budget from ACCA	*Budget from community*	*Budget from government*[1]	*Budget from others*	*Total budgeted cost*
Small ACCA infra-structure projects • in 165 cities • in 19 countries	963 projects	145,990 households	2,046,426 (33%)	1,253,744 (20%)	2,620,083 (42%)	395,145 (5%)	6,284,949 (100%)
Big ACCA housing projects • in 104 cities • in 15 countries	111 projects	8,611 households	3,900,256 (4%)	11,750,344 (13%)	73,025,280 (80%)	2,617,914 (3%)	91,313,674 (100%)
Total small and big ACCA projects	1,074 projects	154,601 households	5,946,682 (6%)	13,004,088 (13%)	75,645,363 (78%)	3,013,059 (3%)	97,598,623 (100%)

1. This includes land, infrastructure, materials and cash. Governments have provided land for the big projects in 63 of the projects (either free, on long-term lease or for sale at subsidised rate on instalments).

Source: ACHR Secretariat, 26 November 2012

Table 4.4 Major projects completed or underway with the financial contributions of different stakeholders

Country	*No. of projects*	*Budget contributions to projects (US$)*					*Beneficiaries*	
		from ACCA	*community*	*government*	*others*	*Total project budget*	*No. of households directly benefiting from ACCA loans or grants*	*No. of households in the whole community*
Bangladesh	1	43,000	30,000	436,875	122,800	632,675	70	346
Burma	7	271,200	44,500	0	0	315,700	927	20,462
Cambodia	12	393,500	707,500	7,933,465	459,145	9,493,610	630	1,953
China	1	39,000	30,000	0	24,000	93,000	10	10
Fiji	5	200,000	20,000	5,885,000	0	6,105,000	170	2,578
India	2	80,000	42,010	8,920,307	0	9,042,317	58	901
Indonesia	7	245,000	137,159	3,941,117	10,000	4,333,276	735	3,448
Lao PDR	9	333,000	232,600	7,259,755	61,000	7,886,355	656	2,010
Mongolia	5	150,767	38,905	207,780	7,900	425,232	149	420
Nepal	11	359,800	409,904	4,599,763	203,727	5,573,194	509	1,989
Pakistan	3	110,000	3,020,500	0	465,435	3,595,935	414	32,816
Philippines	18	679,989	1,776,088	27,905,912	998,907	31,360,896	2,271	8,510
Sri Lanka	11	450,000	61,410	0	165,000	676,410	496	496
Thailand	8	180,000	3,742,362	984,665	0	4,907,027	1,148	2,538
Vietnam	11	365,000	1,457,406	4,950,641	100,000	6,873,047	368	807
	111 project (in 104 cities)	3,900,256 (4%)	11,750,344 (13%)	73,025,280 (80%)	2,617,914 (3%)	91,313,674 (100%)	8,611 HH	79,284 HH

Source: ACCA Third Year Report (in preparation)

ACCA project money. These city funds are linking the community savings groups with ACCA finance – and with other sources of finance – in new and creative ways, with the national-, city- and community-level funds interacting in different ways. Some of the countries have started with national funds (like UPDF in Cambodia (Phonphakdee *et al.* 2009) and CLAF-Net in Sri Lanka), 70 cities have started with city-based funds (as in cities in Nepal, Burma and Vietnam), some have started from strong savings groups on the ground (as in cities in Mongolia and Lao PDR), and one has not started savings as yet (in China). Local governments have contributed to 21 of these city funds, in eight countries. The total of US$200,000 they have invested in these funds works out to only 5 per cent of the total US$3.7 million capital in all 70 city funds thus far, but it is a significant contribution as it represents a public commitment by these city governments to support an ongoing funding mechanism for low-income residents within their cities.

In terms of participatory government, in 63 out of the first 65 big projects (those realised by the end of December 2010), there is now some form of partnership between communities and the government. The joint city development committees that are being set up are platforms that allow low-income communities to work more equally with their local governments and other urban partners. By December 2010, in 91 cities (out of a total of 107) there was some kind of committee formalising this city–community partnership. National-level collaborative mechanisms are also working now in eight countries (Cambodia, Nepal, Vietnam, Sri Lanka, Mongolia, Fiji, Thailand and Lao PDR). This remains at an initial phase and the challenge is to ensure that the partnerships deliver tangible benefits to those in need. In many of the cities in Cambodia, Indonesia, Nepal, Philippines, Vietnam, Fiji, India and Lao PDR, the successful implementation of the ACCA big projects has led local governments to initiate or agree to partner with the community networks and their support NGOs to implement subsequent housing projects and to link with other housing schemes and development projects in their cities.

Another way in which governments are contributing is by adjusting existing planning standards to make them more realistic, lower cost and easier for the urban poor to develop housing which matches their needs. This is happening in several cities, but the most striking example is in Vinh (Vietnam), where the planning standards for redeveloping old social housing have been changed from an expensive, contractor-driven model to a people-driven model as a result of the ACCA project at Cua Nam Ward. In Lao PDR, the government had never previously given land on a long-term lease to a low-income squatter community; the two big projects in Lao are the first cases of the government giving squatter communities long-term leases to the public land they already occupy, to regularise their status. A number of other policy reforms have been secured. In Cambodia these activities build on a long-standing programme of work. Two emerging policies provide a framework for making city-wide upgrading plans for housing all the poor in the city (on site if possible and relocation only when necessary, to land the government provides for free, with full land title) in which the municipality and the local community networks survey and work out the plans together. In Nepal, the joint city development fund concept, which was piloted in Kathmandu (with matching funds from ACHR, SDI and the Kathmandu Municipality), is spreading to other cities where funds are now up and running (in Bharatpur, Birgunj and Dharan), with local governments contributing money. In South Korea, people living in vinyl house communities have won the right to register their addresses (which is necessary to access various government entitlements like schools, health care and basic services), even if they are considered squatters. This breakthrough came after the work of building the new network of vinyl house communities had begun.

In Thailand, the ACCA funding has helped to pilot new city-based development funds in a few cities (which are managed by the community networks, in collaboration with their local governments) and has helped ignite a city-fund movement throughout the whole country. There are now city funds in some 50 cities. The Thai networks are interested in part because funds offer a new source of more locally controlled monies, reducing their dependence on central government.

Conclusions

The programme interventions described in this chapter provide us with diverse experiences about how to achieve greater equity and justice in towns and cities of the Global South. In this section, some key themes related to these interventions are highlighted and the work is contextualised. Chapter 5 examines these experiences in the context of broader discussions about political change and social transformation.

This chapter has reported on evolving modalities of action that take into account the broader changes that are ongoing in towns and cities in which they are operational (these changes were reported on in greater detail in our first volume on understanding urban poverty and are summarised in Chapter 1). The five programme interventions described in this chapter have learnt from their own outcomes, and those of others, and adjusted to other events, observations and initiatives. Given this context, and the ways in which governments and development agencies have responded to urban poverty, what has emerged as effective strategies? Before summarising a response to this question, it may be helpful to elucidate some of the key contextual factors that appear to be particularly influential in defining these strategies.

Agarwala (2006) discusses the increase in informal employment in India with the 'casualisation' of the workforce and 93 per cent of the economically active population now being employed in casual, temporary or non-permanent employment. Her research concludes that these workers are targeting the state rather than their employers and seeking improved access to the basic goods and services they need for well-being rather than higher wages. She argues that

> shifts in production structures have pushed informal workers' organizations to make two strategic changes to their mobilization strategies in order to fit the conditions of informal employment and to retain their membership. The first

> is to target their demands to the state, rather than the employer, and the second is to make demands on welfare benefits (such as health and education), rather than workers' rights (such as minimum wage and job security).
>
> (Agarwala 2006, p. 428)

This reflects a reality in which traditional labour unions represent a labour elite with a declining power, as the proportion of the economically active population in formal employment is usually very low and may be decreasing. Moreover, many of those who are the worst paid and who have the most insecure incomes have never had the possibility to be unionised due to their concentration within the informal sector. Informalisation has brought highly personalised relations in the labour market with low pay and much insecurity, and this makes it difficult for workers to secure a better deal from employers. Others are 'own-account' workers, working under contract without formal employment. The development sector has been aware of this – and there have been many initiatives to assist labour market outcomes that are meant to benefit lower income groups. However, many of these have been concentrated on providing micro-finance to small-scale traders. There has been little to assist the low-paid informal sector employees or 'own-account' workers. A few initiatives have sought to strengthen the power of such enterprises to negotiate better deals. SEWA, it should be remembered, emerged from a trade union to help organise the informal beedi (cheap local cigarette) makers. SEWA's later experience is also informative here and the organisation has incorporated both housing and informal settlement upgrading within its portfolio of activities as it has sought to respond to the needs of the women with which it is working.

Other initiatives in this area of employment and labour have sought to support traders in their work. There have been efforts to support their struggles to remain in central city areas despite beautification and redevelopment. Recent efforts are discussed by Crossa (2009) for Mexico City, Turner and Schoenberger (2012) for Hanoi, and Maharaj (2010) for Durban. There have also been efforts to prevent the formalisation of trades such as waste collection which inhibits or prevents the recycling in which low-income workers have been securing a livelihood. The formalisation of such activities leads to the corporatisation of service provision and consequently small-scale individual waste recyclers are excluded from access to the waste at collection points or dump sites. Here, informal waste recyclers often have to resort to developing a formal organisation through which to protect their livelihoods. In Brazil, the consolidation of organisations of waste pickers has enabled negotiations with the government and the opening up of initiatives that improve working conditions for the informal waste recyclers by training and health services. This can also raise their incomes by introducing cooperatives that mean they can package waste and perhaps be in the processing stages themselves, adding value to their product, and facilitating waste collection through agreements with shops and companies (Fergutz *et al.* 2011).

Trends towards informalisation with continuing urbanisation mean that as collective bargaining in the labour force becomes increasingly difficult due both to more personalised relationships and fewer regulatory controls, so collective consumption of basic services becomes increasingly important. Those groups that have traditionally focused on neighbourhood improvements have found their experience with negotiating for collective consumption goods increasingly relevant, as Agarwala describes above. Collective consumption goods are those that, because they are jointly consumed, have a powerful orientation towards group organising. Standpipes, sewers (or sewer connections), pathways, street lighting and drainage are some of the most obvious. Water connections may be private but if the water pipe network is not there then such connections are not possible (although groundwater may be tapped as an alternative). Less immediately evident but also critical is secure tenure for those living on informal land which will not be obtained individually but which needs to be negotiated as a collective good.

At the same time, the complexity of most urban centres means that the needs and solutions for neighbourhoods and districts are similar, overlapping and increasingly have to be considered together. As described in the programme initiatives above, working at the city level is essential if grassroot activities at the settlement level are to be effective in improving conditions. Without such collaboration, groups will not be sufficiently strong to challenge anti-poor practices by the local authority and other elites; moreover there are few urban services that can be managed only at the neighbourhood level.

Hence the programme interventions described in this chapter share a strong focus on improving conditions at the neighbourhood level as the impetus for city-level collective action and an orientation towards basic services owing to their critical role in health and well-being. The state is the primary focus for their influencing efforts as it is responsible for the provision of many services, the redistribution of resources between citizens, the regulation of urban areas, and the structuring of economic and spatial activities through its investment and management strategies. The successes and limits of these initiatives point to the importance of moving away from defensive positions towards a proactive engagement. It is not enough to simply challenge inefficient urban management and poverty reduction programmes run by the government; these experiences emphasise the importance of being able to propose alternatives based on their practical realisation.

There are two other notable emerging themes: gender and globalisation. The first speaks to the immediacy of human experiences within relations at the household and neighbourhood level, the second to international experiences. Gendered relations are such that women are usually responsible for household reproduction and the care and nurture of children and older members of the family. In many contexts they also have to contribute to household income and work, although this is frequently from the home as they combine these tasks. In addition, there are significant numbers of female-headed households – in some nations' urban populations, they represent more than one-third of all urban households (Chant 2013). Despite all of these realities, in many social contexts women are seen as subservient to men, with a lower social status, fewer legal rights and a more restricted set of opportunities. The weight of these social norms and values creates expectations that make it more difficult for women to realise their potential. Women exist, for the most part, within a context that demands they acknowledge male superiority.

The programme interventions described in this chapter all involve women as core activists in a process of social change. This is least evident in the case of the Orangi Pilot Project; however, although women have a less public face in Pakistan society, the women are acknowledged to have been key to the realisation of the improvements in sanitation. Across all five interventions, the focus on the immediate locality and the very practical orientation of activities encourages women's participation. As women are more actively engaged in activities, in many cases they become local, city and sometimes national community leaders. In many towns and cities of the Global South, the scale of women's oppression is such that this in itself challenges social norms and values. Even local self-help activities such as improved water supplies, digging trenches and building houses deviate from norms and expectations. Why? Because women are meant to remain at home in a private space, while the male role is dominant in the public sphere. Women are expected to be passive, to respond rather than initiate, to follow and not to lead. Women follow men through the door except when invited by men to go first. They sit on the floor when men sit on the table. They take the minutes and serve the tea; they do not chair the meetings. The expectation of reduced capacity is challenged by the active physical role in the public arena. The success of women achieving physical improvements is then reinforced by a demonstrated capacity in financial (savings) management when it is finances that are so key to processes of capital acquisition and economic transactions in the formal world. Moreover, this public activism is in areas that require the acquisition of specific collective capabilities. This is not the individual self-help of a businesswoman. These acts necessarily engage neighbours and their friends in a collective effort.

Such dynamics mean that women's collective act of neighbourhood self-help is profoundly political and it is a step against the most immediate and personal form of subordination and repression. If they can challenge relations within their families and gain an acceptance of a new more equal role, then their own individual and collective expectations of what is possible begin to change. Women's collective self-help is an act of affirmation in themselves and their capabilities (Appadurai 2004).

Finally, such efforts have increasingly incorporated activities that go beyond the city to the nation, and then beyond the nation-state. Such transnational growth reflects three realities. First, there is the spread of ideas from one place to another through migration (including international migration). As communities understand what is happening (and what has been achieved) elsewhere, they are interested in replication. Second, international community exchanges have been an important vector through which the residents of informal settlements have been able to build their knowledge, confidence and abilities. When government officials accompany these exchanges, community groups have realised that peer engagements between government officials and politicians are an effective means of influence. As officials see what is possible elsewhere some of their resistance to approaches pushed by community organisations is reduced. Rather than rely on an amalgam of pressure from below and above, it is a way of illustrating possible solutions to agreed problems through peer legitimation. And experience shows that 'peers' of local urban poor groups are taken more seriously because they come from outside of the particular country. Hence an international network has important additional modalities through which to address local needs.

Finally, the internationalisation of local, city and national processes is useful because solutions to common problems are being replicated by governments. The development of strong transnational networks is key owing to the pace of economic integration and similar experiences such as the clearing of inner city areas to allow for infrastructure installation and the expansion of commercial centres, the sharing of ideas between governments specifically in the context of poverty reduction and social policy, and specific alliances between southern governments. Underlying this logic is the logic of twenty-first-century capitalism, so that exploited, dispossessed and disadvantaged communities have to learn from other experiences how to prepare themselves for processes of impoverishment and adverse stratification. Thus they need to learn about likely responses of the state, and how they can renegotiate such outcomes and responses and replace them with those that are more effective and/or less damaging.

The next chapter looks more broadly at processes of struggle and resistance supported by the interventions described in this chapter. This is to enable us to understand shared principles and lines of action within their wider meanings and significance.

Notes

1 This draws on many visits to India by the authors and on Arputham 2008, 2012; Patel 2004, 2013; Patel and Arputham 2007, 2008; Patel and D'Cruz 1993; Patel and Mitlin 2004; Patel and Sharma 1998; Patel *et al.* 2002.
2 This draws heavily on Arputham 2008.
3 This subsection draws on Orangi Pilot Project: Orangi Pilot Project – Research and Training Institute 1995, 1998, 2002; Rahman 2004; Hasan 2007, 2008; Pervaiz *et al.* 2008 as well as on five or six visits by the authors to Karachi between 1991 and today.
4 This does not include the 2,163 *goths* (villages) identified in the recent OPP-RTI survey. Of these, 610 are now urban settlements known as 'Urban Goths'.
5 Exchange rate in December 2012 P Rps. 97 to the US dollar.
6 See http://www.oppinstitutions.org/.
7 See http://www.oppinstitutions.org/water%20supply.htm.
8 See Boonyabancha 1999, 2005, 2009.
9 This subsection draws on Baumann *et al.* 2004; Bolnick 1993, 1996; Mitlin 2008a, 2008b; Mitlin *et al.* 2011.
10 The board has ten members, of which two are professionals.
11 This subsection draws on ACHR 1989, 1993, 2011a, 2011b; Boonyabancha and Mitlin 2012; Boonyabancha *et al.* 2012; and www.achr.net.
12 This publication, *Selavip News*, is still being produced by Father Anzorena and is widely circulated.
13 see www.achr.net.

5 Understanding pro-poor politics and pro-poor transformation[1]

Introduction

The significance of power and politics in addressing poverty has long been understood, as has the importance of a continued commitment of oppressed citizens to struggle for progressive transformation. The struggles against colonial rule and other forms of authoritarian control have been diverse, multiple and often successful. The waves of social protests in the 1980s against autocratic states renewed global interest in the power of citizenship and citizen action. The themes of democracy and decentralisation have long been a part of the portfolio of activities undertaken by official development assistance agencies albeit, in general, primarily with a technical focus. Non-governmental efforts (including agencies in both the Global North and the Global South) have supported explicit and radical political agendas including the movements for democracy such as those in Latin America and southern Africa, as well as citizen organising to address exploitation and discrimination and voter education.

However, while these measures may have brought certain benefits for some low-income or otherwise disadvantaged people living in towns and cities in the Global South, acute needs remain unaddressed for hundreds of millions of people. As described in Mitlin and Satterthwaite (2013), the scale and depth of urban deprivation (and the multiple forms it takes) remain very considerable for individuals, households and communities. To address this and help citizens engage with politicians, political agencies and political institutions to address their needs and interests, numerous ideas have emerged. Key approaches are described and analysed in Chapter 2 and programme activities discussed in Chapter 3. As experiences have consolidated, a number of programme interventions have developed, and these help us understand the processes that produce a more pro-poor politics. Five of these interventions were described in Chapter 4.

This chapter builds on previous chapters to explore what these approaches, agency actions and key programmatic interventions offer to our understanding of how to achieve a pro-poor politics with just and inclusive urban centres. We analyse their success through six themes which together encapsulate critical interactions between civil society and the political systems and structures with positive outcomes. In summary these themes consider:

- Universalism of provision and the meeting of essential basic needs: the relationships between individuals and groups with the state, other political agencies and community activists.
- Legitimacy of the urban poor as citizens that includes recognising that their multiple contributions to city economies and city development are recognised.
- Strategic responses and approaches that simultaneously deliver services and strengthen representative organisations of the urban poor.
- Women as public agents for change: changing gendered roles and securing their emancipation.
- Linking the city to the nation: the implications of working at the level of the city and the level of the nation-state.
- Capability enhancement: helping activists (and other individuals) improve their collective strategies and hence influence underlying power relationships between the organised urban poor and the state, leading to more inclusive urban politics.

This introduction summarises the themes discussed in this chapter, elaborating on the context in which they are realised, and emphasising their interconnectedness. We are not suggesting that the interventions discussed in Chapter 4 are similar or that their impacts are necessarily the same; on the contrary, we recognise that there are important differences between them. However, we do argue that there are notable similarities in the strategies that they are using – and this is what we expand on here.

Prior to discussing political change, it may be helpful to emphasise the anti-poor nature of many political relations. There is an extensive literature on the outcome of clientelist political relations and the extent to which they do or do not benefit the urban poor (see Chapter 2). Whatever the merits of these relationships and the ways in which they provide an avenue for some resources to reach some low-income households, many development initiatives have sought to shift relations away from clientelism. This shift is towards relations between citizen groups and political leaders that provide a more predictable universalised flow of goods and services to households and neighbourhoods in need. However, this is a serious challenge in a context in which low individual incomes, minimal state budgets and resource scarcity predominate. It is the search for universalism that is the first theme – as a counter to the selectivity and exclusion of clientelist politics.

Universalism requires a rethinking of the relationship between formal approved standard-compliant improvements and alternative designs that challenge current rules and regulations to enable greater scales (up to universal provision) through reduced unit costs. It requires a serious engagement with alternatives to both formality and existing informal realities. In addition, it requires a challenge to the vertical personalised social engagements that characterise patron–client relations. In this context, for most nations in the Global South, current models of urban development based on the formal high-cost models of the Global North offer little. Minor modifications to reduce standards are unlikely to be effective in producing alternatives relevant for the numbers in need. Rather, emphasis is placed on developing new models that can be scaled up by local groups who are able to negotiate changes with the political leaders with whom they engage (i.e. the interest is in new models that spread from below because of their resonance with the ongoing efforts of the urban poor to achieve development). The efforts against clientelism are not because there are no gains that can be secured through this, but rather because clientelism depends on scarcity and a lack of predictability – it only works because it selects and discards, and in so doing creates the context for continuing exclusion for some. It is the search for inclusivity and universalism, for reasons of social justice and as a means to build a mass movement that keeps the state on track. A further advantage is that it produces an agency or agencies able to negotiate outcomes from a position of strength. These are central to our understanding of these strategies. Inclusive solutions require low-cost improvements and this is one of the reasons for co-productive strategies for public provision (see below), with the resources of the state being combined with those of citizens in ways that can increase scale – because they work within the limited (usually local) resources available.

The second theme considers the ways in which the poor represent their interests (or allow them to be represented). We argue that there is a need to focus attention beyond the contentious politics highlighted by Tilly and Tarrow's seminal works on social movements towards the totality of engagement between citizens and the state. The focus on contentious politics in social movement theorising has missed a lot that needs to be understood in any comprehensive analysis of the relations between citizens and the state (Goldstone 2004). The focus on contention detracts from an understanding of the importance of legitimacy for social movements and disadvantaged groups, and the different ways in which women and men respond to conflict because of their gendered roles and responses. The agencies associated with the programme interventions described in Chapter 4 seek to avoid being drawn into contentious politics because they perceive it to be a terrain which is disadvantageous to low-income and vulnerable groups. They recognise that the middle classes do much better because they succeed in having their demands viewed as reasonable and responsible (Goodland 1999; Harriss 2006). This theme recognises that the ways in which issues and groups are viewed as legitimate or illegitimate by social and political elites form an important part of the broader context of engagement. The significance of legitimacy in issues related to representation and participation lies in its importance in building political support and avoiding marginalisation. This is particularly important in the context of seeking to engage the lowest-income and most disadvantaged groups. Simply put, vulnerable groups are less likely to be able to afford to enter into conflict; the threat of violence is too great. In particular, gendered roles are such that strategies that lead to contention will tend to exclude at least some of the groups whose involvement is sought.

The third and related theme is the development of solutions that, simply put, address the need to shift from simple claim-making towards the state to co-production in the broad area of collective consumption (i.e. goods that are consumed and/or produced collectively), and hence which reinforce collective practices, build collective capacities and inculcate solidarity among the urban poor. There is an alignment of reasons favouring co-production, including the need to build strong local organisations able to demonstrate alternatives that have popularity and scale to local politicians, the need to draw in resources and the need to strengthen local organisational capacity. It has long been acknowledged that self-help in informal settlements can be used to reduce insecurity; housing consolidation becomes a way to secure tenure because the settlement is looking more developed and authentically urban (Payne *et al.* 2009). Co-production builds on such self-help activities to deepen and extend the political capability of organised communities. A major impetus is the desire to shift urban poor political strategies from being defensive to being pro-active (Connolly 2004). As discussed below in the case of Mexico, co-production enabled communities to move beyond demands to proposals – and in doing so strengthened their negotiating position. Seen through these experiences, we can recognise that claim-making and co-production may not be so diametrically opposed; co-production enables collectively organised citizens to engage the state more effectively. The shift towards co-production emerges from an analysis of why previous diverse efforts to reduce poverty and inequality across North and South have not worked. Most notably, co-production is used in part to contest the dangers of institutionalism weakening the strength of social movements that Castells (1983) warned against. Co-production requires active citizens and, as they engage, also the building of additional capabilities in residents' associations organising at the local level. Stronger organisations lead to greater capabilities as well as to increased legitimacy for citizen movements in the face of social and political elites. As elaborated below, the core reasons for the use of co-production lie more in the need for fundamental changes in political relations than in cost effectiveness.

Moreover co-production has a synergy with the reconstruction of gendered roles. The activities result in improved services, attracting activists who are concerned with using practice and mobilisation as a means to increase inclusion, as opposed to pure advocacy and lobbying. These are often particularly attractive to women who are uncomfortable with and/or disinterested in a more confrontational advocacy and lobbying approach. All of the first three themes contribute to the fourth theme, that of the successful renegotiation of gendered relations. The interventions provide openings for women to leadership roles to address their needs and those of their families, and in so doing enable them to challenge their subordination within the home and neighbourhood. The local relations and collective processes of co-production reduce the vulnerability that women experience as private individuals within a domestic context that is frequently oppressive. Such collectivity also helps women address the more damaging psychological impacts of gendered discrimination, imposed associations of inadequacy and inferiority, and enables them to become active positive role models for each other and their female children.

The fifth theme returns to the constraints of the city, and the complexities of engaging the nation-state. As Castells (1983) and more recently Heller and Evans

(2010) argue, urban transformation at the city level is necessarily limited as city governments are subsumed within nation-states. As greater success at the city level is achieved, central government becomes more important as a source of finance for extending activities and for linking urban centres to create a national platform for change, and for providing the resources to enable basic needs to be addressed, and for participatory processes to take place. Economic success often means more funding available for national governments; the urbanisation that economic success underpins also brings new needs and demands for investment in infrastructure and services in the urban centres where the economic success is concentrated. Slum/Shack Dwellers International (SDI) and the Indian Alliance, with the support for national federation building, are the programme interventions that most consciously develop a capacity to augment a city-level focus with national work. It is not coincidental that for SDI, national government strategies are most notable in India and South Africa, middle-income countries with ambitious nationally-financed urban development programmes. In addition, as elaborated in Chapter 4, the Orangi Pilot Project works at the national level following demand to expand their activities from Orangi to other informal settlements in Karachi and then also to other towns and cities.

The final theme is the need to strengthen political capabilities at the local level so as to be able to consolidate gains and realise others. All five programme interventions view political parties with some suspicion as their experience is that most politicians are generally self-interested and/or concerned primarily with policies and programmes that benefit elites (if only because of the absence of alternative pressures for democratic and representative decision-making). All the programmes view political ideology and political statements promising commitment and support as unreliable and believe that it is at best only a part of what makes a difference to outcomes. Moreover, as suggested by Tilly (2004, p. ix), they view a democracy reliant on sporadic elections as incomplete, due to its inability to support the kinds of participatory practices required at the local level for democratic institutions to include the voices of low-income and disadvantaged citizens. Hence their goal is to build local associations and local processes that are able to engage the state and associated political agencies from a position of strength whatever its political complexion, nature and scale. Moreover, as noted above, the challenge is not simply to engender a positive response; rather it is to change the nature of relations between citizens and the state to enable the more effective realisation of state support for equity, social justice and the reduction of poverty.

After discussing the processes involved, the conclusion analyses the contribution of these interventions to changing power relations within the urban context, and the challenges that remain.

Securing universalism and addressing clientelism

The nature of the political relations between politicians (and sometimes officials), parties and low-income citizens and the shift from clientelism to an alternative relationship that is less personalised, less partial, more predictable and more inclusive is an objective for all five programme interventions. As discussed in Chapter 2, the ubiquitous nature of clientelist relations is widely recognised, and all five interventions believe that there is a need to shift away from these vertical relations of dependency between political elites and the urban poor in which votes (primarily) are exchanged for access to a range of publicly funded and generally publicly provided essential goods and services. Such vertical relations lie behind an anti-poor politics that maintains multiple forms of disadvantage and poverty. The intention of these interventions is to challenge the political subordination of low-income individuals and groups, and to move towards relations between citizen groups and political leaders that provide a more predictable universalised flow of essential goods and services.

As noted in Chapter 2, there is an extensive literature on the outcome of clientelist political relations and what they offer, or not, to the urban poor (Auyero 2000; Benjamin 2000; Fox 1994; Robins 2008; Wood 2003). Such relations are frequently accompanied by actual and threatened violence and by a lack of accountability of government officials and agencies to organised citizens and the electorate more generally (Chatterjee 2004, pp. 138–139). As Wood describes, this is a Faustian bargain in which '[P]oor urbanites, and especially new migrants, have no option but to gain membership of such networks and patronage. The price for such loyalty is not to challenge the structural conditions, which in turn deny them long-term autonomy and rights' (Wood 2003, p. 23). However, the strategy is not simply to challenge clientelism – rather it is to encourage alternative relations. In part this is because, as described by Auyero (1999) in the example of a low-income settlement in Buenos Aires (Argentina), in many cases these clientelist relations may meet some needs as powerful mediators respond to the desperate situation of some of those that they know. More fundamentally, implicit in the strategies used by these interventions is the recognition that a direct challenge to clientelism is unlikely to be successful primarily because – as the frequency of clientelism indicates – this is a systematic response to common pressures rather than a few self-interested individuals seeking to manipulate outcomes in their own interests (Khan 2005, p. 714).

Those living in informal settlements and other dense low-income neighbourhoods are, in general, communities that are in considerable need. The difficulties of managing without adequate access to water, sanitation, drainage, schools and health clinics are self-evident. Local governments which are generally responsible for such provision lack resources and provision is irregular and partial. At the same time, existing models do not facilitate the provision of infrastructure and services at scale. The need to secure such goods and services means that votes are, to a considerable extent, a tradable commodity. Despite their best efforts, residents frequently do not benefit to any substantive degree from clientelist relations that tend to favour the community leaders and the informal settlements they represent (or claim to represent). As discussed in Chapter 2, the infrastructure and services provided through clientelism are limited (i.e. only some of what is needed) and generally of very low quality, and quickly require repair or replacement. The nature of the clientelist relationship requires this – as otherwise there is no need for a continuing relationship between the giver and receiver. Clientelism relies on scarcity: there is no benefit to these uneven exchanges if everyone can access the goods and services that are 'traded'. One consequence of such selectivity is that it is difficult to build solidarity and challenge for political inclusion on better terms. Looked at from another perspective, there is a vested interest among the elite in perpetuating 'divide and rule': see, for example, the experience of FEGIP, a federation of tenants' organisations in Goiania, Brazil (Barbosa *et al.* 1997). This is not to say that groups cannot benefit; they may, but all of the urban poor cannot.

As is evident from the programme interventions introduced in Chapter 4, alternatives to clientelism require new solutions for urban development if they are likely to be able to address the needs of those who live in informal settlements. Conventional approaches are expensive and hence not able to go to scale and provide for all due to a lack of state resources; moreover, such traditional approaches are generally based on northern models of urban development that are rarely appropriate to the context of the Global South. A core requirement is therefore to rethink modalities for informal settlement upgrading (and sometimes new greenfield developments) such that they can be made more predictable and scaled up with lower unit costs and greater numbers. Rethinking models for producing collective consumption goods and services requires rebuilding local organisations.

One conclusion of these programme interventions is consistent with that of the study of associations by Houtzager *et al.* (2007): in and of themselves regular neighbourhood associations do little or nothing to address political inequalities and inequities in service provision. Each of the programme interventions seeks to rebuild local associations and city-wide associations of local associations, deepening their relations with their members by requiring them to engage with money (a scarce resource for low-income households) and incorporating practical work to upgrade informal areas into their political claims for redistribution. We can differentiate local organisations on the basis of their objectives and the strategies that they use to achieve such objectives. Many, as elaborated in the following subsection on residents' associations, have limited objectives – and these interventions are not unique in this recognition of the need to reconstitute residents' associations that exist. Abers (1998, p. 515) discusses similar themes in the context of participatory budgeting, and explains that there has been 'a new kind of neighbourhood organization emerging in Brazilian cities that refused to play according to clientelist rules'. She explains how these new organisations avoid clientelist negotiations and find new strategies to advance the needs of all low-income settlements.

As the emergence of new models for upgrading has taken place, it has become possible to see just how costly the previous solutions were, embedded within clientelist and/or corrupt politics and state practice. As communities develop their own solutions, it is common for local municipal officers and politicians to be surprised at just how cheaply improvements can be made (Box 5.1 gives one example of this from Vietnam). In some cases, there was prior awareness within the local authorities of just how cheaply low-income groups can provide goods and services but it is hard for officials alone to tackle this situation. More active community groups help create a context in which alternatives are considered. One of the keys to the success of OPP was their demonstration of how to reduce greatly the unit costs of infrastructure provision and support for households and neighbourhoods to realise these alternatives. In Zimbabwe, local authorities are now showing increasing interest in communal sanitation as the long-standing requirement that individual plots have water-borne sanitation and water points is too expensive in the current financial crisis. The policy was always too expensive for the lowest-income households, forcing them to contravene regulations and increasing the difficulties of living in urban areas. Political competition is often important here as the Movement for Democratic Change controls the urban authorities and their politicians are happy to critique the programmes of the past and consider alternatives.[2]

Sometimes information can drive the changes that are needed – as in the community-driven mapping and surveying of informal settlements described in Chapter 4 (*Environment and Urbanization* 2012). Work by the Orangi Pilot Project's Research and Training Institute showed that much of the water produced for the city was being stolen through hydrants and resold through a tanker delivery network. In this case, city engineers were aware of the problem but not the scale of the theft. It was only a civil society organisation that was able to negotiate a path through these relationships, secure the information and publish a document (Rahman 2008).

Box 5.1 The costs of convention or why alternative approaches to urban development make sense

In 2007, the provincial government announced plans to redevelop around social housing areas in Vinh. The plan was to demolish the existing buildings and replace them with lower-density social housing, doubling the sizes of plots and homes. The authorities wanted the families who would be unable to pay higher rents to move to other areas where new social housing was to be built. In Vinh's Cua Nam Ward, a different approach was taken. The people took up the opportunities offered by ACCA. They used ACCA's small-project budget to establish a revolving fund for small infrastructure loans, and by December 2009 they had completed three projects providing 110 households with underground sewers and 40 households with a paved walkway. The total project costs amounted to almost $60,000. The community members' contribution was $39,000 and, together with ACCA support worth $9,000, it leveraged government funding worth $11,000. Such collective action boosted people's confidence, including their own potential to save and mobilise money.

The families in Cua Nam Ward wanted to stay together in their neighbourhood. Aware that they were unable to afford the redevelopment units planned by the government, they decided to propose to improve their housing themselves. The plans they prepared, with help from architects, included widening the lanes, laying drains and rebuilding their homes in an efficient layout of rows of two-storey houses. Earlier, there were only one-storey buildings. The new units would be 47 square meters, compared with 30 square metres previously. The provincial minimum standard was 70 square metres, but the communities argued that this was unnecessary and that it was better for them to remain in the area than to move far away. They used this redevelopment plan, and the availability of housing loans from ACCA, to negotiate with the authorities, who finally agreed to the people's proposal.

Construction began in March 2010, and was completed six months later, providing 29 low-income families with improved housing and setting an important example. For the first time in Vietnam, urban poor households who lived in collective housing had won the right to design and rebuild their own homes on the same site, with the support of both the municipal and provincial governments. This was also the first case of a housing community being permitted to build homes that were considerably smaller and hence more affordable than the provincial government's minimum standards. The municipal government was so impressed that it has officially sanctioned this people's standard. It has agreed to replicate the Cua Nam model in 140 other social housing areas. Five of those projects are underway. The municipal government appreciates the approach because it is cost-effective and people-friendly.

The costs in Cua Nam were compared with Ben Thuy, another ward in Vinh, where the government spent $1,166 per household on infrastructure, $141 per square metre on redeveloped housing and $395 per household to demolish the previous dwellings. Of the 114 families benefiting from 'improved' housing, 40 per cent had to be relocated. No one had to move away from Cua Nam, where infrastructure cost (installed by the community with technical support) was $303 per household, the housing cost $72 per square metre and the demolition costs per household were $103. In the meantime, Cum Nam has become a model beyond Vinh. Low-income communities in Hai Duong, another ACCA city, have already persuaded their local government to allow them to follow the example of Cua Nam.

Source: ACHR (2011a)

The example above shows how community-designed and managed solutions are lower cost and can go to scale (through a multiplication of local initiatives). In addition to lower-cost versions of conventional solutions, another necessary step is to shift to collective, sometimes incremental designs. As noted above it is only with a rethinking of approaches to informal settlement upgrading that larger numbers can be reached. Achieving scale requires moving away from formal approved standard-compliant improvements and using alternative designs that challenge current rules and regulations. The work of the National Slum

Dwellers' Federation, *Mahila Milan* and SPARC (the Indian affiliate of SDI, also called the Indian Alliance; see Chapter 4) to provide sanitation in a context in which clientelist politics were ensuring limited access is described in Burra *et al.* (2003). In this case, the market solution was not possible, as individual toilets were too expensive and housing units were often so small that it was difficult or impossible (and expensive) to fit a toilet into each unit. Communities innovated and created high-quality toilet blocks with separate areas for men, women and children and a flat for the caretaker and their family. (Not only are there separate toilets for women and men, there are also separate queues, as women's experience is that when there is only one queue, men often push in). The success of the initial toilet blocks resulted in the Alliance securing funds for the expansion of provision. By the end of 2011, over 20,000 toilet seats (1,039 blocks) had been provided as a result of various initiatives to support the expansion of this alternative design. An estimated 50 people use each toilet seat, resulting in a total of 1 million more people now with access to sanitation and washing facilities.

Universality requires a central engagement with alternatives to formality that engage with the informal realities (Myers 2011). Regulatory changes alone are unlikely to lead to programmatic reform, and what is required is the development of new models that achieve scale as more and more local groups negotiate changes with political leaders to improve their neighbourhoods at the same time as reform from above. In some cases the innovations are more with respect to process than technology. The work of Muungano W.A. Wanavijiji (the Kenyan Homeless People's Federation and the SDI affiliate in Kenya) demonstrated a complementary and important step towards inclusivity when they persuaded structure owners (not absentee landlords) and tenants to share the land in an informal settlement (Huruma) in return for securing tenure so that the re-blocking provided plots for both (Weru 2004). The situation in Nairobi remains very difficult with unequal land holdings and with informal settlements with large scale landlordism – both absentee land owners and those who rent land from these land owners and build rooms to rent to others. Muungano activists believe that inclusive solutions require the sharing of this land which is a painstakingly difficult negotiation but which has been seen to be possible. It took Muungano and the support NGO a considerable time to persuade the first structure owners and tenants to negotiate together to work out a land-sharing deal (Weru 2004), but once practices such as this become embedded in local expectations and actions, then achieving inclusion becomes a possibility. At the same time, the need for cost reductions favours co-productive approaches, with the resources of the state being combined with those of citizens to increase scale.

Co-production does not emerge just because of the need for financial contributions from citizens. The primary interest in co-production lies in the requirement for active citizens, and especially the building of additional capacities in local residents' associations, in order to engage in the provision of collective consumption goods. The core reasons for co-production lie more in the need for fundamental changes in political relations than in its cost effectiveness. The importance of co-production for a new political capability is returned to below following a discussion of political strategies and the limits of contentious politics.

The concept of universalism in service provision is critical for both social and political reasons. In terms of meeting basic needs, it is clearly not adequate for only some needs or some people's needs to be met. In part this is because each person, each life, needs to have access to essential services. However, there are other reasons for universalism. In the case of sanitation and waste water management, for example, the quality of everyone's neighbourhood is determined by the totality of options that are available and one single household can do little to address the consequences of inadequate sanitation/waste management. In dense urban settlements, neighbours need to act collectively. In terms of political strategy, there is little reason for individuals or neighbourhoods to participate in an exclusive strategy that is intended to assist only a small number of residents unless they can be sure they will be included. In a clientelist context, neighbourhood associations have more to gain by negotiating outside of such arrangements. This offers political agencies the opportunity of a 'success' in providing a neighbourhood with services at a lower cost because it is even more selective. Hence clientelism results in competing groups of low-income residents, rather than a collective process to challenge outcomes. However, this is not the only source of competition, and, as discussed in Chapter 2, welfare regimes may offer selective benefits and in so doing reduce group solidarity.

It is clearly not possible for a bottom-up process alone to either overcome clientelism or to provide the finance for universal solutions to be implemented. One response used by SDI and ACHR is to create funds at the city and/or national level that have clear rules, a public allocation process and transparency in what gets funding. Fund allocations offer subsidised loan finance to community groups that meet the agreed conditions. Savings contributions serve multiple purposes helping to build local groups that can manage finance, improve affordability through reducing the final cost of the improvement, strengthen the fund itself, ration resources based on agreed criteria and build the political momentum leading to an increased state contribution. The joint management of such funds encourages state contributions as public officials and politicians recognise the effectiveness of these allocations. Box 5.4 illustrates the functioning of CDFs in Thailand.

However, at the neighbourhood (and later the city) level, careful positioning is required to negotiate with various political agencies, including political parties. The following subsection expands on particular challenges for this, prior to a discussion of the positioning of these interventions in respect of clientelist politics.

Neighbourhood associations, inclusive cities and political parties

Residents' organisations face competing interests both within and from outside, and the local groups involved in all five of the Chapter 4 programme interventions have learned to mistrust both political intentions and the articulation of political ideology.[3] The political processes are often not as well intentioned as they may at first appear. Within the formal political system, existing parties seek to ensure that they occupy participatory spaces in part because they have a political motivation to do so. Political parties have a vested interest in rewarding their supporters and this can reduce solidarity between organisations. One of us will not easily forget a discussion with a community leader in Rio de Janeiro in 1995 when he explained that two things had (in his opinion) defeated the social movement in Rio, namely drugs and democracy. The problems that those designing these programme interventions have observed are similar to those described by Castells (1983, pp. 258–277) in his discussion of the consequences of democratisation and party political competition on the citizen's movement in Madrid. The Madrid Citizen's Movement was one of a number of city movements in Spain which involved hundreds of thousands of residents during the 1970s in struggles for 'the matters of everyday life, from housing to open spaces, from water supply to popular celebrations' (ibid., p. 215). Castells describes the success of the movement but also its demise with the advent of a democratic national state. Castells argues that the movement did not survive the transition to democracy because the leadership affiliated to political parties and chose to contest politics on this basis. Political parties demand upward loyalty to the party, local accountabilities between residents and leaders were reduced, individual leaders competed for residents' support and vibrant local citizen participation fell away. Without unity between urban citizens, their political position was lost (ibid., p. 274). Castells recognises the contradictions involved in a need for both connection and autonomy (ibid., p. 273):

> This was a continuous tension that characterised Madrid's Citizen Movement; it was based, at once, on the self-organization of residents to foster their urban interests and on the connection of their demands to the political struggle against the urban crisis, whilst keeping their autonomy in relationship to partisan politics. The ambiguity of the situation was both a source of creation and destruction: it was a creative tension because it allowed the Movement to expand, to find powerful allies, to shift from local and piecemeal demands to alternative models of urban policy enabling citizens to have a decisive impact on the political mechanisms that were prerequisites for the transformation of the city. It was also the major source of crisis and, ultimately, of destruction. Partisan goals and the Movements' orientations became increasingly divergent.

Such realities help explain the decline of Mexico's urban popular movement, in a country recognised for the strength of its citizen activism, as this movement thought they had achieved their goals through democratic reforms which they had helped drive (Dávila 1990, pp. 37–38). As Rolnik (2011) exemplifies for Brazil, city authorities may be forced to compromise with political interests at the state and national level. She argues that despite national level commitments by the PT (Workers' Party), outcomes at the city level are not necessarily pro-poor, selective investments along clientelistic lines continue and 'urban planning in Brazil continues to be the domain of highly sectorized, centralized bureaucracies that rely on decision-making processes strongly influenced by the interests of the economic and political actors that depend on them to survive' (ibid., p. 251). However, in this case and in others, political parties may not wish to have such realities acknowledged and may resist local groups that seek greater participation and pro-poor outcomes.

Even if political parties are successful in being both well intentioned and effective in delivering needed resources to the urban poor, there is a danger that policies become associated with particular political positions. Indeed, parties want this association. But when the opposition gains power, the urban poor may find that the favourable policies are reversed. Moreover, in practice, political parties compete for control of local organisations either through party branches or through local government. An example of this is described by Arévalo (1997) in Huaycán (Lima, Peru), the location of a land occupation. Once it was evident that the occupiers would be allowed to stay, two political parties began to register their own members. Several other parties joined them in positioning for control over the residents' organisation. *Sendero Luminoso* (Shining Path) also became active in the settlement and threatened or murdered community leaders that opposed them.

A number of local studies highlight the exclusionary consequences of contesting political parties within a particular locality. Hossain (2012) describes the interaction between residents in one informal settlement in Dhaka and the ways in which political elites nullified a vibrant local association and competition between political parties resulted in stratification and exclusion. Sinwell (2012) discusses the involvement of multiple political parties within the Alexandria Development Forum in Johannesburg (South Africa), a platform established to implement a government-funded redevelopment programme. The dominance of the African National Congress (ANC) resulted in other parties and community groups believing that they were excluded. The Communist Party responded by setting up alternative negotiating opportunities for disadvantaged residents, but, she suggests, the result was that groups of residents ended up competing with each other for scarce resources. The contribution of the African National Congress to addressing the concerns of local residents is viewed more positively by a study by Bénit-Gbaffou (2012), but concerns are also substantiated when she describes how the ANC ensured their control over community policing forums (ibid., p. 184) using violence to maintain their dominance:

> Although some civic organizations have developed specific constitutional or electoral measures to avoid being 'hijacked' by ANC members, in some cases they appear relatively powerless when the ANC has decided either to block their relationship with the state; or to take over the leadership of the organization. When political strategies do not work, and in specific cases of high

> local tensions, physical violence can also be resorted to as a means to weaken or delegitimise oppositional organizations.
>
> (ibid., p. 184)

However, despite potentially negative impacts of political parties on the development of inclusive and pro-poor cities, it is critical to recognise that political relationships between those affiliated and those not affiliated to political parties exist as individuals and groups interact at the local level. Castells (1983, p. 299) notes that

> all over the world conscious people have continued to mobilize collectively to change their lives. … People mobilized, in a variety of historical contexts and social structures, without parties, beyond parties, with parties, against parties and for parties.

The complexity of relations between movements and political agencies (inside and outside of the state) and the difficulties in untangling causality in any particular case are illustrated by alternative perspectives among researchers which are likely to reflect both their conceptual frameworks and their informants.[4] Understanding who has done what and the interrelations between political parties and movement organisations (or support organisations) is made more complex because individuals may be members of both. People choose to hold multiple identities at the same time (as in movement leaders that belong to parties) and shift identities (movement leaders who enter politics, politicians who move into movements and/or other civil society organisations). As Bénit-Gbaffou (2012) explains in a study of five low-income urban neighbourhoods in Johannesburg (South Africa), there are frequently overlapping affiliations between community activists and political party members; and these arise because of the genuine concern of these individuals in engaging with matters of public interest. Such overlaps are not just found at the neighbourhood level: support professionals in many countries include those who come from the same political class and even studied with politicians and senior members of the government, those who have worked for the state in previous employment, and those who have close friendships with members of the government (Lewis 2008). What may be more helpful to understand is why individuals with multiple identities pursue some strategies through civil society associations and others through political parties. What are the consequences of such identities which are ambiguous in at least some contexts, and what are the ways in which ideas emerge, gain legitimacy and influence, and are transferred into policies, programmes and practices?

Despite overlapping affiliations and the difficulties of establishing causal attribution, the importance of political ideology in explaining pro-poor politics continues to be articulated (Moore 2005), as does the central role of the state (Gandy 2006; Houtzager 2005). For example, Lavalle and colleagues (2005) argue that evidence from São Paulo suggests that the participation of the organised urban poor in formal participatory institutions is explained by the extent of their political connections. They believe that such links enable the organised urban poor to participate in and influence such institutions, and conclude:

> [T]he organizations that are most likely to represent the poor in participatory institutions are those well connected to the actors of classic representative democracy – political parties and state agencies. Contrary to the conventional wisdom, these civil organizations are not co-opted but instead are more likely than their poorly connected counterparts to organize public demonstrations and to make demands on the government through multiple channels.
>
> (ibid., p. 951.)

In the competitive political context of São Paulo, they go further and state that: '[A]ctors best able to represent the interest of the people living in poverty are those who establish ties to agents of representative democracy' (ibid, p. 960). Note that the emphasis in the study is on formal structures for participation rather than on the full range of informal engagements and formal interactions. Harriss (2006, p. 446) also considers the responsiveness of parties to local residents when discussing findings from surveys of citizens in Delhi, São Paulo and Mexico City. He points to a diversity of experiences between the cities, noting that in Delhi, 37 per cent of respondents went direct to the government for assistance and a further 29 per cent went to a political party, while in São Paulo and Mexico City the figures for the government are 54 and 33 per cent respectively, and for political parties 4 and 9 per cent (ibid., p. 451). Hence, whatever the earlier experience of local associations in São Paulo, this evidence suggests that it does not translate into individuals using parties to access the governmental system. Harriss (2006) also notes the class affiliations of local associations and the ways in which they engage the local authorities, differentiating between citizen associations in South Chennai which are primarily concerned with middle-class interests (including slum clearance), and those in North Chennai where there is more diversity and overlaps across civil and social organisations through women's groups (including *Mahila Milan* from the Indian Alliance). He provides further evidence of the divisive impact of political parties on the unity of organisations (ibid., p. 460). An alternative experience which also highlights the complexity of engagements between political parties and local collective action is recounted in Auyero *et al.* (2009, pp. 15–19) when they discuss the involvement of Peronist activists in the looting of shops in Argentina during a period of falling real wages, rising unemployment and declining state support. This suggests that while political activists may prefer to draw residents into their activities, they sometimes realise that alternatives are needed and align with the movements themselves, adjusting immediate strategies to maintain their long-term position.

The five programmes described in Chapter 4 have many responses to competing political parties, including the ways in which such parties seek to use affiliated local organisations to their own advantage in the management of opportunities arising from participatory governance. The five programmes avoid any party affiliation for the local process so that community networks and federations can build relations with each and any political party that is interested in working with them. They emphasise the accountability of leaders to the local membership and hence a degree of autonomy from the political system. Looking at the wider literature, this emphasis on the need for autonomy resonates not just with Castells (1983) but also with Bayat's (2000) analysis of the strategies of informal dwellers, and with Fox's (1994) study of the rural low-income communities in Mexico. In the case of the federations that are members of SDI, this means that community leaders should not hold office in a political party (although there is much overlapping membership). SDI affiliates agree that federation leaders should not be active within the leadership of political parties (although they may be members), as they believe this will compromise the ability of federations to negotiate and make political alliances. All five programme interventions also combine this emphasis on positive proposal-making with the avoidance of political alignment. One of their rationales is that this reduces their mobilisation potential because there are inevitably different political interests within their constituencies (i.e. informal settlement residents) and that they are seeking an inclusive collective process to advance the needs and interests of all informal settlement dwellers. A second is that it is difficult to avoid co-option if there is a strong alliance with a political party as allegiances become confused and the more powerful agency, the party, dominates. Furthermore, if there is a strong alliance with one party, the interests of the informal residents are less likely to be taken into consideration when that party is out of power. Hence, local groups make an effort to nurture political plurality. In the case of Zimbabwe, for example, it is taken for granted that federation members will include those who support both the MDC and ZANU-PF, despite the hostility between these parties at national level and despite the very real difficulties that informal settlers have faced from a national government controlled by ZANU-PF both before and after the massive eviction programme it mounted – Operation Murambatsvina (Chitekwe-Biti 2009). Building links across political groups to draw in all residents is a necessity if a process is to reach scale and secure universality. The challenge is to avoid simple political trade-offs and to maintain processes that build a united position on universal access to basic services and other needs throughout negotiations with a range of politicians and their parties.

Negotiating legitimacy: contention, compromise and political positioning

The second theme emerging from a consideration of the programme interventions and the more general literature on pro-poor political change is the shift from a focus on contentious politics between citizens and the state towards the totality of engagement including contentious politics, the exploration of non-contentious issues and the identification of new and emerging issues. While recognising that issues related to redistribution are likely to be contested, the approaches of these interventions seek to minimise contentious responses through multiple tactics designed to deepen the state's understanding of the realities of the urban poor and pre-empt conflict. Chatterjee (2004, pp. 40–41) discusses the emergence of a political society in India which bridges the gap between the state and groups who are living in the city but who are not fully legal (i.e. squatting) and with some degree of informal services for their basic needs. He argues that as the state has introduced programmes to alleviate poverty and provide welfare, so a politics has emerged to manage access to such resources while at the same time denying recognition of equal status and claim to urban citizenship. Chatterjee (2004, pp. 50 and 138–139), in keeping with much of the analysis of this politics, emphasises the level of contestation and often violent nature of relations. The programme interventions in Chapter 4 are seeking to challenge this politics and to establish the legitimacy of the urban poor – and hence the inclusion of the organised urban poor in state planning as an agent of urban development. These interventions are not alone in these tactics although there has been relatively little exploration of such political strategies within the social movement literature related to the Global South.[5]

The literature on social movements has, in general, focused on contentious politics (Crossley 2002) and in so doing has given relatively little attention to the scope and complexity of interactions taking place between organised citizens and the state. Tarrow summarises his contribution thus (1998, p. 19):

> The most forceful argument of this study will be that people engage in contentious politics when patterns of political opportunities and constraints change and then, by strategically employing a repertoire of collective action, create new opportunities, which are used by others in widening cycles of contention. When these struggles revolve around broad cleavages in society, when they bring people together around inherited cultural symbols, and when they can build on or construct dense social networks and connective structures, then these episodes of contention result in sustained interactions with opponents – specifically, in social movements.

Tarrow (1998, p. 5) recognises both that movements do more than simply involve contention (and he recognises negotiation with the state and challenges to cultural codes), and that contention is not limited to movements (citing political parties, voluntary associations and individuals), but he concludes that 'the most characteristic actions of social movements continue to be contentious challenges'. This, he suggests, is because they lack other resources such as money, organisation and access to the state, and hence 'contention is the only resource movements control' (ibid., p. 5).[6] However, social movement theorists also recognise that many middle classes do much better because they succeed in having their demands viewed as reasonable and responsible, and negotiate a positive response from the government before the issues become contentious (Goodland 1999). Harriss (2006) provides an example of this when he describes how middle-income groups in India are using associational life and emerging public spaces to put forward their views on the desirable direction for urban development. Drawing on research in Delhi, Chennai and Bangalore he concludes that the middle-classes use such spaces to good effect, quoting one of his interviewees who explains that 'the rich *operate* while the poor *agitate*' (ibid., p. 461, original emphasis).

All five programme interventions in Chapter 4 view it as important to avoid being drawn into contentious politics because in their experience this position is unlikely to be advantageous to low-income and disadvantaged groups. In their experience, contention provokes an antagonistic reaction from the state as it involves a challenge to the state's legitimacy (i.e. their right to make decisions) as well as a challenge to the content of state decision-making. This does not mean that they always avoid it; on occasion it may be used – however, it is used with care. This is because the antagonistic reaction that confrontation provokes means that it is more difficult to secure state resources and state collaboration in changing laws, regulations and approaches. The federations and networks described in Chapter 4 believe themselves to be a force for positive changes in their towns and cities, and they ask that what they are doing is recognised as a legitimate contribution to progressive transformation that offers benefits to all. Prapart Sangpradap, a community leader from the Bang Bua Canal Network in Bangkok, explained his perspective on these struggles and successes thus:[7]

> In Thailand, we have been fighting for a slum law for ten years. We mobilized all the communities to support this bill, because those of us who live in squatter settlements and slums should have rights too. But we never got those rights, and we never got that bill. The way we got our land and housing and security only happened when we made concrete change and showed the possibility by people, showed a new way. We are the ones who have to make that change, according to our way. And that change becomes its own law.

The urban poor are frequently excluded from political debate and are often even represented by those in power as a violent mob or rabble; the federations and networks recognise that such a stereotype is used to restrict their access to political influence and seek to challenge this negative image at every opportunity. In this task, women's leadership creates new possibilities for challenges that may still be 'forceful' but with a lower risk to the reputation of the organisation and its legitimacy. For example, women from *Mahila Milan* in India campaigned for access to water by bringing their children with them to 'sit in' government offices that had refused them access to water. In so doing, they put out the officials but reinforced their reputation as parents caring for the needs of their children.

A second reason for questioning the effectiveness of antagonistic confrontation (in addition to the reduction in legitimacy) is that the negative response of the state makes it harder for a pro-poor movement to go to scale as it reduces the participation of the most vulnerable groups who cannot afford the risks associated with such antagonism. Citizens campaigning against the evictions are at risk of violence, and even rape and murder (UN-Habitat 2007). Simply put, in many instances, low-income and disadvantaged people cannot afford to enter into conflict because the likelihood of (often violent) retribution is too great: a problem that Tarrow recognises (1998, p. 101): '[O]rdinary people are more likely to participate in forms of collective action that they know about than risk the uncertainty and potential violence of direct action.' The avoidance of confrontation is even more important in the context of nurturing women's involvement and leadership. Patel and Mitlin (2009, 2010) explain how the emerging women's leadership in India developed a new strategy to challenge evictions (see Box 5.2).

Box 5.2 A gendered approach to anti-eviction struggles

In India, the National Slum Dwellers Federation (NSDF) was set up in 1976 to support slum dwellers in fighting eviction. It was established by male community leaders who were protesting against the threat of eviction in their neighbourhoods, and these leaders used women's anger and outrage to fill protest marches and confront the authorities. Male activists evaded the police by placing women in front. The police also found it more difficult to enact violent responses against women protesters. Hence, women were involved but the male leadership tended to see women as passive contributors to mass rallies and protest marches.

Then the male leadership recognised that their movement was not progressing beyond marches and demonstrations. The leadership observed that men were generally comfortable with fighting aggressively against eviction and wrongdoing by the state, whereas women – while they passionately sought security of tenure and basic amenities – felt less compelled by confrontational strategies. During the mid-1980s there were many eviction threats in Mumbai and street battles in defence of the rights of pavement dwellers to reside there. SPARC, the NSDF's support NGO, asked women living on pavements in the Byculla area and members of *Mahila Milan* what they wanted to do. The women replied: 'we don't want to fight and we don't want to stay on the pavements either! Go and speak to the municipality and to the state government and see if you can explain to them our situation.' The Alliance forged between NSDF, *Mahila Milan* and SPARC began to work on longer term strategies, but meanwhile they needed a solution to the continued evictions.

The pavement-dwelling women agreed that the confrontational strategy had done little to stop evictions or address the damage caused by evictions, including the loss of their belongings and the trauma to their children. They had developed their collective strength through savings and learnt both how to help each other with loans, and how to negotiate with the state to secure benefits such as ration cards. They developed another response to evictions. Rather than confronting the police, they decided to outwit them. When the police next came to evict them, the women offered to take down their dwellings. They dismantled their shacks and stacked their belongings and building materials neatly on the pavement. This left only rubbish on the site where their shacks had stood, which they invited the police to take away. The police were willing to do this, as they could then go back and report that the dwellings had been dismantled. Once the police had left, the women replaced their dwellings. As a result, they kept their material possessions, they and their families were not traumatised by the experience, and the police began to see that they could negotiate with the poor. This showed the group that, when the poor are in a vulnerable position, a collective demonstration of strategic resistance is as powerful as confrontation, and more effective psychologically. Their slogan was '*todna tumhara kam, ghar bandhna hamara kaam*' ('it's your job to break our house, it's ours to rebuild it').

Source: Patel and Mitlin (2009)

A third reason for seeking a positive position with regard to the state is the recognition of the potential value in alliance-building with middle-income groups. The federations and networks are aware that their case is more likely to succeed if they can build support among middle-income groups and particularly interested professionals. The importance of alliances that enable low-income disadvantaged citizens to benefit from better relations between social movements and professional groups and other elites is acknowledged by Tarrow (1998, p. 79). Goldstone, reflecting on movement activities in the US, reinforces the importance of strategies to extend relations and build networks; he argues that it is not simply movement activities that seem to be important in securing success but rather the complex alignments between citizens, the actions and impacts on voting, and hence the ways in which politicians recognise these underlying interests, and adjust accordingly (2004, p. 354).

The federations that are SDI affiliates consciously build platforms similar to that of the Urban Resource Centre (see Box 3.5) where interested professionals and community groups come together to share perspectives about the problems and discuss possible strategies for improvements. City-level federations and networks build strong links with politicians, enhanced by community-to-community exchanges in which politicians are invited to participate. Professional staff from support NGOs do likewise with officials. There are frequent disagreements and much debate but the focus is on dialogue. The ACHR tradition evidenced in ACCA is to build a platform at the city scale which brings together key groups and individuals. This platform may be organised around a particular intervention that emphasises a commonality of interests, such as the settlements on sites belonging to a common landowner, as in Thailand, where networks address common issues arising from living on State Railway Authority land or in other cases from living on canal-side land.

Fourth, experience has shown these interventions that contention may be hard to maintain in the longer run, provoking apparent successes that turn out to be insubstantial. The experience of many activists is that significant activities take place and concessions are won but it is very hard to realise these gains in the longer term. Once the movement has disbanded, it can be hard to remobilise in part because of the fluidity and immediacy of many of the relationships. Hence it is important to build around something that is maintained, which is unlikely to be contentious. Ruby Papeleras, a community leader of the Philippine Homeless People's Federation, explains how confrontations are often provoked by external political interests who

> nurture anger against injustice in the community people, stir them up and get them to fight. But after the protests and the barricades, those outsiders go back to their homes, while we are still here, still living with these problems, without any solution. Political factions also use poor people in this way, as cannon fodder when the opposition wants to bash the current administration.
>
> (Papeleras *et al.* 2012, p. 464)

All the networks and federations have groups that become frustrated – some of which then undertake the kinds of contentious actions discussed by Tarrow, Tilly and others. Groups invade land, illegally connect services, stage demonstrations and pickets, and occupy council offices; such events are catalysed by the scale of political exclusion and deprivation. They are used by the federations and networks tasked with negotiations to demonstrate the difficulties members face and the intensity of their needs. But at the level of the organisation, there is an emphasis on pragmatic solutions and an engagement that suggests how the state (local, city, national) can respond positively to this situation.

These programme interventions are far from unique in using these strategies. Many other movements also negotiate alongside their more public challenges to the state,[8] and other movements have recognised the need to avoid party political alignment. While the residents involved in the five programme interventions may have practised contentious policies from former and/or ongoing participation in political parties and/or the trade union movement and/or other movement organisations, they also have considerable experience in what are, in most cases, long-standing practices of engagement with local councillors and in some cases local government officers. By choosing to invest and enhance these links, the programme interventions bring to the forefront the potentiality of legitimacy.

This is encouraged by the reality that the boundary between civil society and political society is hazy and consistently crossed by some. The section above discussed multiple political identities but there are also other links. For example, residents in informal settlements include current or ex-local authority workers who may be (or have been) junior within the bureaucracy but who have connections. Political activists also living in informal settlements may find work in national government agencies and utilities. Benjamin (2004, p. 183), in a study from India, discusses the ways in which lower-level government officials are responsive to the needs of the urban poor, in part because they and/or their relatives live in informal settlements. Peattie (1990) describes how the organiser of a large-scale invasion of land in Lima in 1971 was a driver in the Ministry of Housing.

The focus on contention means that such strategies have been under-researched and inadequately recognised. Goodwin *et al.* (1999) argue that there is a risk of an over-concentration on explicit political engagement for social movements at the expense of understanding effective strategies and/or simply ways of being that challenge convention; however, this perspective is somewhat exceptional to the body of social movement literature. In terms of literature drawing on research in the Global South, Auyero *et al.* (1999) also argue in favour of a more rounded and nuanced analysis of the political strategies of the urban poor, although in this case it is across the dual strategies of clientelism and contentious politics. Some alternative perspectives have been presented, such as the work of Bayat (2000), who highlights the ways in which residents who face severe constraints on contesting adverse outcomes develop pro-active strategies to challenge their lack of options. He describes the ways in which the informal urban poor struggle to secure their livelihoods within modern cities dominated by political elites. Their response has been to 'encroach', finding ways to acquire urban informal space, secure illegal services and earn informal incomes. Bayat himself argues that 'these very simple and seemingly mundane practices tend to shift them into the realm of contentious politics' (ibid., p. 547), but there is much that he describes that is not contentious: rather it is low key and very far from the explicit politics of trade unions and political parties (ibid., p. 548). Benjamin (2004, p. 185) recognises similar strategies in his research in Bangalore: he argues that the urban poor are most successful when they use 'politics by stealth', negotiating to advance their claims and interests while avoiding a public challenge, in this case to secure land. As Bayat argues, this can become a collectivised protest, although one in which advancement is less by direct confrontation and more by scale, i.e. large numbers of people trading on the streets to the point where the state cannot afford to act against them all. In this context, governments are forced to compromise or they risk damaging their own legitimacy by being seen to be ineffectual. While Bayat sees conflict as inevitable, those designing these interventions are conscious of the costs of such conflict.

The responses that Bayat (2000, p. 553) describes are primarily individual and informal responses; while individuals build collectives these are neither resourced nor sustained: indeed he terms them 'non-movements'. What characterises the interventions discussed in Chapter 4 is the way in which they build collective institutions able to learn from experiences and to incorporate improving practice within their modalities of action. The core objective is the effective influence of political processes and agencies from below through strengthening social relations and encouraging informal interaction around areas of common interest and concern within and between the neighbours in low-income settlements. Investments in alternative approaches are concretised through precedents and enable a grounded sharing of experience with the state from an informed position. This is all about the framing of pro-poor interventions as win-win situations for citizens and the state – and this requires both strong relations and the ability to strategise to ensure that the needs and interests of low-income and disadvantaged groups remain central and are addressed.

Important to a comprehensive political strategy is the use of precedent-setting actions to drive an alternative development agenda grounded in practical examples of actions that both deliver essential goods and services, and strengthen the organisational capabilities of the urban poor. The solutions that emerge are, in every case, examples of co-production.

Co-production, politics and building a base

The third theme common to the five interventions is their use of co-production with the joint planning, resourcing and delivery of solutions to improve conditions in informal low-income settlements. All five take solutions developed by and/or with low-income residents and upscale them with the active involvement of both residents' associations and the state. The emphasis on collective consumption reflects residents' needs and offers a potential for securing political processes that work at the city level to secure greater justice and equity (see McFarlane (2008) for this discussion in the context of Mumbai, and Gandy (2006) for Lagos). As discussed in Chapter 2, within the focus on the urban management approach, there is a recognition that the only way to resolve competition between commercial and residential needs for basic services, between discourses of modernity and neglect of informal settlements, and between specific neighbourhood improvements and city scale is to expand and upgrade provision.

While different models are being used, as well as differing levels of coordination and formalisation, the pattern of co-production is consistent across these interventions. The use of co-production builds on earlier experiences, including those in Latin America. Castells (1983, p. 197–198) analyses the strategies of the squatter movement in Monterrey (Mexico) in the 1970s. He explains how members either stole or negotiated for building materials for their schools, health and civic centres. Families collectively planned their settlement providing water, sewerage and electricity for themselves through illegal connections to the city services. After negotiations, the state paid for health and education staff but the community retained management of the health clinics and schools. The community negotiated the conditions under which they received support. As in this example from Monterrey, the five programme interventions believe that the terms of engagement with the state are critical to the effectiveness of any resource allocations, and they do not want to make simple claims for state delivery. Although Harvey (2012, p. 87) does not use the term co-production, for the interventions described in Chapter 4 this strategy is part of the answer to his question about how to use the 'creative powers of labour for the common good': securing 'more public goods for public purposes' from the state while simultaneously strengthening the self-organisation of the population.

In the ACCA programme described in Chapter 4, initiatives are planned and undertaken by the residents of informal settlements as collective processes:

collective information collection (settlement mapping, city-wide surveys), collective definition of problems and search for shared solutions, bringing together networks of savings groups to establish collective funding systems that they manage (the City Development Funds), building collectively a platform for negotiation and partnerships at city level with local government and other key stakeholders, and collective claiming as citizens. In addition, as land is obtained or tenure negotiated for land already occupied, collective land tenure. All these have importance for poverty reduction, yet so few funding agencies have recognised this. They also provide the means through which the urban poor see their capacities to address their development needs – the needs that their societies have never provided for them (Boonyabancha *et al.* 2012).

Co-production approaches were introduced in Chapter 2. Those described in Chapter 4 are a small subset of examples of co-production: they focus on citizen-led co-production, rather than that led by government agencies to reduce costs and improve local ownership. For the groups that are engaged in such co-production, its use is very strategic. A first benefit is that co-production helps establish a positive engagement between informal neighbourhoods and the government. Organisations build links between neighbours with common interests, and engage with the state around those interests. For example, the national federations that are affiliates of SDI strengthen local organisations with savings, a process requiring regular day-to-day contact between neighbours, and then link the savings groups through federations. But not all residents wish to participate in savings. Co-production, the joint management of local services, diversifies the ways in which local relations are formed and strengthened. Jockin Arputham, leader of the National Slum Dwellers' Federation in India, reflecting on the unsuccessful nature of past political practice, when grassroots organisations followed the tactics of trade unions, recognised that they were easily dismissed in part because they did not have a strong enough activist base (D'Cruz and Mitlin 2007). The practical nature of local organising helps to build strong links between residents, resulting in significant demonstrations of popular support, most typically through the scale of financial contributions and participation in events (some of which include politicians). This itself instigates a positive political response. For example, when *Mahila Milan* and the National Slum Dwellers' Federation designed toilets for their apartment blocks in Mumbai, they provided shared toilets in part because the management of these would bring women together over the collective task and hence equip them to respond to other needs. While to outsiders it may appear that the community groups have been co-opted by the state and persuaded to take on government's responsibilities, for the groups themselves this is the way in which they build the political processes and organisations necessary for a mass base and to strengthen their relations with the state so as to be able to influence policy, programmes and practices more effectively.

Second, the concrete engagement with the state consolidates relations between urban social movements and middle- and lower-ranking government officials and politicians. These relations as well as the co-productive activities help enhance what the state does to improve infrastructure in informal settlements (i.e. the solutions that are considered) and the quality of infrastructure improvements when they are installed. Shared financing ensures that residents are active in both participation and in observing and analysing the contribution of the government, helping to strengthen accountability and improve state performance. Involvement in policy and programme design offers an opportunity for social movement leaders to understand some of the dilemmas of urban development at the city level, and enables movements to negotiate for components within programmes that strengthen their own practice. Intense and regular exposure to government processes through collaboration with officials, state workers and politicians on agreed and joint tasks helps strengthen the political competence of community leaders (and increases the capacity of the state), and nurtures the social relations required for both informal and formal politics to be more inclusive. Some of these experiences are illustrated in Box 5.3 with work in Gobabis, a small town in Namibia: this account also demonstrates that rather than simply co-producing housing, the goal is the co-production of urban neighbourhoods.

Box 5.3 The co-production of low-income neighbourhoods in Gobabis (Namibia)

The challenge to address shelter needs is particularly significant in the smaller towns and villages in Namibia due to the scarce resources available for development of these needs. About 50 per cent of households in Gobabis (population 13,000) live in informal shelters, where 85 per cent lack access to toilets. There have been no options for low-income groups to secure affordable land for house construction, as the only available housing was on individually developed plots that were too expensive. The Shack Dwellers Federation of Namibia has been organising in Gobabis for some years, and by 2006 had established 16 savings groups. These groups had developed relations with the local authority so as to win their support.

In January 2002, these groups secured a block of land for 50 members from the government, and the development of infrastructure and services together with housing construction and savings for house loans started immediately. In May 2004, following on the initial stages of this incremental development, the municipal officials, councillors, community and NGO established a land team to prepare for further community land development (tenure and services) in the township. This is not a conventional project with comprehensive budget and annual plans, but an evolving process that has addressed emerging needs and combined different resources in agreed strategies. Thirty-two of the families took loan finance provided by the government's Build Together programme which provides capital for housing construction with subsidised interest rates.

The community made a significant contribution through sweat equity. Estimated construction costs were reduced by 25 per cent through the preparation of all excavations and block-making by community members. The members also did their own bookkeeping and record keeping with their support NGO, the Namibia Housing Action Group, providing technical support to the process.

The municipality agreed to finance a community information centre and this has been constructed within the block of land and services as a meeting place for the residents and town federation groups. Records and reports are updated monthly and are displayed in the community centre.

The national government provided financial assistance to the Federation through two routes: directly to households, with the provision of housing loans through the Build Together programme; and through regular donations to the Twahangana Fund, whose capital has contributed to servicing loans to Federation members. But perhaps of greater importance for the local group is that the municipality continued to be supportive of their efforts and activities. In 2004, an application for a further 100 plots was submitted to the municipality and was approved, but land could not be made available to the groups as it was already occupied by shacks whose occupants were waiting to be relocated. When the savings scheme realised that the municipality was doing little about the necessary relocation, they persuaded the officials to establish a land committee made up of municipal officials and savings scheme members. The committee meets monthly at the Federation's information centre, with community members chairing the meeting. As a result of this greater cooperation, the resettlement has taken place and the land has been made available to the Federation. Now, further options to improve existing areas are being discussed.

Source: Muller and Mitlin (2007)

This relationship-building (together with the federating and networking activities) helps to address the critique made by Houtzager (2005, p. 9) that the efforts of local organisations lack coordination and hence are not effective in their relations with the state. It offers an alternative to the state coordination that Houtzager (2005, p. 11) suggests as a solution – and is one which protects the autonomy of urban poor organisations from a partisan political process. The goal is to build relations with government and political agencies from a position of strength with the knowledge of service delivery, organisational strategy and mass organisation. The willingness of citizens to participate meaningfully is critical to the credibility of the social movement. One politician said to an NGO activist at a meeting of more than 5,000 members of the Indian Alliance: 'I can see this is not a rent-a-mob' (Mitlin 2007, p. 354). In other words, he could see that this was not an organisation that would be easily swayed by promises. The collective engagement between groups of the urban poor helps groups to see new options. Ruby Papeleras (2012, p. 472) from the Philippine Homeless People's Federation explains how the practical engagement around the 'small projects' in ACCA helps to bring the different community groups together, and shows them how they can work together with different approaches to create a positive contribution to urban development, and present political agencies with a united front:

> One of the real breakthroughs of the ACCA process is that urban poor alliances are becoming strong in several cities now, which link our federation with other federations and urban poor organizations in the city. We are all realizing that when we do things in a scattered way, as separate organizations, we have no strength, and we cannot accomplish anything. In most cities, the only time all the urban poor unite is when some politician comes along and says, '*We'll give you rice if you go to my rally!*' They only unite because of the rice! Here, we are trying to unite around common issues and to appreciate each other's strengths: the federation is good in savings and we do housing and upgrading, while the others may be good at advocacy and policy reform. We are going to make this strategy of uniting all these scattered and sometimes-competing community organizations in each city one of the goals of our new Urban Poor Coalition Asia.

The nature and depth of activism is likely to provoke a positive response from politicians (once they accept that they cannot control the process), because such groups are not politically aligned and therefore have votes that can be secured, and because they address real urban management problems. At the same time as engaging the government, the emphasis on city-wide systems and solutions that can be scaled helps to reinforce the work of the federations and networks at the city scale. The iteration between local neighbourhood negotiations and pressure on the city government for reform highlights the work that needs to be completed. The importance of these approaches has been recognised in abstract: Gandy (2006, p. 390) elaborates the potential when he argues that infrastructure networks in Lagos provide an opportunity to create a strong collective process at the city level able to address urban poverty, strengthen governance and provide 'new and more legitimate modes of public administration'. The programme interventions offer a demonstration of this.

Third, as shown in Chapter 4, the strong focus on practical activities increases the scale and diversifies the nature of community activists, and provides for a common learning platform to reconcile different perspectives between residents allowing for a stronger collective approach. Both the interventions in Chapter 4 and the more general literature recognise that these grounded participatory and co-productive processes bring a new kind of citizen into these political activities, one who was not previously active in part due to their disassociation from politics. Bovaird (2007, p. 856) argues that co-production draws in those who 'want to deal with common concerns at the "small politics"' level, concretely and personally, but who distrust political parties and old grassroots organization and do not wish to become 'expert activists'. Abers (1998, p. 526) in the discussion of participatory budgeting comes to similar conclusions: 'Few of the participants in the *Extremo Sul* forum had previously been activists – for the most part, they were just ordinary people hoping to improve their neighbourhoods.' This is seen too in interventions described in Chapter 4. ACHR members and the federations that are SDI affiliates are aware that the pragmatic focus of co-production draws in many women who have not had previous experience of political involvement (this is discussed further in the next section). In some cases, the existing community leadership chooses to become involved and to practise a different kind of politics.

As well as drawing in a new kind of activist, co-production helps to reconcile the different perspectives held by residents and enable them to move forward with greater common understanding and hence unity. Auyero (1999) highlights the different perspectives held by residents when asked about politics and improvements to their neighbourhoods. Such differences often pervade grassroots organisations. When the Orangi Pilot Project began working on sanitation issues, they found much support for the existing system of lobbying political parties for improvements. After several years of encouraging different community groups to try alternative approaches to sanitation, perspectives about effective strategies for local upgrading began to change. In local neighbourhoods, levels of community involvement and political attitudes vary based on individual and collective experience; and co-production activities build local residents' grounded engagement with each other, enabling them to learn more about what works and to move forward together.

A fourth benefit is that communities can come together to make some progress in addressing their needs even if it takes time to bring the government into this work. In Karachi, the Orangi Pilot Project recognised the potential of Orangi's hill location and hence the ability of the model to use the stream gullies (*nalas*) for secondary drains long before the state was willing to support them. The ability of communities to 'get started' has helped to achieve scale – as local groups self--replicate functional practices they attract interest and attention from political well-connected groups who begin to ask why the government is not assisting these communities. As visibility increases and local effectiveness is demonstrated, communities make implicit and sometimes explicit demands on the state to respond to their efforts: trunk infrastructure in dense urban areas cannot be provided through self-help but requires some degree of state investment. ACCA's design of small and large projects is to facilitate this process. Communities are encouraged to work with the local authority in the first stage (small projects) but if they face difficulties in doing so, they are also encouraged to proceed on their own to demonstrate their potential. The larger projects are designed to require local authority involvement.

A fifth benefit is that the co-production of public goods in informal settlements fits with the nature of the upgrading process which necessarily involves a combination of government and household monies. Local groups need to be actively involved to maximise the effectiveness with which infrastructure improvements can be planned alongside investments in houses (for the higher-income residents able to afford the additional investments). Collective participation in basic services facilitates later stages of the upgrading process.

Hence the use of co-production is linked to the belief that the nature of grassroots civil society groups matters in terms of their ability to negotiate political outcomes that are favourable to the poor. And, more specifically, that the nature of groups arising from a co-production process (practical action, engagement with the state, networking of neighbourhood groups) offers particular benefits to low-income citizens and others living in informal settlements. As they engage the state more effectively and build a consensus about a way to improve their lives, they can negotiate for more from the government. The strategies described above extend political practice through drawing in new groups, building the individual and collective capabilities of the urban poor. As a result, political outcomes are more likely to favour low-income or otherwise disadvantaged urban citizens.

This use of co-production by these interventions is different from its practice elsewhere where it may be used to reduce costs through a labour contribution, to improve the information available to a project management enabling a top-down project to be realised more sensitively, and to improve the accountability of state agencies through participatory planning (Brudney and England 1983). Progressive state programmes are essential but in themselves they may do little to change political relations and reduce poverty and inequality (see Chapter 3 and the discussions of state programmes). In this context, the specific approaches to co-production matter. These programme interventions are using co-production to catalyse a shift in power and improved outcomes for the urban poor. This has parallels with the analysis by Castells (1983) as to what is needed to make a city more just, inclusive and accessible for the urban poor. He emphasises the need to change the vision as to what a city can be and is, enabling new understandings and visions of the urban future to emerge. Central to this new vision, promoted by these programme interventions, is the recognition of interdependency between individuals, households and communities, and local organisations that are downwardly accountable and which are able to interact with more powerful political agencies but at the same time remain responsive to local realities and mediate between different needs and competing priorities. Through the organisational processes of co-production, and in some cases savings-based organisations, and networking and federating links, these interventions seek to enable neighbourhood collective action to connect to the larger scale while still retaining their embeddedness within their localities. The next-but-one section explores this process. First, we elaborate on the interface of these processes with local gender relations.

Gender: addressing disadvantage and securing women's leadership[9]

As noted above, the relations developed through savings groups together with the processes of co-production are more likely to engage women than traditional confrontational community organising. Pragmatic collective actions are a better fit with their gendered roles and responsibilities than demonstrations and protests.

In particular, savings-based organising draws women into a collective process that simultaneously raises their aspirations, nurtures the confidence to achieve these aspirations and develops the skills they need to do it. These new roles are not those of the entrepreneurial businesswoman; rather they involve new public identities for women centred on their leadership in the acquisition of essential goods and services for their families.

Women's central engagement in strategies to address their own disadvantage and develop their roles as political activists is a critical component of these interventions. The disadvantages that women face are well known. Women endure relationships within which they are treated as inferior and akin to being a child. They are rendered incapable, in part through being viewed by men as being less intelligent and secondary to them in many other aspects. They exist, for the most part, within a context that demands they acknowledge male superiority. One aspect of this is that women are kept in the private sphere, i.e. the home. In some places that rule is explicit. In others it is implicit and enforced through women's greater responsibility for children and other dependants. Alongside this private setting for their daily activities, women are expected to be passive, to respond rather than to initiate, to follow and not to lead, to be selfless rather than self-interested (Moser 2009, p. 73). In short, gender relations mean not only that women and men do different things but that women have a lower social value than men. If they are allowed to participate, they sit on the floor when men sit on the table. They take the minutes and serve the tea; they do not chair the meetings.

The triple burden of income-earning work, reproduction and community work that women face is widely acknowledged (Moser 1993). The experiences point to an emancipatory potential within this third role. It is important not to overstate this issue; women's interests are not always advanced through them taking on an active role within the local community. Sometimes there is a real burden as they struggle with additional responsibilities that impose considerable cash and time penalties. However, the experiences within the programme initiatives described in Chapter 4 point to this potential and arguably the gains they have achieved have been because they provide a positive opportunity for women to become active leaders within their communities, and through women's leadership a different kind of politics has emerged.[10] As summarised below and elaborated in Mitlin *et al.* (2011), such community activities enable women to take on new public roles and realise some of the previously hidden potentials; they also help build supportive relations between women and in so doing help ease difficulties they may face within the household. Our argument here is that these interventions provide openings for women to make a public and active contribution towards addressing their needs and those of their families, and in so doing to challenge their subordination within domestic and neighbourhood power structures. How do they do this? The collective processes of savings create new relations between women who are neighbours and reduce the vulnerability that women experience as private individuals within a domestic context that is frequently oppressive. When savings activities are combined with the realisation of neighbourhood improvements, this leads to new roles for women that challenge the norms into which they have been socialised.

Savings-based organising provides an alternative to the home; these meetings occupy a semi-public and semi-private space as they take place in the homes of their members. Such savings rarely attract much male interest. In savings groups within the federations that are members of SDI, men typically make up only about 10 per cent of the members. Women are allowed to create informal groups to save their spare change, and together they develop new skills. As women are encouraged to visit other neighbourhoods through exchanges between savings groups, they have the opportunity to gain a new perspective on the problems they face and the reasons for them. Together they begin to plan how to address their need for tenure security and basic services. However, while many women participate to address their financial and/or residential development needs, the benefits extend beyond these two dimensions and include the opportunity both to address their emotional needs and build their political agency. It is hard to separate these different needs: improving material conditions and improving the sense of well-being and happiness are related. One of the women from a savings scheme in Malawi belonging to the Malawi Homeless People's Federation and affiliated to SDI describes her experience:

> We came to town for a better life. But it is a troubled life. We are alone. We rent this broken house, without it we are homeless. Because we only have little money sometimes, we are not secure in our living. We are prisoners of our poverty. There is no one to talk to, no one to share my troubles with, no one to discuss solutions with. I joined the federation and learnt to save and loan. The savings group women all know each other. We all help each other in our troubles. We sing and dance! In our group, we share ideas, so many ideas. We are rich with ideas.[11]

Participating in savings groups is generally viewed as complementary to domestic tasks by powerful figures within the family. The fact that achievements reinforce women's existing reproductive role (through activities to provide family members with health care, improve basic services and address housing improvements) reduces the likelihood of a patriarchal challenge to this alternative non-familial space. The altruistic element (as savings schemes emphasise their contribution to helping the weakest and most vulnerable members of the community) is seen as 'morally legitimate', and this enables women to defend their participation if it is criticised. At the same time, this social process has a number of characteristics that directly contribute to the quality of women's lives. Local meetings are easy to get to as women are allowed to move around the neighbourhood and this can extend women's horizons. Many women have few alternative forms of support open to them, although some may be active in faith-based organisations. A savings group can offer specific assistance that reduces vulnerability and insecurity, with some or all of the following benefits:

- practical support in childcare and food provision;
- support to confront or leave abusive relations, or at least challenge the accompanying and damaging sense of guilt;
- a social space to share ambitions and then be supported and challenged by their peers to realise them;
- physical and emotional space to get away from the ongoing demands of household tasks even for a short period – a space for women-to-women dialogue that reduces the isolation that women may experience within the household;
- enhancement of women's social status and sense of pride through their involvement in support for those who are more vulnerable than themselves;
- self-organisation, which in this case does not require women to negotiate difficult relations with external groups until the scheme is more established;
- as the group consolidates and networks with other groups, the opportunity and support to engage with more powerful groups to secure basic goods and services (and it is not possible to address most material needs without this).

It is the immediate engagement in local development activities, running meetings, managing finance, constructing homes, surveying their neighbours and digging sewers that enables women to discover their own capabilities and future potential. The women's nutrition organisations in Peru (discussed in Chapter 2) draw women into a range of practical activities around securing, processing and distributing food to improve the nutrition of children and others: this also enables women community leaders to emerge who are able to challenge existing practice and negotiate alternatives with men (Barrig 1991; Bebbington *et al.* 2011). Such activities, undertaken with each other, help women address the more damaging psychological impacts of gendered disadvantage and overcome any sense of inadequacy and inferiority. Collective self-help challenges a conception of women as incapable, physically weaker and less intelligent. In this way, these activities undermine powerful processes of repression. As a result of this new understanding of their capability, women change their relations with the men in their families, and persuade them to accept new roles and responsibilities (Patel and Mitlin 2010). Women's 'self-help' in addressing the need for basic services is an act of affirmation, an act that says they do not have to wait at home while men fail to address their families' needs; an act that engages them in the accumulation of useful skills, engineering, surveying, building and financial management. With this new consciousness, it is easier for women to challenge the fiction of their inferiority.

These processes of public activism are necessarily collective, preventing the vulnerability that women experience if they are isolated. A public presence of a women's collective exemplifies and then amplifies their potential contribution. Community savings groups, in short, help women both to address their gendered needs and to take on a public role. A new member of a Shack Dwellers' Federation of Namibia savings group captured these dual benefits, both personal and political. 'After independence … ' she said, ' … I did not feel the independence.

But now with the federation, I feel the independence as I also have a say in things. If I want to talk, there is someone from the federation who passes by. If there is a problem, one can call on a sister from the federation.'[12] This member went on to explain how if the local neighbourhood faced a problem such as a lack of sanitation, they would go together to talk to the town clerk or the representative on the regional parliament.

Challenging traditional roles is not an easy process and savings schemes have to provide support. One of the leaders in the Philippines, for instance, spoke about how, when she first began to take a leadership role in the federation, her husband locked her out of their house and sometimes burned her clothes in his anger. Over time, he came to accept her choices and she became a national and international leader.[13] Discussions with other women leaders suggest that this experience is far from unique. In some cases, men resist savings-based organising at the neighbourhood level. In Zambia, for example, the savings schemes faced such opposition to women's participation from local men that they wrote and then performed a short theatre sketch to demonstrate how both men and women could benefit from activities. Two years after they began, according to the national leadership, they rarely faced this problem and no longer needed to perform the sketch.[14]

The activities they prioritise may be those particularly affecting women but there is a holistic approach to family needs. In Mumbai, for example, *Mahila Milan* groups work with the police to set up community police stations in informal settlements in which they play active roles, supported by police officers appointed to work in their settlement. Once established, one of their tasks is to close down many illegal drinking places, helping to reduce alcohol abuse and domestic violence (Patel and Mitlin 2010). The groups may also offer those involved in making and selling drinks illegally support to develop alternative livelihoods (Roy *et al.* 2004). At the same time, SDI federations may be dominated by women, but they are conscious that they have to bring their husbands with them, and that this requires addressing their families' needs and those of the community. Women from a savings group in Paradeep (Orissa), who had designed and built their own community toilet, suggested that what they needed next was a loan programme for their husbands – mostly fishermen who needed credit to help mend or replace their nets. However, the activities that are more frequently prioritised are those that address the needs of all residents for greater tenure security, access to basic services and improvements to housing.

From the city to the state

All five programme interventions believe that national politics is important, and each of them is conscious that the state agencies that make the most difference to the quality of life of their constituency are local (city or municipal) governments. Major resource flows (that are redistributional) may be more likely to take place at the level of the national state. However, these programme interventions all choose to locate the critical node of change at the city level – with the belief that solutions have to emerge from communities located in informal settlements, and be tested, improved and validated as they spread through a city process. As Mohan and Stokke (2000) argue, social movements have a potential for radical and redistributive change, but the challenge is to use decentralisation and devolution processes offered by government and prevent governments from splintering communities into individuals negotiating access to basic services through market mechanisms. Rather the need is to use such processes to build a strong collective at the grassroots level that is clear about its autonomous position, and which is able to negotiate and make the linkages at different political scales (ibid., p. 262). In so doing, movements are more likely to secure the required changes at both national and local government levels. The danger of participation if it is not linked into institutions able to negotiate with the political system (at appropriate levels) is that it does little to redistribute power and resources. Thus the goal is to build a strong local base at the settlement level, build networks between settlements in the city to be able to negotiate with local authorities and other relevant agencies, and to create a presence at the national level.

The programme interventions concur with the more broadly based social movement literature about the importance of securing appropriate support from national government to enable progressive urban change (Bayat 2000; Heller and Evans 2010; Tarrow 1998, pp. 60–66). Each of the programme interventions has had some impact at the national level, reflecting both the opportunities that are available and their own maturity. The UCDO/CODI case is in a slightly different position as it is itself a government programme operating nationally. With respect to the other programme interventions, the work of the National Slum Dwellers' Federation and the Indian Alliance is now used as a reference point for national policy with the participation of senior figures from the Alliance in the preparation for the twelfth National Plan (from 2012–2017), in the design of national funding programmes for informal settlements/'slums' (RAY[15]) and an engagement in national as well as state policies. In the case of Pakistan, as the effectiveness of the Orangi Pilot Project's design for Karachi became evident, the model became national policy. Federations affiliated to SDI have engaged with the national governments consistently as the federation model creates a national platform for previously excluded informal settlement dwellers. By the end of 2011, there were 15 federations with agreements with national governments (UPFI 2012). Across the network there has been considerable variation in the responsiveness of national governments within three domains: regulatory changes, provision of non-financial resources and provision of financial resources. The Asian Coalition for Housing Rights' ACCA programme is, in part, drawing on the experiences of the Community Organization Development Institute (CODI) and its work to encourage and capitalise savings groups. Some of the ACCA processes have begun to influence national policy. In some cases, most notably in Fiji, this has moved ahead of the savings groups and their networks, and has been done in response to government interest. In other cases, such as in Cambodia, this has emerged from concerted action at the local and provincial level as groups have patiently built up their relations alongside their activities (ACHR 2011a).

ACCA's work in Thailand has combined with CODI to develop the autonomy of the community processes with the instigation of city-level funds (see Box 5.4). This innovation for Thailand reflects on the citizen networks' own understanding about what is needed to maintain their autonomy in the context of an unreliable state. Castells (1983) in his historical analysis of urban social movements emphasises the importance of such autonomy if movements are to successfully manage a political engagement. The experiences show how Thai network leaders, aware of their vulnerability in the context of changing state commitments and an unstable political situation, have sought to secure their ability to allocate investment funds according to their own assessment of their needs and interests.

Box 5.4 Community Development Funds (CDFs) and ACCA

The ACCA programme in Thailand has supported the formation of community development funds (CDFs) at the city level, initially in Bang Khen district of Bangkok, and Chum Pae town in the northeast. As community networks observed the operation of these funds, members came to realise that CDFs gave them the freedom to continue *Baan Mankong* upgrading activities without having to follow the procedures required to obtain a government loan through CODI. While eight communities in Bang Khen are in the process of, or have completed, *Baan Mankong* upgrading, other communities have seen delays in obtaining government loans for upgrading. These households have turned to the CDF for a loan. The success of these first two funds and the initiatives they have enabled led to a scaling-up process facilitated by the national network of low-income community organisations (NULICO), with the eventual goal of ensuring that every city has a CDF.

By April 2012, there were 62 CDFs operating and 243 in the process of formation across the country. While not all groups have succeeded in securing a financial contribution from their local authorities, these authorities support the CDF, providing advice and meeting facilities.

Community members find that the process of obtaining a loan from the CDF is more rapid than through CODI, as fewer procedures are required. Some CDFs act as 'buffer funds' providing housing loans while the community waits for larger-scale *Baan Mankong* funding from CODI, or as a buffer margin in case of delays in loan repayments. In the first round of loan-granting in Bang Khen, seven households from three communities benefited from a loan directly, and in the second and third round this number increased to 69 beneficiaries from seven communities within the CDF. While the scale is smaller than *Baan Mankong*, it plays an important role in keeping community processes active through the implementation of housing projects and demonstrating to other stakeholders the community's desire to press on with housing improvements, while also opening up possibilities for further development through welfare, income generation and other small initiatives. Loan conditions are similar to the *Baan Mankong* programme.

The value added by such funds has led to a substantial increase in membership as ACCA seed capital has increased the revolving capital of local funds and allowed loans to be awarded for a wider range of uses, including housing and livelihoods. For example, the Bang Khen CDF is supplemented by a welfare fund, and an insurance fund, to which each household makes monthly contributions. In Chum Pae, the network decided to use the fund to buy a rice field, and this type of unconventional solution to ensure food security is possible because the city-level fund is theirs to use. Most CDFs are structured with a number of different funds, from loans for housing repairs to disaster funds as insurance against housing damage.

The CDFs have the flexibility to play a role broader than that of the ACCA project by supporting welfare, disaster rehabilitation and providing insurance. The types of funds within a CDF are established according to the needs of the community network within that particular city. By pooling together various smaller funds and savings groups, there is scope for larger-scale loans and welfare support within a city, thus allowing for city-wide upgrading and community development to occur.

Source: Archer (2012)

Among the other four programme initiatives there are efforts to build up a position at the national level that is the authentic representation of locally grounded activities. Such a position emerges organically as both civil society groups and city governments seek additional monies to scale up local successes. There is a strong belief that if the process does not emerge in this way it will not address needs; nor will it have the capacity to build a political will and make a lasting contribution. At the same time it can be a difficult position to maintain in the context of a national government that has its own political interests and policy agenda.

Governments in middle-income countries are more likely to have housing programmes in place to support upgrading and greenfield development. Reflecting this, SDI affiliates in Brazil, India, Namibia and South Africa have all accessed central government monies from existing programmes (in addition to limited specific support). In some cases, lower-income governments are also willing to support housing. Some of the members of the Asian Coalition for Housing Rights have secured a national profile and an influence on policy. For example, the Urban Poor Development Fund (UPDF) in Cambodia has been active for many years prior to the commencement of ACCA and has established new community-led options for improving low-income shelter options (Phonphakdee *et al.* 2009). By 2010, the community groups recognised that there had been two policy breakthroughs at national and city levels (ACHR 2011a). First, the 'Circular No. 3' policy directive (approved in May 2011a) is very closely based on the city-wide community upgrading strategies and procedures that have been developed by the UPDF and the National Community Savings Network; and second, the new national housing policy (which the Asian Coalition for Housing Rights is helping to draft). These two policies provide a framework for making city-wide upgrading plans for housing all low-income families in the city (on site if possible and relocation only when necessary, to land the government provides for free, with full land title) in which the municipality and the local community networks survey and work out the plans together.

Some of the complexities of national housing programmes are evidenced by South Africa. Here, the South African Homeless People's Federation emerged at the time of democratisation (Bradlow *et al.* 2011). The national government had developed a housing policy centring on redistribution through a capital subsidy. Construction interests had had a major influence on the programme and the subsidy was mainly to be delivered through projects managed by commercial contractors (Gilbert 2002b; Huchzermeyer 2003). The SDI affiliate responded by pioneering precedents in community-managed housing construction. As a result, they attracted the attention of government and were able to catalyse the introduction of a new sub-programme nested within the capital subsidy, the People's Housing Process (PHP) (Baumann 2003; Khan and Pieterse 2006). The PHP financed community self-build but over time the option became increasingly regulated as the government responded to criticisms of the programme by raising both standards and the unit value of the subsidy. The increasing levels of regulations made it hard for the community to drive a process that was constantly interrupted by inspections and subsequent delays in phased payments (Baumann 2007). Moreover, despite the considerable scale of the main housing subsidy programme with two million units within 15 years, the housing backlog increased due to natural growth, smaller family sizes and migration. There is the acknowledgement that significant numbers have left the subsidy-financed homes that have been provided for them (Tomlinson 2003, p. 84; Zack and Charlton 2003).

The social movement faces an increasingly professionalised set of solutions combined with no capacity to release monies even with an explicit ministerial commitment (Sisulu 2006). Families continued to live in informal settlements, or increasingly rent informally in formal areas. The Federation has found itself constantly seeking to influence an inappropriate policy to be a little better. While influence has been sought and secured at the city level, with a generous nationally financed housing programme, South African city governments are themselves orientated towards subsidy acquisition. The challenge faced by the federation is twofold. It has to engage with and use the existing policy framework, showing how it can improve its relevance to government by engaging with empowered active communities and their associations. It then has to articulate an alternative vision more conducive to a community process that can be achieved through a progressive process of policy and programme reform. In recent years, the evident constraints of the capital subsidy have become more evident, most notably related to the adverse location of many of the new dwellings. The emphasis has switched to informal settlement upgrading and the affiliate is developing new high-density incremental development options.

The experiences in South Africa and elsewhere illustrate the highly contested nature of urban development processes in towns and cities of the Global South. The problems are not simply related to a lack of understanding about what is needed to nurture a bottom-up process and competing political interests. There are considerable financial gains to be made in urban development, and contestation to control land and service provision is acute. These problems are alluded to in Rolnik's (2011, p. 251) analysis of the limited power of local authorities in Brazil; she concludes that planning decisions continue to be 'strongly influenced' by political and economic elites. Even more exclusionary and politically partial urban development processes are described in the case of Mumbai by Weinstein's analysis (2008). Municipal governments that wish to promote a more inclusive equitable model seem to struggle. Klink and Denaldi (2012) elaborate the problems faced in Curitiba where the metropolitan authority is limited in its ability to influence real estate investments and reduce the housing backlog. They also explain how, within neoliberal economic policies, local planning authorities face a difficult context even for relatively simple urban management tasks, let alone the more complex process of building an inclusive city.

There are several reasons why there is a need to develop city programmes of pro-poor urban change prior to a national programme. First, the alternatives may be too radical to be understood and supported at the national level if there are no demonstrations of how the required policy changes improve the situation. Moreover, effective city-level demonstrations of improvement build up the political support and legitimacy that is required for change. Second, these models have to be tested out locally such that community groups can be involved in modifying designs as they emerge. Third, there is a strong belief that national programming is not going to be effective unless opportunities at the national level are matched with pressure from below as local groups seek to use resources in ways that address their needs. Hence residents' associations need to know and understand what is available, and pressure their local government to take up these opportunities.

As support moves to the national level, with an emphasis on universal provision and co-production, the challenge is to build a link with others in some need, including lower-middle-income groups. The strategy broadly resonates with Nelson's (2005, p. 120) analysis about the importance of building a political alliance between low-income and disadvantaged groups and what she terms the middle strata (not the middle class) who are 'somewhat better off and less insecure materially than the poor' (ibid., p. 121). This group may include those with incomes below the poverty line. However, while their incomes may be low, they may be less politically excluded than the very poor, with more confidence, better connections and a greater propensity to join groups. Nelson argues that only those programmes that help a significant proportion of citizens are likely to be maintained in the longer term (ibid., p. 125). Moreover, she suggests, their greater level of social inclusion increases their effectiveness as they are associated with less social stigma than more targeted programmes (ibid., p. 127). Castells suggests that there is an ability of urban movements that seek collective consumption goods and services or which seek greater cultural identity or a decentralisation of political power to transcend class differences (1983, p. 32), although he also points out some of the difficulties in doing this in his discussion of the Mission Coalition Organization in San Francisco. More generally, there is an acknowledgement within the co-production literature that lower-middle and middle-income groups also participate, although the ways in which they can dominate openings is also recognised in Chapter 2. Boonyabancha (2009) describes the ways in which the savings groups in Thailand develop an open city-wide process with many agencies to ensure that there are mechanisms to provide checks and balances. The experiences of SDI and ACHR are that if solutions are not found for these groups, they occupy the space and support offered to the lowest-income groups. Hence strategies have to be identified that favour the lowest-income most disadvantaged groups but which also provide possibilities of advancement for others. This challenge is a difficult one, and arguably for these interventions it remains work in progress.

Political capability

Appadurai (2001) entitles a paper about the Indian Alliance (SDI) as 'deep democracy' in reference to the ways in which SDI seeks to reconstruct political relationships and secure more pro-poor forms of governance and governmentality. Each of the five interventions seeks to change political relationships, introducing new opportunities for a formal engagement between the organised urban poor and the formal political system while also seeking to ensure that the engagement is on terms that favour the urban poor – and this necessarily involves an informal political process and recognition of the informality of their everyday lives (Myers 2011).[16] Favouring the urban poor means, as Fox (1994) suggests for rural Mexico, facilitating the building of autonomous organisations that have a representative practice and a leadership with enhanced capabilities to undertake development including the management of finance. It includes, as argued by Devine (2007, p. 303), the ability to create local organisations with organic supportive social relations between members and leaders that are able to contest the politics of the powerful. Underlying the strategy of all five programme interventions is a shared perspective that progressive urban politics requires the active engagement of organised citizens in the processes of urban development. These programmes share key values and believe that people should be able to access basic services, live in secure shelter and be able to address their shelter needs, and that governments should respond to the needs of all citizens, including low-income or otherwise disadvantaged groups.

Underlying their approaches is the understanding that community organisations need to be able to participate in political processes at various levels so as to secure a more inclusive city. Arguably their conception shares a lot with a

Lefebvrian concept of the 'right to the city'.[17] This is not a formal legally secured right – rather it is a vision and a practice that is defined by a sense of equity and of belonging. It is a right to participate at multiple levels in multiple domains, it is the right to belong to groups and communities, to have a home that is safe and secure, and to have an identity that enables each individual to realise their potential, and to have lives with safety, security and aspiration. This right recognises that the formal city involves social processes alien and exclusionary to many low-income households, and so it includes the right to work with the collective of the urban poor to present an alternative imagery of the urban poor as a vibrant force for positive development. These are rights defined and realised in the lanes and alleys within informal settlements as people come together to work out how to improve their water, sanitation facilities, drainage, access to schools, health care, security. … and prevent eviction. This is the right to remake public and private space through collective practices that provide for basic needs and which reduce the vulnerability of the household as they build the collective. This is the right to a sisterhood that supports women outside of the home, and a right to join youth organisations able to help children and young people managing the difficulties they face. This is the right to re-define the commodified city, transforming it into a living, vibrant space offering choices to its residents. This right includes access to a set of inclusive horizontal political relationships that nurture rather than dictate and which are more concerned with opportunities than with constraints. It is a right to the urban life that Simone describes (2011, p. 268) when he articulates urban interactions that are: 'an ordinary part of the city, demanding neither exception nor special development' and which include 'the ways in which residents enter and exit each other's lives, conversational spaces and social transactions without an overarching sense of propriety, eligibility or relevance – commenting, joking, berating, and congratulating about everyday performances of all kinds'.

As noted above, this is very far from a conception of a right as defined by a legal rule or practice. It is also not a right that is given by a benevolent state according to an abstract argument made by public intellectuals or a right won through a process contested in the courts. Rights as a legal instrument can be important but they can also be profoundly disempowering to those for whom legal processes are socially distant and inaccessible. They involve both a language and behaviour that is not familiar to the realities of the urban poor. Moreover, instruments of the law in the form of the police and courts may be corrupt. When Perlman (2010, p. 189) asked 1,200 people randomly selected from low-income formal and informal neighbourhoods in Rio who committed more acts of violence against the community, 22 per cent said the police, 11 per cent said the traffickers and 48 per cent said both (i.e. that it was not possible to distinguish between the two). In Msunduzi (KwaZulu-Natal, South Africa), one study reports that over 50 per cent of respondents believe that judges are corrupt, 50 per cent believe some police are corrupt and 20 per cent believe that most police are corrupt (Piper and Africa 2012, p. 226). Such examples are indicative of the suspicion felt by many of the urban poor towards the formal processes of safety and security. For approaches to urban poverty reduction that place faith in the state and see it as a neutral space that can be secured for and by the poor, the legal institution is there to be gained for the people as well. But for those whose realities are grounded in ongoing struggle against more powerful vested interests, legal rights are sterile, inflexible positions that are essentially inaccessible and which may never be attainable. Low-income movements have been promised rights as the solution to protests, only to find out that the rights are simply on paper. This points to a reality in which the state is not neutral, and where rights with their emphasis on formalised professionalised processes favour higher-income and highly educated groups (Chapman *et al.* 2009).

But even though the formal processes of rights as entitlements may be a difficult terrain, the idea of rights has a powerful symbolic meaning because the language of rights speaks to the sense of injustice that provides a counterpoint to experiences of prejudice, discrimination and disadvantage. If the discourse of elites is related to the undeserving poor whose neighbourhoods are places of mob violence and criminal activities, then the language of rights challenges these perspectives.

If not an approach based on rights, then what? The five programme interventions believe that a core challenge is to build local organisations able to negotiate their way towards an alternative urban reality that works for the urban poor. The goal is to build local associations able to engage from a position of strength whatever the nature and scale of the state. The complexity of the process and its component parts including the interest in the local organisations and the collective input into the co-production of goods and services becomes clearer when analysed through this perspective. In summary, self-organisation and self-help builds a positive identity. Federating and networking, together with mapping and enumerations, build a conscious identity of the urban poor as structurally disadvantaged by the development of the city and needing to organise in ways that complement party policies (ideology), ethnic affiliations and other allegiances to secure development and justice. This self-identity is used to strengthen links between informal settlement dwellers across the city building solidarity such that groups are able to negotiate with both formal and informal political interests in the city. The focus on savings and improvements in basic services builds a process that helps women step into public roles. The use of their own household resources to catalyse a process, either through savings or through small-scale loan capital that can be rapidly repaid and used elsewhere, gives them the power to undertake development without being dependent on the state. These actions, together with a willingness to work with local government in the co-production of basic services, build up public legitimacy alongside greater self-confidence. The groups form alliances with a range of interested groups including other community groups and professional agencies such as NGOs, university departments and elite individuals who are concerned with the future of the city. Such alliances help them position their work and further legitimate people-centred development. Core in the strategy of each is to present opportunities to work with the state on an agenda for improving low-income settlements in the city. Of course, this is not always the agenda of the government. There are powerful groups hostile to the needs and interests of the urban poor (see, for example, the discussion about Delhi in Bhan 2009). But there is always the possibility to shift the agenda of the state by presenting politicians and political agencies with a political opportunity. Rather than movements using the political opportunities that are available, they create their own – and persuade the state to respond positively.

Tools and modalities of action are chosen to build mass organisations. In the case of OPP, this is something that they take up through organisations that are networked through the URC. For example, when evictions of informal settlements were threatened over the circular railway upgrading in Karachi, groups negotiated with other, more powerful informal political groups. In this city, it is not clear that it is possible for the urban poor to gain significant amounts of their own power and influence due to the high levels of political violence. However, in other cases, the scale of organisation can emerge as a more explicit threat to the politicians if they ignore the process. Large-scale mobilisations of thousands of residents, such as those related to the opening of a housing exhibition (Appadurai 2001), are suggestive of the power of such movements to mobilise large numbers of citizens and potential voters.

Central to this process, therefore, is the development of new experiences, skills and eventually capabilities, both individual and collective, at the local level. Whitehead and Gray-Molina define political capabilities as 'the institutional and organizational resources as well as the collective ideas available for effective political action' (2005, p. 32). Political capabilities, they suggest, are all about effective collective action, and successful engagement in policy-making and implementation (ibid., p. 34). These are the capabilities to agree priorities, strategies and tactics, realise them and learn from the experience to improve performance. Appadurai (2004) also identifies the ways in which the organising processes of SDI's Indian affiliate help to inculcate a 'capacity to aspire' among the individual participants, increasing both their individual and then their collective ambition.

Capabilities emerge and are further honed through practice. As suggested by Appadurai (2001) in the context of the Indian Alliance, there is a willingness within these programme interventions to take on the mechanisms of the state, deliberately transforming their tools and techniques (the processes of governmentality) to strengthen members' capabilities and specifically their knowledge base and relational capital. For example, rather than conducting state-sponsored surveys to 'find out about' the urban poor, communities gather information about themselves both to address self-interested manipulations of other data collectors and also because this will inform their own priority setting. Despite this, informal settlements have long been some kind of 'black hole' on the urban landscape ignored in maps and censuses because of a fear that this would validate people's claim over the land. SDI and OPP and more recently ACHR have developed their capacity to map and document informal settlements to a considerable degree (see *Environment and Urbanization* 2012; Karanja 2010; OPP-RTI 2002). The purposes are multiple and entwined: they are both to mobilise residents (and increase numerical strength) and to provide material to negotiate with government, establishing residents' credibility through the provision of such information and setting the terms for a dialogue with the state. Surveys are identified by Foucault as one of the mechanisms through which the modern state establishes itself as the prime power within the area that is mapped as its territory. Is there then a risk that communities themselves will become co-opted by the state if they take on this role? Appadurai (2001, p. 33) in his observations of the federation in Mumbai is sanguine, and argues:

> my own view is that this sort of governmentality from below, in the world of the urban poor, is a kind of counter-governmentality, animated by the social relations of shared poverty, by the excitement of active participation in the politics of knowledge, and by its openness to correction through other forms of intimate knowledge and spontaneous everyday politics.

The work of the savings groups in Cambodia demonstrates how surveying and mapping can be a tool that local residents can use to challenge the government's policy of eviction in Phnom Penh. As they gathered information about which settlements could be upgraded and which may need to share the land with the legal owner or relocate, then it became possible to work with local government on a city-wide plan and to build a momentum for pragmatic and pro-poor improvements (ACHR 2004, pp. 20–24). By creating this information base, people owned it and could use it to address their needs and interests.

Mapping is not the only tool of the state that these interventions seek to overturn. Rather than the government setting standards that define self-help solutions as inadequate and add illegal contravention of building regulations to the problems that shack owners face, communities seek to establish these standards through an internal negotiation and dialogue with the state. Rather than government setting the rules to access state programmes, federations and networks argue that the rules should be jointly set with communities playing an active role in the design, learning and redesign alongside state officials and politicians. Each of the five interventions has examples of this – OPP-RTI's sewer and drain mains designs that built on to existing drains, the Indian Alliance rethinking the design and management of community toilets, the Malawi Homeless People's Federation showing the value of using adobe (this was officially illegal when they began to use it), the Namibian and Zimbabwean Federations showing good-quality housing on plot sizes below the official minimum size, the hundreds of community initiatives supported by ACCA that did not follow official standards but showed improvements that worked, the families in Cua Nam Ward showing their own designs for upgrading with housing sizes below the official minimum standard, and so on.

Capabilities to manage relationships are also critical. The programme interventions seek to build intuitive organic relations between low-income residents rooted in the local realities of being disadvantaged on gendered, ethnic and/or spatial grounds. From this, they develop a set of networked institutions able to strategise from a position of advantage and to negotiate with government agencies. This they do using tacit and professional knowledge to further a set of agendas that have to be negotiated with elites. There may be resultant tensions and the danger is that working with the state and securing scale places at risk the processes that these interventions seek to nurture. As Appadurai (2001) notes, in creating their own tools of governmentality, community organisations may run the risk of replicating the oppressive tendencies of the state. The paradox is that such organisations may become disciplining agents in their own right; for example, managing access to basic services through communal water points and meters, and/or policing access to land. As Castells (1983) discussed, success leads to institutionalisation and, even with community-managed processes of citizen governmentality, this shifts community-led processes towards state formalities.

As noted above, Appadurai argues that SDI's Indian affiliate has been able to avoid the risks of institutionalisation. Here we can make first a conceptual and then a pragmatic response. The conceptual response is to emphasise the importance of collective practice. Both Foucault and Castells highlight the ways in which the state individualises its relations with citizens, weakening their collective practices to reinforce its own power (Castells 1983; Dreyfus and Rabinow 1982, pp. 139–140; Rabinow 1991). As the examples above demonstrate, the process of co-production for collective consumption (public services) resists individualisation, both increasing the density of working relations between groups and strengthening their consciousness about the benefits of such collaboration. Multiple forms of collective practice are encouraged, including women's networks, savings groups, city and national federations and networks, construction teams and in some cases enterprise groups. Such collective processes help strengthen the ability to negotiate successfully with the state, and hold the leadership to account. The importance of co-production in basic services is emphasised by Castells' own analysis which is that 'the management of urban services by state institutions, while demanded by the labour movement as a part of the social contract reached through class struggle, has been one of the most powerful and subtle mechanisms of social control and institutional power over everyday life in our societies' (1983, p. 317). Co-production which strengthens multiple capabilities of the organised urban poor enables movements to undermine these processes of social control.

Pragmatically, there is a very evident strategy used by four interventions (excluding OPP) which is to set up city- and national-level funds to provide a mechanism through which communities can engage the state but maintain a level of autonomy. Rather than being seen as investment capital, funds are a tool through which these programme interventions seek to maintain their own ability to make decisions and manage development with the state, and hence continue to challenge underlying structures of power (Mitlin 2008b). Funds are jointly managed accumulations of capital. Although initially capitalised by community contributions and donor funds, they seek additional monies from the state. The objective is to provide a source of investment capital for local improvements *and* to allow the local communities to develop their own management processes for such improvements. Joint local government and community funds share some broad principles with participatory budgeting, i.e. that local communities should determine investment priorities. Funds enable communities to develop other political capabilities, including financial management skills. While many individuals have the skills they need to manage small businesses, community funds help them develop further capacities, including those of public accountability, interest payments, construction investments, and the blending of state, donor and community finance. These financial and managerial skills help them interact with officials and politicians from a position of strength. In the case of the Orangi Pilot Project, similar levels of financial autonomy are secured through the component-sharing design which means that financial responsibilities are clearly distinct.

Hence these interventions use both knowledge and finance to strengthen their ability to articulate and advance solutions which, they believe, work for the urban poor. However, this does not mean that these processes are considered to be free from problems. One major concern is related to the ability of the model to formalise successfully without losing the engagement of local groups that are embedded in the 'informal' and its component parts. Developing hybrids will require that the process is led by leaders of the informal, or at least that they make a major contribution towards design and implementation. As described in Chapter 4, these are, in each case, alliances between organised communities and professionals who seek to support the emergence of new models, engage with officials and other professional groups, and raise funds to allow for the development of community-designed precedents that the state may wish to follow. However, we are not naïve about the very real power dynamics that are involved in these relationships and associated processes (Wilson 2006). Such powers are reduced by the creation of funds as they ensure that NGOs no longer control decisions about the allocation of development finance. As Bolnick (2008, p. 324) argues, in the context of conventional development assistance 'it is not possible to talk of real people's participation or equal partnership when the decision to keep power and resources within the hands of professionals and out of the hands of the communities is one of the preconditions of the engagement'. The answer lies in the capacity of these and similar interventions to establish accountabilities that balance the privileged social status of professionals (Mitlin 2013b). Such accountabilities need to work against their ability to dominate decision-making and open space for a more broadly based negotiation about the direction to be taken.

A further concern is that, whatever the merits of the collective process at the city level, local groups may find it difficult to continue to participate either because they secure housing improvements and hence no longer understand the purpose in participation, or because they fail to secure housing and give up the struggle. Simone and Rao (2012, p. 327) suggest that experiences in one CODI-supported neighbourhood in Bangkok, Klong Toey, are varied. In this case residents have been able to improve their material and social conditions and secure housing, attaining greater security and establishing effective local organisations and networks. But the authors are concerned whether or not the collective politics required to protect these gains will continue, as there will be new aspirations and activities 'that existing collective organizations may find difficult to engage, monitor and influence' (ibid., p. 327). Debts and increasing demands from their political partnerships may, Simone and Rao suggest, lead to organisations returning to political contestation, even if there is no likelihood of success. Equally, organisations need to have strong enough processes to hold leaders to account. Devine (2007, pp. 308–309) discusses the ways in which membership organisations can perpetuate hierarchies and the emergence of new elites. Since leadership of community-level collective organisations is voluntary, the leader becomes the bridge to outside actors, such as landowning agencies and government officials, and thus acts as the gatekeeper, controlling channels of information in and out of the community. Whether this position ends up being exploited depends on community organisational structure – in many instances, a subgroup system helps limit the hierarchy of informational flows, while membership of wider city or national networks of the urban poor also acts as a mechanism for accountability, through wider participation by community residents.

A third issue is the type of urban development. Shatkin (2007) emphasises the importance of national-level political support for the expansion of CODI's support for housing improvements. However, although in this case efforts have been made to keep down housing costs, there is evidence that the process may encourage a set of urban development solutions that are expensive enough to limit the scale and reach all the lowest income groups. Politicians favour complete housing over incremental housing.

In practice this means pressure to construct units of more than 20 square metres as a first step (i.e. a two-room house or more): these improvements fit within existing regulations and offer 'modern' dwellings but at higher (and sometimes much higher) unit costs. This pushes the community-driven process away from scale. Moreover such investments merely confirm for the majority of those living in shacks that progress requires a substantive housing improvement. However, once a project is being designed with a definite set of beneficiaries and product, then communities are reluctant to argue for less. Those who cannot afford this product tend to withdraw rather than contest, in part because they are ashamed of their lack of investment capacity. This issue is raised as an example of the challenges that emerged, are recognised and which need to be resolved if city-wide inclusive urban development is to be secured.

The objective is to change urban development outcomes towards greater inclusion in access to housing with secure tenure and basic services, and to change political relations towards those that favour decision-making by the urban poor themselves. But the practice remains difficult. For the five interventions, such difficulties are not problematic as the emphasis on institution-building and capability enhancement centres on collective learning including experimentation and analysis. *The point is less to find a solution, and more to find a process that enables the urban poor themselves to learn from experiences, and be challenged to find new and better ways of addressing needs at the global scale.*

Conclusions

In the five programme interventions described in Chapter 4, the challenge to existing political processes may be subtle and non-confrontational but it is comprehensive and multi-faceted through elements that reinforce each other. All five focus directly on the consolidation of financial and physical assets within urban poor groups – both collective and individual – through savings and financial accumulation, through the use of loan finance and through direct investment in basic services (individual, neighbourhood and city systems). All five also seek to influence the nature of local associations with an emphasis on practical work to improve the local neighbourhood in conjunction with negotiating for increased state investments. That is, these interventions seek to change local associations away from simply making claims on the state to being active self-help agencies with a direct involvement in local improvements. This involvement is seen as essential to strengthening local associational practice, and to enabling local groups to negotiate with the state with knowledge and a degree of permanence. These associations are then networked at the city and (sometimes) national level to enable them to support each other, learn about their own challenges and options through observations of others, and pressure the state for a substantive inclusive response to the needs of informal settlement residents. The interventions all build links between these networked and/or federated communities and interested professionals. The professionals advise them on technical aspects (at the settlement and city level), build contacts with interested officials and sometimes politicians, and legitimate people's approaches. Over time this builds into coherent approaches to both local upgrading and policy interventions. The community process deepens as the associations or federations negotiate with politicians, learn from their successes and failures, and understand the impact of particular programme and policy approaches on both the needs and interests of the urban poor.

These programme interventions seek to change the distribution of power between organised citizens and the state such that the needs of residents of informal settlements are met. That is, they seek to redistribute power in order to redistribute resources, reform practices and change values. There is a very strong sense that meeting the needs of informal residents cannot be done by simply convincing elites of the efficacy and/or justice of these investment. Rather it can only be achieved if low-income citizens have representative and accountable organisations able to propose alternative forms of urban development and negotiate for them to receive support. Moreover, these needs are not viewed as one-off but rather meeting these needs has to involve a strategy to contest ongoing exclusionary social and political processes. Hence they propose ways to meet the needs for basic services that have synergies with strategies to strengthen neighbourhood associations and networks of such associations. The weaknesses of civil society to provide or coordinate basic services at scale are fully acknowledged. But equally acknowledged is the demonstrated inability of the state to act in the common interest of low-income and disadvantaged citizens. Hence the need for an alternative politics, embedded within the lives and practices of these citizens, and reaching across the city and beyond defines new urban practices. This, in turn, creates the demand for these practices at the same time as negotiating their supply, and also holds the state to account for its actions.

Rowlands (1997, p. 13) develops a fourfold categorisation of the dimensions of power relevant to an understanding of how groups secure advantage or have disadvantage imposed upon them. She identifies:

- power within (i.e. the power gathered through internal strength, self-acceptance and self-respect);
- power to (i.e. productive power);
- power with (i.e. relational power);
- power over (i.e. controlling power).

Table 5.1 maps these programme interventions within this analysis of power to summarise the ways in which they contribute to pro-poor and more equal cities.

As emphasised at the beginning of this chapter, the five very different programme interventions (shaped by context and orientation) have certain characteristics in common in their design. One is how much local residents' associations can and should be reconstructed, and how much they should draw on what exists already. A second is the relative importance of national political grassroots structures in addition to city networks, and the balance between national and city-based efforts. A third is the more effective relationships between professionals and the organised urban poor, and the subsequent roles and responsibilities of each together with the accountability systems that are put in place. A fourth is the complexity of the intervention and the number of sub-programmes in different areas that might be included. A fifth is the way in which subsidy finance is blended with community monies, the degree of separation between state and community funds, and the balance between grants, savings and loan finance.

Whatever the similarities and differences between these initiatives, they offer insights into the key challenges of urban development in the twenty-first century. In many urban centres, economic development has taken place without a corresponding increase in state capacity and/or recognition of the need for provision of essential goods and services. In many locations, organised citizens have sought to address their needs for these basic services. In other cases, a more capable state has sought to deliver a top-down solution as described in Chapters 2 and 3, but many of these did not deliver the hoped-for benefits. In addition, most were difficult to access for the lowest-income and most disadvantaged citizens. Collective struggles (and successes and failures) in the provision of serviced and secure residential land have provided a rich learning opportunity.

The five programme interventions all seek to move forward within a context where the state's involvement is essential for both financial redistribution and collective management at the level of the city, but historically the state has been unable and/or unwilling to perform these roles effectively enough to result in inclusive solutions. If this has happened, it has been a temporary political

Table 5.1 Understanding how these interventions change the nature and distribution of power

	OPP	*UCDO/CODI*	*ACHR/ACCA*	*SDI including Indian Alliance*
Power within	Strengthening of lane organisers in informal settlements and linking of these individuals. Development of Urban Resource Centres with the remit to network groups (across Karachi and other cities).	Savings groups (mostly among residents of informal settlements) and the networks that emerged. Creation of processes favouring participatory democracy.	Savings practices among residents of informal settlements. Emphasis on the importance of networking across all grassroots groups in the city. Using resources to include all.	Savings practices, mapping and enumeration. Claiming the label of 'slum'.[1] Strong emphasis on solidarity across all informal settlement dwellers. Women encouraged to be leaders – identities to challenge gendered discrimination.
Power to	Install and fund sanitation and drainage. Show up the lack of a response from the state and then show the benefits of working together.	Design and implement neighbourhood improvements and housing supported by loans to strengthen community level revolving funds. Show scale of what could be achieved through networking.	Secure tenure, undertake a range of small infrastructure and housing initiatives. Demonstrate the power of people's own actions and request support. Network at city level to bring in local government.	Secure tenure, small infrastructure and improved housing conditions. Suggest directly and by example that the state could do more if it works with low-income groups and their organisations.
Power with	Collaboration with government over drainage and sanitation. Development of relationships between state and NGO professionals. URC reaching out to other professionals and strengthening community networks.	Each other – through city networks of savings groups – and with local government and other groups in the city as platforms are formed at the city level. Links between rural and urban groups.	Building up of city networks of active communities. Links to academic institutions concerned with urban issues. Collaborative work with a range of NGOs and local governments. Many different approaches. Alliances with regional groups and networking, and peer assessments across all nations.	Creation of an international network able to assist local and national federations with negotiations. Alliances with other citizens' groups at city level. Partnerships with local universities. Alliances with other regional and international groups.

Table 5.1 (continued)

	OPP	*UCDO/CODI*	*ACHR/ACCA*	*SDI including Indian Alliance*
Power over	Challenging corrupt or over-costly practices in construction contracts.	Challenging attempts to remove informal settlements from city centres and other well-located sites.	Challenging the marginalisation/isolation of informal settlements and their residents within city development processes.	Challenging the conventional models of state engagement with informal settlements; mass mobilisations of residents to demonstrate the scale of organisation.

[1]There are parallels here with the discussion in Venkatesh (2009, p. 16) when he described the identification of young men in Chicago with the term 'nigger' and associates this identification with their own analysis of the scale and nature of their social disadvantage.

commitment associated with particular regimes. The interventions respond to this with complex strategies offering complementary pressures at multiple points. Successes have been achieved (although there have also been setbacks), but each intervention recognises that success is limited, simply enabling the process to move to another level, with greater political engagement but a further set of challenges.

For each of the programme interventions, there are a set of significant questions to address as they move forward. These include some that are shared across the five interventions. There is a continuing need to understand more about the processes of institutionalisation as these strategies are scaled up by the state. Within this institutionalisation, there is an opportunity to learn about the contribution of external agencies (NGOs, governments, international development assistance agencies) in nurturing the initiatives of collective citizen groups prior to and throughout such institutionalisation. Each of these programme interventions seeks an autonomous space for communities within their schema for state support. For OPP, this includes component-sharing, with the clear delineation of state and civil society contributions. In the case of both ACCA and SDI, it is community savings funds and the networks of such funds. Both of these strategies provide for an organisational form that has identity and actions irrespective of the state. Is this sufficient? And what happens to these processes as community-led practices are scaled up with state or international support?

What changes take place in broader state and citizen relations? In the discussion above, the following all emerged as important: changes in material redistribution and resource provision with improved access to land and basic services; changes in decentralised authority to local government and/or to joint programmes run by civil society and the state; changes in expectations/aspirations and consciousness on both sides; changes in citizenship with rights, entitlements, obligations and responsibilities; and changes in the levels of political inclusion (formal and informal). How are such relations changing, and how can programme interventions such as these contribute to progressive transformation?

What are the impacts on the bigger scheme of things and broader relations between citizens and the state? What seems to be critical in this case is both the level of centralisation within the political system (i.e. how power and resource availability are distributed between districts, urban centres and higher levels of government) and who are considered to be legitimate political agencies at each level. Should the state configure and realise a vision of development itself? Or should it inspire a developmental society? A very different but equally important set of ideas, values and discourses are the kinds of equity and justice that are considered important. For example, is political equity (one vote each) more important than resource or service equity (all with access to treated piped water)? What levels of income inequality and/or social mobility are thought to be acceptable or not? A third set of attitudes (related to those of equity and justice) are those related to inclusive identities: for example, do people consider themselves to be members of one social group (e.g. Indians) or are their class and/or caste identities too strong for this to be the case?

It is fitting that we end this chapter on questions rather than answers. The continuing scale and depth of need requires that we acknowledge what has not worked as we look at practice and theory to map the way to a more just and equitable urban future.

Notes

1 This chapter draws on a paper presented by Diana Mitlin to an workshop in Delhi (November 2011) at the Effective States and Inclusive Development Research Centre, University of Manchester.

2 The Movement for Democratic Change (MDC) is a political party in Zimbabwe. It emerged in 1999 in opposition to the Zimbabwe African National Union – Patriotic Front (ZANU-PF) organised from the civic movement that had mobilised to a 'no' vote to the constitutional referendum that would have given new immunities to the state and which sought to take land from white farmers without compensation.

3 The willingness of community leaders to align with different political interests suggests an underlying analysis that is close to Khan (2005, p. 719) (i.e. that each political faction is out for itself), although as argued in this chapter the activists behind the programme interventions described in Chapter 4 appear to be more optimistic than Khan in that they do seek to ensure that political factions deliver on social goals.

4 For example, Houtzager (2005, p. 8) attributes participatory budgeting to a political party, the Workers Party (PT), while Abers (1998, p. 43) argues that its origins lie with UAMPA (Union of Neighbourhood Associations of Porto Alegre). Our point is that alternative interpretations are part of the mix and reflect both political strategies and claims that contributions be recognised, and multiple interpretations of events.

5 Suh (2011) is one exception. She discusses the complexity of strategies used by the women's movement in South Korea and suggest that movements can engage the state and be autonomous, avoiding co-option and absorption (ibid., p. 464).

6 McAdam *et al.* (2001) focus explicitly on this frame of action in a volume entitled the *Dynamics of Contention* which defines contentious policies as: '[E]pisodic, public collective interaction among makers of claims and their objects when (a) at least one government

is a claimant, an object of claims, or a party to the claims and (b) the claims would, if realized, affect the interests of at least one of the claimants.'

7 Speaking at the ACCA Committee Meeting in Bangkok, 14 December 2011.

8 Connolly (2002) discusses how FONHAPO in Mexico enabled a positive engagement between community groups and the state as the strategies of the urban poor shifted from being defensive to being pro-active in housing improvement and development.

9 This section draws on Mitlin *et al.* (2011).

10 There is relatively little discussion in the development literature of the significance of such local relations for well-being. One exception is Devine (2007) who highlights the importance of shared identities and material improvements for members of self-help groups in rural Bangladesh.

11 SDI (2007) *Voices from the Slums*, Shack/Slum Dwellers International, Cape Town, p. 71.

12 Interview with members of Oruuo Pomwe savings schemes in Omaruru with Diana Mitlin, 26 May 2010.

13 Presentation to Rajandapur Conversation, Dhaka, January 2009.

14 Discussions on a field visit, August 2007.

15 Rajiv Awas Yojana (RAY) is a programme set up by the Government of India Ministry of Housing and Urban Poverty Alleviation.

16 Such hybridity is considered essential. As Bayat (2000, pp. 548–549) argues: '[I]n a quest for an informal life, the poor tend to function as much as possible outside the boundaries of the state and modern bureaucratic institutions, basing their relationships on reciprocity, trust and negotiation rather than on the modern notions of individual self-interest, fixed rules and contracts.'

17 Our understanding of Lefebvre owes a lot to discussions between one of the authors and Andy Merrifield; this was extremely helpful in interpreting Lefebvre's conceptual thinking and applying it to this context.

6 A future that low-income urban dwellers want – and can help secure

Introduction

The most important conclusion from our work is that in most urban contexts in the Global South, poverty can only be reduced significantly when urban poor groups and their organisations can influence what is done by the local and national government agencies that are tasked to support them, *and when* they have the space to design and implement their own initiatives and then scale up with government support. It is the learning from their own work and from each other and the demonstration to local government of what they can do that enables creative co-production with the state and larger-scale programmes to develop. For the networks or federations of slum or shack dwellers or homeless people, co-production enables them to secure legitimacy and to gain more political influence, improved policies and a greater share of state resources. Much of what is discussed in previous chapters is about when and how this is possible.

This chapter looks forward to what international, national and local development agencies and governments can do to support urban poverty reduction. The Rio plus 20 (UN Conference on Sustainable Development) in 2012 approved an outcome document entitled 'The Future We Want'.[1] We make the case that this has to be a future that urban poor groups want, and are allowed to articulate and develop themselves.

This book has described many initiatives that helped reduce one or more aspect of urban poverty. Taken together, in all their diversity in what was done and who was involved, they show that progress is possible. They remind us of how much the innovation in this was catalysed and supported by community-driven processes. What is also notable is how many of these initiatives received no or very limited support from international aid agencies and development banks. This suggests that there are institutional constraints upon such international agencies and banks that limit their contribution to community-driven processes. What is equally notable is that national and local governments have offered limited support and have not always been constructive in their efforts. This all points to the critical contribution of the urban poor themselves. If they are not organised, able to represent themselves, articulate and negotiate for what makes sense in terms of contributions to their own efforts, then progressive development is unlikely to take place. But the limitations of purely local citizen contributions are also evident. If the urban poor are not organised at the city level, are not experienced in financial management and political negotiations, are not in structures that require them to think broadly rather than parochially and to be accountable to those they claim to represent, then development is likely to remain selective and exclusionary for at least some of those most in need.

It is difficult to draw general conclusions on what initiatives might be considered 'best practice' or even more modestly 'good practice', since so much of what was done was influenced (and often limited) by the particulars of each location and its political economy. Professionals and researchers get excited by particular experiences that they think can be replicated – for instance, the community-driven upgrading programme in Thailand supported by the Community Organizations Development Institute – without recognising the particular political circumstances and cultural factors that made it possible. But many of the initiatives described in earlier chapters that contributed to reducing poverty have some 'good principles' in common.[2] A first such principle is explicit provision for more voice for low-income groups and more voice at the city scale, and usually also for supporting their active engagement in developing solutions (although in very different ways). Most initiatives built on the power and ingenuity of grassroots organisations and their collective capacities. Within this, most encouraged and supported the active engagement of women; for some, this was one of their defining features.

A second shared principle is that all recognised a need to change relationships between urban poor groups (or informal settlement residents) and local government, and many have developed this into co-production. All included a strong focus on local initiatives on housing, land tenure and basic services. All sought a larger scale and impact through a multiplicity of local initiatives with this multiplicity (and groups working on them) building to effect political change. For urban poor groups, changing relations with local government – and other state agencies – requires strong collective autonomous organisations for the reasons explored in Chapter 5.

Most initiatives included great care in how money was used – to make the money they could raise (through savings and direct community contributions) and other support negotiated from outside go further. Where possible, this used loans so that the repayments allowed the funds to revolve. Working at any scale above the household requires collective financial capability from the neighbourhood level up; without this, local groups will not be able to participate meaningfully in the development projects and programmes that take place to address their poverty. As shown in Chapter 4, building that capability is possible, and has many other benefits.

In the companion volume to this book, we emphasised nine different deprivations associated with urban poverty; the figure listing these and highlighting some of their immediate causes is reproduced here (see Figure 6.1). Exploring the effective interventions in Chapter 4 has shown us that it is not possible to address these deprivations individually – effective programmes deal simultaneously with most if not all of them.

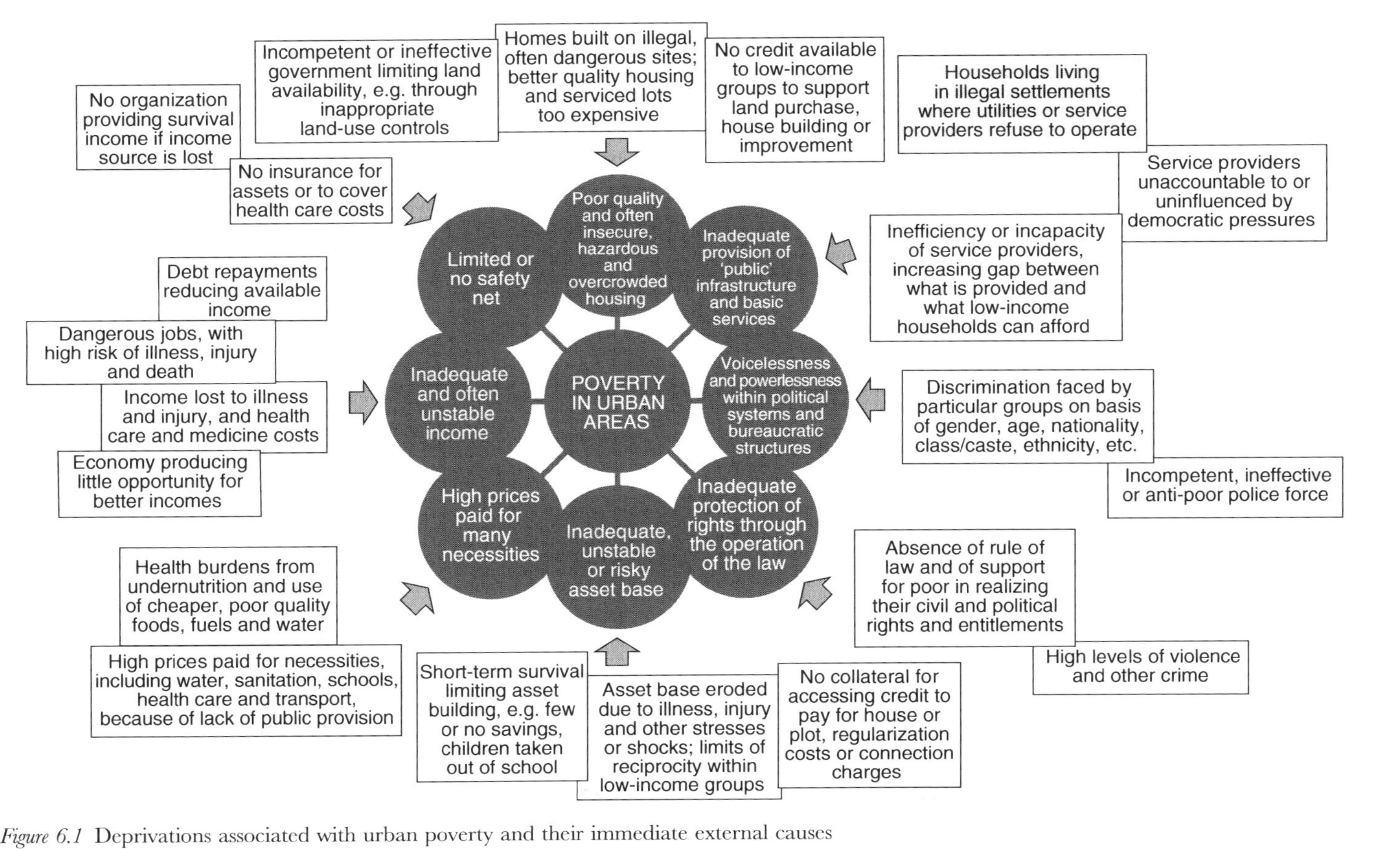

Figure 6.1 Deprivations associated with urban poverty and their immediate external causes
Source: Mitlin and Satterthwaite (2013, p. 281)

After reviewing what is being done (and not done) with regard to reducing urban poverty in earlier chapters, perhaps we need to add to the list of deprivations associated with urban poverty the lack of constructive relationships between urban poor groups and local government. This book has pointed to many examples of where the development of such a constructive relationship brought many benefits to low-income groups and also to the local government and to the city. Chapter 4 suggests that simply to claim political voice is not enough – and specific skills and capabilities are required. To be effective, urban poor groups need to be able to develop their own representative organisations and to develop relationships with other such organisations and groups in their city. Such organisations need to be able to challenge the incapacity of professionals and politicians to come up with realistic solutions – that meet needs and address urban poor groups' priorities at scale using available resources. Their political voice needs to be strong enough to challenge the institutional weaknesses of the bilateral aid agencies and multilateral development banks and their lack of support for effective urban poverty reduction. Thus here we have an expansion in our concern for urban poverty from seeing it as different aspects of material deprivation to also seeing it as the result of political and institutional inadequacies – at all scales from small informal settlements and wards through municipalities to cities, metropolitan areas, states, national governments and international agencies. From this also comes an interest in how governments (and occasionally international agencies) have sought to address these inadequacies (the focus of Chapters 2 and 3) and how urban poor groups and their organisations see and address these inadequacies (the focus of Chapters 4 and 5).

What is understood by urban poverty reduction?

There is little disagreement about some aspects of poverty reduction – for instance, reducing hunger and deficiencies in the provision of some 'basic services' (such as schools and health care, water and sanitation) – although there are disagreements as to how best these are provided and paid for (and, for water and sanitation, what should be provided). There is also little disagreement that an important part of poverty reduction is reducing or removing the large preventable disease and injury burdens and premature death (for instance, for infants, children, youth and mothers). The Millennium Development Goals may be seen as targeting a range of deprivations on which there is general agreement.

Then there is the priority given to economic growth that is still held up as critical to poverty reduction. There is general agreement that low- and middle-income nations need stronger, more successful economies, although there is disagreement on the extent to which (and the mechanisms by which) this reduces poverty. The companion volume to this noted the lack of evidence for income and other benefits among most of the urban poor from economic growth. In summary, the discussion concluded that economic growth is important, but its benefits are all too infrequently shared with those who need them most. Once again, organised, representative urban poor groups are needed to change this.

There is general agreement as to the validity of addressing the deprivations associated with 'living on poverty' and, as the Millennium Development Goals state, seeking 'significant improvements' in the lives of 'slum dwellers', although there is less agreement as to what this should entail and by whom. A large part of the health burden associated with poverty mentioned above and summarised in Chapter 1 comes from very poor housing and living conditions, although this often appears to be forgotten. For instance, the focus of external funding is often on addressing one or more particular diseases and not on addressing the housing and living conditions that underpin risks from these and from other diseases and injuries. Chapter 3 described the nations and cities where serious attempts have been made to reduce the number of urban dwellers living in poverty (and the importance of comprehensive upgrading to this) but also how few national governments and international agencies have seen this as a priority. Chapter 3 also discussed how, when there are efforts to improve housing and basic services, these rarely include those with the lowest incomes (and the greatest needs). Such experiences highlight the difficulties in reaching those who are the most disadvantaged with sectoral interventions planned from above.

In much of the above, the priorities and agency of 'the poor' or those facing deprivations that can be considered part of poverty are ignored. This can be seen in so many development frameworks that fail to engage them (and in most to engage with urban issues at all). This may be seen in the formulation of the MDGs too – and in the discussions underway on the post-2015 development framework.

'Success' in poverty reduction is measured by changes in indicators chosen and measured by 'experts' even as the massive deficiencies and inaccuracies in the data to do so have long been evident (see the companion volume to this for details). Indicators such as the number of persons living on less than a dollar a day (now usually adjusted to $1.25) still get repeated (and used to apparently show a dramatic fall in the proportion of poor people), even as it is known to be a very inadequate indicator of whether someone has or does not have the income needed to avoid hunger and other deprivations. These are also used at the highest levels – for instance, in the background papers prepared for the 'High-level Panel of Eminent Persons' appointed by the UN Secretary-General to advise him on the post-2015 process.[3] Applying the dollar-a-day poverty line (whether or not this is adjusted a little) shows that there is virtually no urban poverty in most low- and middle-income nations.[4] Set a poverty line unrealistically low and poverty can be made to disappear.

In this book and our previous volume, we stress how much the scale and depth of urban poverty is underestimated by this measure. Faulty data or inappropriate definitions also lead to large underestimations for many other deprivations. The official UN figures on improvements in provision for water and sanitation are usually presented as if these showed increases in the proportion of people with their needs for water and sanitation actually being met – when again in urban contexts, the indicators on who has 'improved' provision for water and sanitation are known to massively understate the proportion with provision to a standard that cuts down health risks and ensures convenient and affordable access. For many nations, national indicators on access to schooling and health care seem at odds with the deficiencies documented on the ground in informal settlements.

The only data on housing conditions used to monitor the MDG target that seems to have global coverage (as there are statistics for most nations) is the number (or proportion) of urban dwellers living in 'slums'. But there are serious doubts as to the accuracy of these statistics for many nations. First, there are the criteria used for defining 'slum' households. A household is defined as a slum household if it lacks one or more of 'improved' water, 'improved' sanitation, durable housing or sufficient living area (UN-Habitat 2012). But as noted above, a large proportion of households with 'improved' water or 'improved' sanitation still lack provision to a standard that meets health needs (or, for water, what is specified in the Millennium Development Goals as sustainable access to safe drinking water). If there were the data available to apply a definition for who has provision for water and sanitation to a standard that cuts down health risks and ensures convenient and affordable access, the number of 'slum' dwellers would rise considerably in many nations.

Second, the 'slum population' statistics show very large drops in the proportion of urban dwellers living in 'slums' in some nations (see UN-Habitat 2012) for which there is so little supporting evidence. For instance, the proportion of the urban population living in 'slums' in India is said to have fallen from 54.9 per cent in 1990 to 29.4 per cent in 2009 (UN-Habitat 2012). For Bangladesh, the proportion is said to have fallen from 87.3 to 61.6 per cent during this same period. Where is the supporting evidence for this? Certainly, in India, there was some (official and academic) bewilderment as to the accuracy or validity of these figures. It may be that most of the apparent fall in the slum population was simply the result of a change in definitions – as a wider range of (inadequate) sanitation provision was classified as 'improved' (IFRC 2010). Third, one can only wonder at the basis for the statistics on the proportion of the urban population living in 'slum areas' that are provided for many nations for 1990, 1995, 2000, 2005, 2007 and 2009 – including nations for which there is little or no census data for the last 20–30 years.

The MDG's desire to achieve quantitative targets and perhaps the desperate need for all the agencies involved in international development to show success mean that critical issues of quality are forgotten. What proportion of urban children that according to official statistics are at primary schools are at schools where the quality of teaching is poor, classroom sizes are very large, the availability of basic books is limited and teachers often do not turn up? What proportion of the population living in informal settlements are having to pay to send their children to cheap[5] but usually very poor-quality private schools because they cannot get them into government schools? What proportion of the urban population said to have access to health care services have to put up with poor-quality services that can only be accessed with difficulty (and often long queues) and which often do not provide much-needed treatments?

Some discussions of urban poverty are extended to include some consideration of the rule of law. But this is generally seen in rather abstract discussions for nations rather than in the specifics of providing a just and effective rule of law, including policing for low-income urban dwellers (especially those living in informal settlements) that also addresses issues such as the discrimination which some or all of them face in access to services and employment.[6] As Chapter 5 discusses, research has recorded the low opinion of low-income residents with regard to formal institutions of law that have failed them on many occasions. Some discussions of poverty recognise a longer list of services to which low-income groups should have access including emergency services (ambulances and fire services) and disaster risk management. As Chapter 3 describes, cash transfers that actually reach low-income households with small increments to their income have been effective at reducing hunger and some aspects of extreme poverty, and in some nations also at considerable scale.

Some discussions of poverty include in their definition a lack of voice, although here too this is rarely specific in the sense of the means by which low-income groups get more voice – for instance, in elections (large sections of the urban poor population may lack the documents or legal address needed to get on the voters' register) or in holding politicians to account, and in demanding and getting better services from providers (obviously impossible if unconnected – you cannot hold the water or sanitation utility or health care facility to account for poor services if you get no service).

In all this discussion of poverty reduction, the very people whose needs are the justification for international development assistance (and national 'poverty reduction' programmes) are almost never consulted. Aid agencies and development banks do not engage them in discussions of effective measures and priorities. Chapters 2 and 3 report on how some national and local governments have done so but these examples are rare. Whether or not a particular person or household is poor is defined by external experts. It is also 'expert' judgement that defines which particular 'needs' may or may not be addressed – and may or may not be in ways that actually help urban poor groups avoid deprivation. How many of the poverty reduction strategies (and papers) have been published in the languages spoken by urban poor groups? How many of their 'civil society' consultations took the trouble to include representatives of urban poor groups and to listen to them and then to work with them and act on what they said?

Another aspect of this is that development assistance so often comes tied to particular 'expert' views of what 'the poor' need – improved cooking stoves, micro-finance, eco-san toilets, immunisation, mobile phones, advice on urban or peri-urban agriculture and so on – and perhaps those who work on these needs will be disappointed that there is little discussion of them in this volume. This is not to say that these cannot be important contributors to reduced deprivation or better health, but what we have sought to highlight is a need to allow urban poor groups and their organisations and federations to define their priority needs, as outlined in the next section.

Another way to reduce urban poverty

This book has sought to present the evidence for another kind of development assistance; one that works with urban poor women and men and their organisations. One that is oriented to their needs and priorities because they can influence what is chosen – and its design and implementation. One that is accountable to them as well as to governments (and, when they are involved, to international agencies). Above all, one that recognises their knowledge, skills and agency. This involves 'expert' staff from local and national governments and international agencies engaging with them, listening to them and being influenced by what they hear. Seeking interventions that respond to their priorities and their observations of what works for them – and what serves them in getting more effective responses, especially from local governments. This book has described some approaches that have sought to do this and presented experiences that collectively show that there is another way to address urban poverty. That does not involve massive sums (although the trunk infrastructure needed to ensure adequate provision for water, sanitation, drainage and road access may, because of very large backlogs).

Here, we also draw on our own personal experiences to add to the evidence presented in earlier chapters about the effectiveness of alternative approaches – approaches that worked with urban poor women and men and supported wherever possible their productive engagement with local governments. We have both seen at first hand the competence and capacity of urban poor groups – especially those rooted in savings groups in which women dominate both as savers and savings group managers. Sitting with these savings groups in many different cities and nations as they plan some new collective measure, discuss visits to other savings groups that want to learn from their experience, reflect on how initiatives are progressing and what more needs to be done (and who will do it), examine who has particular needs that require new measures (for instance, help with loan repayments). … Plan and implement surveys and enumerations of informal settlements[7] (generating the data that really does serve planning for poverty reduction and that they own). … Accompanying them on visits to key people in the local government to present their priorities and discuss how local government support can help address these. Seeing how new perspectives on almost all aspects of poverty reduction are raised by their comments and analyses. In addition, experiencing their generosity in their willingness to share their experiences with us and allow us to sit in on their discussions. Seeing also their humour and the rich cultures from which they come. Perhaps, above all, seeing how their own savings groups and the organisations and federations or networks of which these are part have allowed them to contribute more to addressing their collective needs and empowering them to do so.

It is difficult to envisage sustained poverty reduction in urban areas without sustained pressure from below from those who suffer poverty. But this raises the issue of how to build and then sustain this pressure in ways that get positive responses. Even the larger national federations of slum or shack dwellers recognise that they lack the political strength to actually secure the scale of resources that they need from local government and to persuade local government and other state agencies to adopt development solutions that work for them. Thus their strategy is to develop local precedents that address their needs – building new housing on land negotiated or sometimes purchased from local government, upgrading, community toilets, a detailed survey and mapping of informal settlements – and with which to engage local government. When they can take the city engineer or town planner or a local politician or civil servant to see what they have built or improved and then also produce a detailed costing to show how much has been done with limited resources, it helps change the way they are viewed by professionals and politicians. Meanwhile, the federations or networks are also encouraging other member groups to try out their own initiatives. All such initiatives provide learning opportunities for those engaged in them – and for other groups who visit them. They (and what they do) need to be seen as legitimate by politicians and civil servants. Thus, what they are seeking in this is not just a positive response but a change in their relationships with the state: to be seen as legitimate citizens with the same rights as middle-income groups in formal settlements. But also to be continuously engaged in co-production with the state also means community influence in what is done and help in sustaining the federations and networks. Chapter 5 discusses the depth and sophistication of the strategies that have emerged from the urban poor as they have sought to learn from past failures and develop new, more effective development options.

Here, there is a deliberate choice by most networks or federations of slum/shack dwellers to avoid contentious politics; as noted in Chapter 5, this is a terrain that is disadvantageous to low-income groups. The networks or federations are certainly capable of contention where needed – for instance, where an informal settlement is suddenly marked for demolition. But this is a last resort because from this point of confrontation, collaborative relations have to be rebuilt, negative stereotypes of the 'urban rabble' challenged, and new positive identities for the urban poor reinforced.

There is also a conscious choice to avoid being drawn into political parties. This may bring short-term disadvantages but it avoids their agendas being seen as associated with one party and thus only considered by one party. It also protects their bottom-up focus and horizontal structure; political parties demand loyalty from the bottom up, and so often want to control local organisations such as residents' associations. As they receive support from some local politicians or civil servants, so this draws in more resources – and if local government supports their work, this increases the scale and scope of what can be done. Thus it is building on self-help activities but in ways that develop their capabilities and their relationships with local governments. Women also find that this process works for them, providing openings for their skills and the resources they can mobilise, addressing their needs and priorities and opening up leadership positions for them.

Chapters 4 and 5 gave many examples of how this multiplication of local initiatives and engagement with local government has produced changes on a more substantive scale – for districts or municipalities within cities, and even at city government level. This included examples of city governments supporting their city-wide strategies (for instance, in the mapping and enumerating of informal settlements) and institutions (for instance, the City Development Funds). It also included the ways in which the strong local initiatives link together and build networks or federations able to influence higher levels of government – for instance, metropolitan government, state/provincial government and national government.

Might not this process also produce city neighbourhoods that are more convivial and more equal? Where the limited resources available to local governments support collective priorities with and through strong local savings groups and, through this, also ensure that women's needs and priorities are fully included. This allows a constant process that addresses the multiple needs in informal settlements, and where each success encourages further action and collaboration. Where residents in each locality have the capacities to address particular local needs, including the needs of the poorest or most vulnerable individuals.

There are examples of initiatives undertaken by grassroots organisations that did not work very well. Sometimes, this is simply because local government was incapable of providing much-needed support. Sometimes it was because the grassroots organisations took on more than they could manage. Sometimes because of a particularly powerful and ruthless local community leader that would not give up or share power. There is also the need to be seen and recognised as a grassroots organisation or movement; as co-production develops with local government or as local government appreciates these organisations' capacities, they can begin to see them as contractors or NGOs and treat them as such. But given that these groups are the ones with the least resources, the lowest educational levels and the least influence on government, what they have achieved is remarkable.

The following sections look at what the international and national and local development organisations can contribute to these efforts.

Universal access to *good-quality* basic services

The Millennium Development Goals (MDGs) were set up to make governments and international agencies focus on some aspects of reducing poverty and meeting needs for some basic services such as water, sanitation and health care. But many of the MDG targets leave the job half done – halving the proportion of people who have inadequate incomes, suffer from hunger, unsafe water and inadequate sanitation between 1990 and 2015. The MDG target for achieving significant improvements in the lives of slum dwellers was for just 100 million, which is around 10 per cent of those in need when the target was set – and for reasons that remain unclear this only has to be achieved by 2020, not 2015.[8]

Would the global leaders who set up the MDGs offer piped water to only one child in a family with two children? Or if extending piped water to an informal settlement, offer it only to half the residents – so that those who are to the left of the water mains get water and those to the right do not? Of course, with global targets, offering to reach only half of those in need is less acute and less personal. It does not involve a parent selecting which of their children should get preference in access to water or sanitation. But the need for water and sanitation that is good quality and not too costly is such that its absence means that many will die young or live with disease burdens that blight their lives and their development. In addition, a focus on halving the population without access to some service will generally mean that this focuses on reaching the easier-to-reach groups and not those most in need (see Waage *et al.* 2010). Thus getting half-way to a target does not mean that half the investment and effort has been made. It also means a politics in which the powerful decide who is included and who is not – and the consequences are an exclusionary culture in which some entitlements are legitimated and others are not.

So can we really set goals and targets that leave the job half done or less than half done? Shouldn't the target be universal access to water and sanitation that is safe, convenient and affordable – universal access to good-quality, affordable health care and to safety nets that really do provide safety? Memories on this seem to be short. In the 1970s, within UN processes, governments formally committed to providing safe drinking water and sanitation to everyone by 1990 and even designated the 1980s as The International Drinking Water Supply and Sanitation Decade. There were other formal commitments made by governments to universal provision for education and health care in the 1970s too.

The scale of the deficiencies in provision for water and sanitation in rural and urban areas is astonishing – despite all the promises, commitments and declarations made by governments and international agencies over four decades. As the companion volume to this documented in some detail, in most sub-Saharan African nations and many Asian nations, less than a quarter of their urban population has water piped to their premises. Most cities in sub-Saharan Africa and many in Asia have no sewers and no covered storm drains. This includes many large cities with more than a million inhabitants. For many of the large cities that do have sewers and storm drains, these only serve 5–20 per cent of their population.

As global institutions and official development assistance agencies reflect on what should follow the Millennium Development Goals, there is the opportunity to recognise this flaw. What is development if it does not involve the acceptance that every woman, man and child is able to secure the basic needs required for their good health? It is extraordinary that, in this age of prosperity, such basic values seem to have been forgotten. As the scale of the huge inequalities in housing conditions and access to basic services is documented and discussed among both government and development agency staff alike, surely a critical first step is a universal standard of basic provision – for safe, sufficient, accessible, affordable water, for accessible and affordable sanitation and drainage that reduces the risk of faecal contamination, for accessible, good-quality health care and emergency services for all rural and urban dwellers. With such a commitment, the interventions described in Chapter 4 can be considerably more effective, sharing ideas and planning and implementing programmes at scale in ways that are able to bring lasting change.

How are low-income households going to access basic services if these are just another market opportunity?

Almost all the conventional responses to urban poverty face the contradictions between the cost of what is needed, the funding required and the very limited capacity of low-income groups to pay.[9] But getting full cost recovery for any form of infrastructure or service from low-income households for which they choose to pay brings great advantages. The number of households reached is not limited by the lack of subsidy. It also avoids the need for external funding. The same is true for loan finance. The expansion and extension of good-quality sewers and drains to which all households connect supported by the Orangi Pilot Project – Research and Training Institute was permitted by the fact that unit costs were lowered to the point that the inhabitants of each lane could raise the funding that this needed. The community toilets that *Mahila Milan* helped design and build and now manage have charges that seek to ensure that everyone can afford to use them while providing sufficient income to cover running and maintenance costs (the capital costs were covered by local authorities). The community savings groups formed by low-income groups (including those that are at the base of the national slum/shack dweller federations) are sustained because they do not need external subsidies to function. Many of the programmes described in Chapters 3 and 4 have included provision of small loans that are affordable and taken up by low-income households to help with improving or extending their homes. Loans have also been widely used within CODI in Thailand and in many of the initiatives supported by ACCA so that loan repayments extend and expand what these can support – including funding for local or city-wide development funds.

There are also the services provided by utilities or entrepreneurs in informal settlements that get full cost recovery – and may even generate substantial profits. In settlements where few if any households have toilets in their homes, 'public' toilets can prove very profitable, especially if there is little competition. It is difficult to determine where and when these are valuable to low-income households and where they are exploitative. Perhaps many water kiosk operators or water vendors or enterprises providing toilets and washing facilities have aspects of both; some families may be able to afford these services but in other cases families cannot afford to purchase on a sufficient scale and have to suffer the consequences. Certainly, the refusal of local government or private sector utilities to work in informal settlements means many market opportunities for private provision for water, sanitation, health care, schools, solid waste collection – and often also for electricity and security.

In many cities, utilities and municipal authorities alike have come to realise that their interests are better served by a formalisation of service delivery, irrespective of whether or not tenure security is offered. Utilities (whether public or private) are now providing piped water supplies and electricity to those living in informal settlements but charging prices that households struggle to meet – or cannot afford. Rather than 'accumulation by dispossession' (to use David Harvey's phrase) with the eviction of low-income residents from informal settlement to enable government officials and politicians (both legally and illegally) to make money from redeveloping their site, people are being integrated into the market for the benefit of the same elites. Of course, utilities should be managed in the interests of all citizens, with profits being spent on extending services and ensuring that basic services are provided to all – but, as discussed in Chapter 3, there is a growing body of evidence from a range of cities that low-income households cannot afford these services.

These experiences suggest that some local authorities and utilities are increasingly seeing the public provision of services as income-generating – rather than being concerned to manage these services so as to improve household well-being and provide for the public good. The companion volume to this showed the limitations to this approach. With these costs, such services are not affordable to the lowest-income households. As a result, these households still use surface water (shallow wells and watercourses), they practise open defecation if there is no affordable alternative, and they risk the dangers of illegal electricity or manage without.

Entrepreneurs have long made money from investments in industries and commercial services. This system has provided increased opportunities for many. It has helped urban centres to grow strong and in many cases prosperous. In recent years, the informalisation in labour markets has reduced risks for businesses and increased some of these profits. Growing informalisation in both employment and housing means that not much of this income is spent on formally produced goods and services. Service payments are a way in which the formal system 'captures' the incomes of people living and working informally. On the one hand, it is fair that people pay for services, and most households want legal connections. But on the other, prices have to be affordable for most people most of the time. The interventions described in Chapter 4 have developed local organisations and institutionalised practices able to work with local government and utilities to reconcile some of these contradictions. There are no simple, easily replicable mechanisms but there are social processes that are able to test out solutions, improve their functioning and take them to scale.

Rethinking finance for development

This book has included many examples of using finance in different ways and drawing it from different sources. This includes the money mobilised by informal savings groups and the many functions this has – providing loans, supporting group capacity to manage finance and initiate their own initiatives.

Chapter 4 described how the Asian Coalition for Community Action has provided small grants to 950 community initiatives to upgrade 'slums' or informal settlements in 165 cities in 19 nations but with each community choosing what to do. It also supported all the initiatives in each city coming together to assess what they could do at the city scale – and engage the city government. Additional funds were available as loans to support larger initiatives. In many cities, this not only supported grassroots organisation–local government collaboration but also led to the development of a City Development Fund in which all the active community organisations have a stake by pooling financial resources and through which larger scale initiatives can receive support. ACCA supported the setting up of 107 City Development Funds.

A report on the first two years explains the principle of 'insufficiency', because there is not enough development funding to 'sufficiently' meet the needs of informal settlements.

> The US$3,000 for small upgrading projects and the US$40,000 for big housing projects that the ACCA programme offers community groups is pretty small money but it is available money, it comes with very few strings attached and it's big enough to make it possible for communities to think big and to start doing something actual: the drainage line, the paved walkway, the first 50 new houses. It will not be sufficient to resolve all the needs or to reach everyone. But the idea isn't for communities to be too content with that small walkway they've just built, even though it may be a very big improvement. Even after the new walkway, the people in that community will still be living in conditions that are filled with all kinds of 'insufficiencies' – insufficient basic services, insufficient houses, insufficient land tenure security and insufficient money [...] the ACCA money is small but it goes to as many cities and groups as possible, where it generates more possibilities, builds more partnerships, unlocks more local resources and creates a much larger field of learning and a much larger pool of new strategies and unexpected outcomes.
>
> (ACHR 2011a, p. 9)

For a civil society initiative, this was large – US$11 million over four years. But this is very small in relation to the cost of most conventional donor-funded initiatives. For an official aid agency or development bank to manage this would have been their worst nightmare; imagine the staff and administration costs that would have been involved if each of the 950 initiatives had had to develop a conventional project proposal that had to go through all the stages needed for official approval. And what about the need to monitor and evaluate each of these initiatives? In addition, since many of the initiatives were loan-funded, the institutional costs of accepting and managing repayments in 19 different national currencies would have been prohibitive.

The same is true for the Urban Poor Fund International, also described in Chapter 4, which has supported over 100 initiatives that were chosen, assessed and implemented by grassroots organisations and federations. This too is small in comparison to conventional development assistance projects – some US$17 million from 2002 to 2012. But as Chapter 4 describes, this supported a great range of initiatives in a great range of nations with many of these initiatives also leveraging substantial contributions from local governments. These also strengthened many federations with activities in over 400 cities and many small projects that have achieved full tenure for over 200,000 households and partial improvements for many more. Again, for any official development assistance agency, managing such a fund would be a nightmare. The key point here is how to obtain funding that serves on-the-ground development to the people and institutions that can use it well and be guided by and accountable to urban poor households? ACCA was possible because it could work with and through institutions in each nation and city. The upgrading programme supported by CODI described in Chapter 4 was possible because the funding was available to and managed by networks of grassroots organisations. The Urban Poor Fund International could work because each initiative that received funding was managed by their national and city federation.

These each show us a working finance system in which urban poor organisations have the power to decide what is funded with decisions made being accountable to them as well as to external funders. They also show the development of local or national funds which are also accountable to grassroots organisations. The model used by OPP-RTI on sanitation in informal settlements was different – but this was possible because the inhabitants of each lane where the sanitation was to be installed organised and helped manage the costing, the implementation and the funding. Thus the key issue is – where are the (mostly local) organisations on the ground that can make best use of money for poverty reduction in ways that are accountable to the urban poor, that bring in other funding sources (including the savings of urban poor groups) and that encourage and support collaboration with local authorities? The importance of the national federations or networks of slum and shack dwellers or homeless people and the local NGOs that support them is precisely that they provide this. In addition, with their national organisation, they can also push for supportive national changes (as many have done, as described in Chapter 4 for the reasons elaborated in Chapter 5).

Is this a new paradigm for development funding?

Perhaps it is too early to suggest that this shows a new trend or (to use a much overused word) even a new paradigm in development. It has certain features shared with other initiatives. First, it makes funding available direct to low-income groups; but many forms of social protection now do this at the household level. Second, unlike social protection initiatives that provide income supplements to individuals, it funds collective initiatives chosen by grassroots organisations. In so doing, it encourages them to plan and act collectively and, as noted above, to bring this to the city level and engage local governments; collective action helps to ensure that the urban poor are strong enough to challenge and overcome more powerful groups who do not act in the interests of all urban citizens. Then, to go further in setting up city or national funds that can continue, widen and increase support for community initiatives. The Urban Poor Fund International that has supported hundreds of community initiatives and where the use of funding is determined by agreements made by federations of slum or shack dwellers also has these two features.[10] Such funds are not alternatives to social protection – rather they are a complementary mechanism designed to improve tenure security, access to basic infrastructure and services, and enhance political voice. They address components of urban poverty that are untouched by conventional approaches to social protection and welfare provision.

Can aid agencies and development banks support this?

Official aid agencies and development banks were not set up to work directly with low-income communities. They were set up to work with and fund national governments. Aid agencies have to be accountable to the government that funds them (and beyond this to the voters who put the government into office). Multilateral development banks such as the World Bank and the Asian, African and Inter-American

Development Banks have to be accountable to the governments that sit on their boards – especially those that provide them with funding. These funding agencies have no direct accountability to low-income groups, although these groups' unfulfilled needs are what justifies their work and the funding they receive. Initially, it was assumed that international funding agencies would support national (recipient) governments to address these unfulfilled needs. It was also hoped that this would support stronger economies that in turn would also help address unfulfilled needs through increased incomes and larger government capacity to provide the basics – secure housing, water, sanitation, health care, schools, rule of law and provision for voice.

However, as this book and its companion volume has described, this has not happened for a large and in many nations a growing number of people. Around one in seven of the world's population lives in informal settlements in urban areas. In the absence of support from governments, aid agencies and development banks, individuals, households and communities have had to manage themselves. City economies would collapse without their labour and informal enterprise activities; yet city governments often ignore them or see them only as a problem. Most aid agencies have also ignored them. There has been some progress where and when low-income urban households and their own organisations and federations had some political influence – as in several Latin American nations when they returned to or strengthened democracies with political changes that included city governments that were more accountable and better funded. This provides us with a reminder that to be sustained, pro-poor policies and practices at national and local levels need urban poor groups to be organized. In many nations, there are now urban poor organizations and federations that actually want to work with local governments – and that bring their own knowledge and capacities that are so valuable in actually reducing urban poverty. These are also bringing the knowledge and capacities of women into these initiatives and in so doing supporting their empowerment. If large, centralised aid agencies and development banks cannot work directly with urban poor groups and their community organisations, can they learn to work with and through intermediary institutions that are on the ground in each city, and that finance, work with and are accountable to urban poor groups? As, for instance in the city development funds and the national funds organised and managed by the slum/shack/homeless people's federations? Are development assistance agencies prepared to give up sole control of the decisions and work in collaboration with these slum/shack/homeless networks and federations to reach those most in need, at scale, with integrated programmes that bring effective development to the urban poor?

Will the agenda of slum and shack dwellers ever be considered?

Global discussions and urban poverty

Many discussions are now underway on the development framework that will replace the Millennium Development Goals post-2015. As mentioned earlier, this includes a 'High-level Panel of Eminent Persons' appointed by the Secretary-General to advise him on the post-2015 process. There are the many UN agencies involved in developing thematic papers for the post-2015 discussions. There are also the international discussions on sustainable development goals coming out of Rio +20 (the UN Conference on Sustainable Development in 2012) – seeking to reinvigorate a concern for local and global environmental issues within development. There are also the evolving discussions on aid effectiveness within a 'High-level Forum'. This is generating lots of discussion and material. *But in all of these discussions, so little attention is given to the local – to local contexts, to local government, to local organisations of urban poor groups and other local civil society groups, to local finance, to local resources, to the local data needed to inform action, to the accountability of national governments and international agencies to the residents of each locality. In addition, so little attention is given to urban populations.*

Looking at the make-up of the 26 persons who make up the High-level Panel, most are working for (or used to work for) national governments – including former or current ministers, prime ministers and presidents. Several are working in or have worked in the large official development assistance agencies. Two represent private enterprises. The Panel is said to include 'representatives of governments, the private sector, academia, civil society and youth, with the appropriate geographical and gender balance'.[11] But no one on the panel is a representative of the urban poor and their organisations and federations. Only one is from local government. Where are the grassroots leaders and the local NGOs and local governments they work with who really have contributed to meeting many development and environmental goals within their localities?

Development in urban areas depends on local institutions

Almost all development interventions in urban areas are local in the sense that they depend on local institutions – for water, sanitation, electricity, piped gas (where this is available), solid waste collection, schools, street cleaning, daycare centres, playgrounds and public spaces, health care clinics, emergency services, public transport systems, policing, bank branches … (see Chapter 3 for more details). These may be government agencies, private sector enterprises or civil society organisations, or some mixture of these (including co-production). Where some of these fall under the responsibility of national or state/provincial governments, their realisation relies on local offices of national governments or collaborative arrangements between national agencies and local governments. So it is the performance of local (state, civil society and sometimes private sector) institutions that is so critical for meeting MDGs and most other development or environment goals.

Chapter 3 discussed the very considerable diversity between nations in the allocation of responsibility for the goods and services mentioned above between different levels of government – and of course, also in the form of local government. But in almost all nations, local government plays a significant role in this provision. Wherever living standards are high, local governments play a major role in this achievement – often the primary role. This may be seen in the wide range of responsibilities they have for provision, maintenance and, where needed, expansion of infrastructure and services that usually includes provision for water, sanitation, drainage, streets, emergency services, parks and public spaces. Their responsibilities often extend to health care services and schools, and many include social protection measures (although usually with national government). They play key roles in ensuring health and safety – for instance, through building standards, land-use planning and management, and environmental, occupational and public health services.[12] They usually play key roles in disaster prevention and preparedness (UN-ISDR 2012). Thus clearly local governments can have a major influence on performance towards meeting most of the MDGs and their targets. Good local governance is also central to democratic participation, civic dialogue, economic success and facilitating outcomes that enrich the quality of life of residents (Shah 2006). For most sectors, policy, standards and oversight are often national responsibilities while actual provision and administration are local. As Nigeria's National Planning Commission noted, 'Without state and local governments, federal programmes alone would amount to attempting to clap with one hand' (Nigerian National Planning Commission 2004, p. vii). In many nations where urban poverty has been reduced, it is the increased competence, capacity and accountability of local governments that has contributed much to this – and to meeting many of the MDG targets.

National governments and international agencies are only as effective as the local institutions through which they implement their policies and programmes. Even where interventions are the responsibility of national ministries, or infrastructure or services are delivered through private enterprises or international NGOs, their effectiveness usually depends on local government support, coordination and oversight – and effective local government usually depends on representative organisations of the urban poor able to manage the local politics and ensure that scarce resources are not used in ways that perpetuate exclusion, disadvantage and inequity. This may be seen in many of the examples given in earlier chapters. In the past, these local representative organisations have been missing. But what has changed is that these groups are now in place in many towns and cities in the Global South – and in many of these they have established collaborative relations with local government.

The UN system and the official aid agencies and development banks fail routinely to support the contributions of local governments and local civil society organisations (including representative organisations of the urban poor) or even to acknowledge them as stakeholders. If low-income urban dwellers are considered in their discussions (and usually they are not), they are simply targets to be reached. Or occasionally their leaders are invited to official conferences to legitimate the agenda of the organising agencies.

Who will address international goals and targets?

The discussions on the post-2015 framework for development need to pay more attention to who has to act to meet goals and targets and how they are to be resourced and supported. Here, local governments and local civil societies have great importance.

The MDGs are very clear about what they want to achieve but say very little about who needs to act to meet the goals. Goals and targets are constructed by 'experts' who seem to give little thought to who needs resourcing to ensure that these goals and targets are met. The MDG agenda is a set of technical, sectoral, macro-economic undertakings that overlook the very local and integrated nature of social transformation (Vandermoortele 2011). Most of its goals are allocated to one sectoral ministry or agency with a technical fix rather than building local competence and capacity to address goals together. Waage and colleagues note the need to avoid goals that are 'compartmentalised into responsibilities of different line ministries nationally, subnationally, and locally, which means that the potential for simultaneous actions in the same location, working with the same communities and households, is unlikely' (Waage *et al.* 2010, p. 999).

The people whose needs are meant to be addressed are at best passive recipients to be targeted; there is no mention of their roles or their rights to set targets and to receive support to address these issues. Nowhere is there any recognition of the agency and capacities of grassroots groups to address the goals themselves. Yet we know from experience that these groups are often the most effective agents for their own development, that they can catalyse action from others, and that this agency is essential for these targets to be met. Perhaps more to the point, their political influence is needed to make sure that local governments also address the targets.

Can international frameworks support local pro-poor agendas?

Despite the key role played by local institutions in implementing and 'localising', internationally agreed development and environmental agendas, their roles remain under-recognised and under-supported. Those who are discussing and determining the post-2015 agenda tend to be at a vast distance from local realities. When they talk about 'localising' the MDGs, they mean at national level, not within local government or civil society. When they discuss good governance, they refer to the activities of national governments, not to the vital relationships between citizens and their local administrations. When they measure progress, they use nationally representative datasets, relying on aggregate data to demonstrate success, but failing to reveal who is being left out and where they live.

In the 20 'thematic think pieces' compiled by experts from various Task Team members, mostly UN agencies, there is so little discussion of local institutions – even in papers which cover issues that depend on local institutions such as health, disasters, inequalities, employment and governance. There is also no mention of local governments in many of these and other papers discussing the post-2015 development agenda. Thus it is not surprising that many national governments fail to take local organisations seriously in addressing the MDGs.

Local governments may also determine whether citizens have access to entitlements provided by national government. Especially in urban areas, where so many residents live in informal settlements, a lack of documentation (for instance, a legal address) may prevent them from getting on the voter register, access to basic services, getting their children into a school, or gaining access to government-supported health care. Local authorities may be reluctant to provide these services to those living in informal settlements because they feel that this encourages the development of even more such settlements. Or high-density settlements and narrow lanes may simply make it inconvenient to provide those living in such settlements with services like piped water or waste removal. Access to these services determines whether many of the MDG targets are met for urban populations.

Lack of attention to urban areas and their low-income residents

This book and its companion volume has documented in some detail how most local governments in urban areas have failed to ensure provision of even minimum basic services and tenure security to much of their population. But this type of exclusionary partial politics is not inevitable. The programme interventions we discuss in this volume have developed the tools and mechanisms that enable local communities to challenge outcomes and find new and more collaborative approaches. As partnerships are built and new, more effective modes of urban development are identified, improving conditions in informal settlements is possible at the city-wide scale. But just as effective local government requires strong citizen groups able to hold local politicians to account and assist with the planning and implementation of improvements, it also requires finance and supportive policy frameworks at the national level.

A review of the papers and discussions that are part of the post-2015 processes shows an astonishing lack of attention to urban poverty. Urban issues are not even mentioned in most of the UN-led thematic papers. Where they are mentioned, it is mainly in the context of urbanisation and economic growth. Many of the MDG targets and indicators are designed for rural contexts and so under-report the scale of deprivation in urban areas. Most documents do not cover the implications for poverty of living in informal settlements – the insecurity, the lack of services, the lack of access to entitlements, the high infant, child and maternal mortality rates. ... There is also the lack of attention to urban poverty among most aid agencies whose work is described in Chapter 3. Perhaps this is changing; there seems to be more discussion of urban poverty now, including more conferences and seminars and more institutions developing urban programmes (in part because there is or may be more funding here?). But how much does this involve the urban poor?

Take one example.[13] The sixth World Urban Forum was held in Naples in September 2012. Its theme was The Urban Future. Organised by the United Nations Human Settlements Programme (UN-Habitat), it included this UN agency's official programme for over 150 'networking' and other events organised mostly by international NGOs and academic institutions. One innovation that UN-Habitat has pioneered within the UN system is supporting a strong focus on the importance of local governments. This is never easy in that all UN agencies are accountable to national governments which may not support the policies and practices of some local governments. But in the official events, many of the speakers were city mayors, along with representatives from national governments and international funding agencies, academics and a few NGOs.

What was absent from almost all the official events in this Urban Forum was any representation of the networks and federations of slum or shack dwellers. It is as if they have no role in defining the urban future. One possible excuse could be that they were not present (although this would raise the issue as to why they were not invited and supported to come). But there were plenty of representatives and leaders of national federations or networks of slum/shack dwellers from many nations taking part in the side events that external agencies were allowed to organise, even though there had been no UN support to get them there. In addition, perhaps surprisingly, these side events often included presentations not only by federation leaders but also by local government or national government staff that work with them. Thus in a session on alternatives to evictions organised by Shack/Slum Dwellers International, the Mayor of Iloilo in the Philippines (Jed Patrick Mabilog) talked about the importance of his government's partnership with the Philippines Homeless People's Federation – which was then confirmed by Sonia Fadrigo from the Philippines Homeless People's Federation. The Mayor of Harare (Muchadei Masunda) spoke of his commitment to stopping evictions and the value of the partnership between the city government and the Zimbabwe Homeless People's Federation and its support NGO, Dialogue on Shelter. This was confirmed by Davious Muvindi from the Zimbabwe Federation. A session on city-wide upgrading organised by the Asian Coalition for Housing Rights included short presentations by many community leaders and local government politicians and civil servants about their partnerships. What was notable about this presentation was the scale of the city-wide upgrading initiatives all over Asia driven by community organisation and action.

It is amazing that after decades of discussion on participation and permitting 'voice' to urban poor groups, very large forums and conferences on urban issues can still be organised without engaging the urban poor – even as these events are justified by their apparent importance for addressing the needs of the urban poor. Anyone who actually listened to the presentations of these and other federation members and leaders during the World Urban Forum were reminded of how clear they are about their needs and priorities, and the challenges they face in getting these issues addressed, as well as how often these differ from our assumptions about their needs. The Mayor of Iloilo ensures that there are representatives of the Philippines Homeless People's Federation on key committees within his government, including those allocating funds and those determining infrastructure priorities. Why weren't representatives of urban poor organisations, federations and networks on the committees that organised this and previous World Urban Forums? Why are the powerful global institutions that are now developing a post-2015 development agenda so reluctant to engage the urban poor directly? The formulation of the Millennium Development Goals did not consult them; if it had, it would have had a much more ambitious and relevant target for improving conditions in slums.

The preparations for the post-2015 development framework will probably forget to involve the representative organisations of slum dwellers – as on the Eminent Persons panel. Those who set up this panel probably think that because there are one or two representatives from NGOs, these represent 'the poor'. Or it will assume that their priorities will be represented by other professional groups (experts). This has to change. As Adnan Aliani from UN ESCAP commented at the 2012 World Urban Forum, in so many countries it is no longer an issue of people needing to participate in government programmes; it is an issue of government learning to participate in and support people's programmes.

Urban poverty reduction and climate change

This book has not reviewed the contribution to urban poverty reduction of climate change adaptation – although the companion volume to this included a discussion about the potential contribution to poverty of different direct and indirect impacts of climate change. Human-induced climate change will certainly increase risks to large sections of the urban poor – and continue increasing risks until global warming stops. As the previous book noted, hundreds of millions of low-income urban dwellers who are so at risk now from extreme weather, sea-level rise and disruptions to food and water supplies will see these and other risks increase. The global discussions on this are still far from producing the agreement needed to reduce greenhouse gas emissions and avoid dangerous climate change. In addition, the scale and scope of international funding for poverty reduction will be influenced by the large and probably increasing share of development assistance (or other forms of financial aid) that will be allocated to climate change adaptation – or compensation for climate change impacts (what is now termed loss and damage). Will funding for climate change adaptation contribute to reducing risks and vulnerabilities among urban poor groups (including those living in informal settlements that governments regard as illegal)? Or might it even contribute to increasing poverty as adaptation measures displace them from their homes and livelihoods? Table 6.1 outlines the implications for climate change adaptation and mitigation of different approaches to poverty reduction in urban areas.

Will the relations between low-income households, their associations and local government be strong enough to withstand the difficulties of more intense or more frequent storms, flooding, landslides and heatwaves in informal settlements? Will their residents be fully involved in determining the best course of action – for instance, upgrade in situ or relocate? Will the programmes to relocate those living on sites most at risk allow those to be relocated the influence they need in choosing relocation sites, organising and managing the move and developing the new settlements? Or do we face increasing tensions, conflict and even violence as residents and local governments struggle to cope, where more powerful groups get adaptation that serves them (or they move) – and where predictions of areas at risk are found to be wrong and there is loss and repeated relocation?

Table 6.1 Approaches to poverty reduction in urban areas in the Global South and their implications for climate change adaptation and mitigation

Approach to urban poverty reduction	*Direct impacts*	*Indirect impacts*	*Issues*	*Implications for adaptation*	*Implications for greenhouse gas emissions*
1: Economic growth	More employment or income-earning opportunities/higher incomes for some	Rising incomes and more people with adequate incomes increasing demand for services and generating more taxes and other revenues	Government facilitating this & removing barriers for private enterprise success; less "success" in poverty reduction than often hoped for; impact exaggerated by inappropriate measures of poverty (as described in this volume). More difficult for external agencies to successfully support enhanced livelihoods in urban areas?	May allow those whose incomes rise to adapt but of itself, does not address risk (eg build risk reducing infrastructure) or increase resilience of urban areas	If successful, rising per capita GHGs from expanded production and consumption. Possibilities for combining mitigation and adaptation eg recycling groups, densification of urban centres.
2: Pro-poor/inclusive economic growth	Minor adjustments to the above that are meant to help ensure fall in poverty. May include support to informal sector and financial services that meet urban poor's needs. May include urban management efforts to improve provision of energy, transport and other basic services for economic growth and/or basic needs.				

Table 6.1 (continued)

Approach to urban poverty reduction	*Direct impacts*	*Indirect impacts*	*Issues*	*Implications for adaptation*	*Implications for greenhouse gas emissions*
3: "Meeting basic needs" – tenure, provision for water, sanitation, drainage, health care, schools, electricity…	If done well, eg in effective 'slum' upgrading, reduces many aspects of poverty. Usually state-led but some successful examples of community-led and larger local government-community organization partnerships.	Can produce major health and time benefits; can support more successful household enterprises (as provision for water, electricity, roads… improve)	Success depends on capacity and competence of (local) government and relations with urban poor; also on whether this resolves difficult issues eg tenure, good quality well-maintained infrastructure and services	Most of this should reduce disaster and climate change risk. Climate change adaptation can be integrated into this	If successful, some minor increases in GHGs from larger, better quality buildings and infrastructure. Improved public transport may lower emissions, especially with densification.
4: Support for Housing	Assists households to get safe, secure homes with infrastructure & services	Improved social status. Perhaps more secure incomes	Relatively expensive; incremental housing cheaper but may contravene regulations. Housing units supported may go to non-poor	Depends on quality of housing and chosen site	Depends on housing design, construction quality & if attention is given to energy efficiency

Table 6.1 (continued)

Approach to urban poverty reduction	*Direct impacts*	*Indirect impacts*	*Issues*	*Implications for adaptation*	*Implications for greenhouse gas emissions*
5: Social protection, safety nets, measures for food security	Providing or subsidizing food or particular services or funds (eg conditional cash transfers) for 'poor'	Can produce significant improvements in nutritional status and health	Needs effective government structures to be able to reach 'the poor'?	Provides some low-income groups with some aspects of increased resilience but does not address large gaps in protective infrastructure	If successful, increases in consumption of lowest-income groups but scale of increase has very small implications for GHG growth; shifts to cleaner fuels may reducing GHGs from energy use. Very considerable possibilities for employment generation for urban centres that seriously address mitigation (and adaptation)
6: Livelihoods and household assets	Microfinance and market access for small scale/ informal enterprises	Often needs change in attitude by government on informal economy	Many urban poor dependent on wage labour that provides very poor returns and not served by this.		
7: Rights based approaches	May address lack of rights but problems related to realization of rights often remain	May lead to improved access to basic services, and/or improved security	Assumes a strong legal process that works for low-income groups (often not there)	This can combine poverty reduction with risk reduction from disasters/ climate change but	

Table 6.1 (continued)

Approach to urban poverty reduction	*Direct impacts*	*Indirect impacts*	*Issues*	*Implications for adaptation*	*Implications for greenhouse gas emissions*
8: Urban poor led initiatives – voice, services, tenure, rule of law … often supported by social movements	Many different kinds of urban poor led initiatives have reduced many aspects of poverty. In some nations, this has produced more competent pro-poor local governments	Organizations of the urban poor with more capacity to negotiate and act	Limits in scale and scope if government is hostile to urban poor.	scale and scope depends on supportive local governments and national government finances, and may also require support from external funders.	
9: Urban poor groups working as federations with local government	As above but scale/scope of what can be done increasing considerably. Urban poor's direct involvement improves state policies and programmes and improves access to tenure and basic services	Empowered local groups & their federations to negotiate other benefits eg community-police partnerships for policing informal settlements	Poverty reduction agenda widens as urban poor organizations develop partnerships with local governments eg city-wide surveys and maps of settlements at risk to help prioritize and support action		

But this book cannot assess the implications for urban poverty reduction of climate change adaptation in the Global South because, as yet, there are too few experiences to review. There can be powerful synergies between reducing everyday environmental risks faced by low-income groups (with major contributions to poverty reduction), reducing disaster risk (with major contributions to poverty reduction) and building resilience to climate change impacts. Here, there is the potential for climate change adaptation to contribute to poverty reduction and often to mitigation too. But this is unlikely to happen unless urban poor groups have political influence and good working relationships with local governments. Almost all the poverty reduction measures described in this book that worked also help build the resilience of urban poor groups to climate change. It is also unlikely to happen unless the international funds for climate change adaptation can learn to support and work with urban poor groups and local governments – and, if they cannot, at least to support the local funds that can. Of course, as we learn more about the specifics of what risks are changing in each location, so development and disaster risk reduction can adjust to these. But it is households who no longer face the deprivations discussed in these two volumes that are generally far more resilient to climate change.

MDGs, post-MDGs and development assistance in an urbanising world

Here are eight points for development assistance agencies to consider if they really want to reduce urban poverty:

1 *Don't just set targets; be clear about how they can be met and by whom.* The MDGs and their various targets are clear about what they want to achieve (and by when) but say nothing about how. They don't set out who is responsible and capable of meeting the targets and who needs their capacity to act enhanced. Most goals and targets will not be met unless grassroots organisations and their federations and networks as well as local governments and the agendas they develop together are supported.
2 *Go back to universal targets* that include universal provision for: safe, sufficient water (which in urban areas is measured by the proportion of households with regular supplies of treated water piped to their premises); sanitation (which in urban areas is measured by the proportion of households with good-quality toilets in their home or immediate neighbourhood); primary health care, schools and emergency services accessible to all (with more attention paid to ensuring good quality provision).
3 When considering finance to support the achievement of goals, consider *where finance is needed, available to whom and accountable to whom.* There is a danger that the post-MDG discussions will just generate a new list of goals without considering the financial and other mechanisms that are needed by local government and civil society to support their achievement. There is a need for local financial institutions in each urban centre that work with and are accountable

to urban poor groups. There are already many of these functioning from which to learn. The federations and networks of slum/shack/homeless people's organisations and the local governments that work with them are critical allies in this.

4 *Have indicators that actually match goals and targets.* Measurements are needed to assess whether targets are met. However, as this book and its companion volume has described, some of the indicators being used to measure progress on MDG achievements are flawed for urban areas – the dollar-a-day poverty line (and its adjustment to $1.25 a day at 2005 prices), the statistics on provision for water and sanitation and on slum populations. If poverty lines were set in each nation at levels that match the costs of food and non-food essentials and adjusted for where such costs are particularly high (for instance, in larger and more prosperous cities) it is very unlikely that the poverty reduction target has been met – or will be met by 2015. This would also produce a very different picture of global trends in poverty.

5 *Support local processes to generate the data needed for setting priorities and benchmarks and monitoring progress.* This means radically changing the very basis for generating data – no longer relying on national sample surveys that provide so little useful data for local actors about where needs are concentrated. There is also a need to consider how to provide data on some key qualitative issues – the extent to which there is a constructive relationship between urban poor groups and local governments, what constrains the development of representative organisations of the urban poor, the availability of funds to support the work of grassroots organisations, and so on.

6 *Encourage and support local governments and civil society organisations to develop their own goals and targets and to recognise their roles and responsibilities within the post-2015 development process.* Agenda 21 coming out of the UN Earth Summit in 1992 had a short section on Local Agenda 21s. This is one of the few times that the key role of local governments in meeting environment and development goals was recognised. Perhaps surprisingly, the agenda for change coming out of Rio plus 20 is one of the only examples of global discussions on development and environment that actually takes local governments' roles seriously.

7 *Avoid vague and ambiguous statements.* Sadly, a commitment to sustainable development means nothing today unless it specifies what is meant. The term 'sustainable development' is used to mean so many different things, including even sustainable economic growth. The term 'sustainable urbanisation' has also come to be widely used, but it is not clear what this seeks to sustain (and it is even less clear what it hopes to develop). What is needed is for the term 'sustainable development' to be used to highlight the two priorities emphasised by the Brundtland Commission in 1987 – meeting the needs of the present (i.e. ending poverty) without compromising the ability of future generations to meet their needs.

8 *And what about climate change?* Somehow the issue of climate change was left out of the MDGs and their targets. Oddly enough, building resilience in urban areas to the impacts of climate change is dependent on points 1 and 2; this

needs local competence and capacity, partnerships between those most at risk, and local governments and basic infrastructure and services reaching everyone. It also requires finance systems that support on-the-ground knowledge and capacity to act (points 3 and 5).

Some of the discussions around the post-2015 development agenda are entitled 'The future we want'. It would be nice if it actually was the future that those who currently suffer hunger and other forms of deprivation want.

Notes

1 See http://www.uncsd2012.org/content/documents/814UNCSD%20REPORT%20final %20revs.pdf.
2 See the comment by John F.C. Turner on this – whose work, writing and teaching did so much to point to more effective ways of addressing housing issues for low-income groups (see in particular Turner 1976). He noted that few of the case studies said to be 'best practice' had the level of detail needed on procedural software or technical hardware to assess their applicability in different circumstances, let alone in different fields. 'Recognizing the differences between practices, tools and principles is a necessary if insufficient step toward understanding what is "best" or, at least, "better than" and, therefore, toward knowledge of common ground defined by particular values' (Turner 1996, p. 199).
3 See *Poverty by the Numbers*, Working Paper Series Drafted by the High-level Panel Secretariat, Prepared for London High Level Panel Meeting, 1–2 November 2012.
4 See the figures on this presented in Ravallion *et al.* (2007) and in World Bank 2013.
5 These may be cheap but they can still represent a significant proportion of the income of a low-income household. When visiting Kibera, the large, informal settlement in Nairobi in 2010, we were shown a large private school and were told that it cost 100 shillings a month to send a child there. *The Economist* in an article about Kibera pubished in December 2012 reported on a better-quality private school than the one we visited where the cost of a place was 7,500 shillings a year, although the teacher interviewed also reported that they do not expel children who cannot afford to pay these fees (*The Economist* 2012).
6 Although there are some exceptions (see, for instance, Moser 2004; Roy *et al.* 2004).
7 See the special article on this topic in the April 2012 issue of *Environment and Urbanization.*
8 This section draws on ideas first developed in a blog by Diana Mitlin – http://environmentandurbanization.org/real-issue-universal-access-affordable-basic-services.
9 This draws on ideas first developed in a blog by Diana Mitlin on this topic – see http://environmentandurbanization.org/basic-service-provision-shouldn%E2%80%99t-just-be-money-maker.
10 See http://www.sdinet.org/ for more details.
11 http://www.un.org/sg/management/pdf/ToRpost2015.pdf.
12 For most of these, responsibilities are shared with national government or different aspects of the responsibilities are divided – for instance, national government sets regulations and standards, and local government enforces them.
13 This draws on ideas first developed in a blog by David Satterthwaite – http://environmentandurbanization.org/6th-world-urban-forum-will-agenda-slum-and-shack-dwellers-ever-get-considered.

References

Abers, Rebecca (1998) 'Learning through democratic practice: distributing government resources through popular participation in Porto Alegre', in Mike Douglass and John Friedmann (eds) *Cities for Citizens*, Chichester, UK: John Wiley and Sons, pp. 39–66.

Abrams, Charles (1964) *Man's Struggle for Shelter in an Urbanizing World*, Cambridge, MA: MIT Press.

ACHR (Asian Coalition for Housing Rights) (1989) 'Evictions in Seoul, South Korea', *Environment and Urbanization* 1:1, 89–94.

——(1993) 'The Asian Coalition for Housing Rights: NGO profile', *Environment and Urbanization* 5:2, 153–165.

——(2004) 'Negotiating the right to stay in the city', *Environment and Urbanization* 16:1, 9–26.

——(2011a) *107 Cities in Asia: Second Yearly Report of the Asian Coalition for Community Action Program*, Bangkok: Asian Coalition for Housing Rights.

——(2011b) *Assessing ACCA in Sri Lanka*, Bangkok: Asian Coalition for Housing Rights.

Archer, Diane (2012) 'Finance as the key to unlocking community potential: savings, funds and the ACCA programme', *Environment and Urbanization* 24:2, 423–440.

Adato, M., D. Coady and M. Ruel (2000) *An Operations Evaluation of PROGRESA from the Perspective of Beneficiaries, Promotoras, School Directors, and Health Staff*, Washington, DC: International Food Policy Research Institute (IFPRI).

Agarwala, R. (2006) 'From work to welfare', *Critical Asian Studies* 38:4, 419–444.

Albee, A. and N. Gamage (1996) *Our Money, Our Movement: Building a Poor People's Credit Union*, London: Intermediate Technology Publishing.

Almansi, Florencia (2009) 'Rosario's development; interview with Miguel Lifschitz, mayor of Rosario, Argentina', *Environment and Urbanization* 21:1, 19–35.

Almansi, Florencia, Ana Hardoy and Jorgelina Hardoy (2010) *Improving Water and Sanitation Provision in Buenos Aires. What can a Research-oriented NGO do?* Human Settlements Working Paper Series, London: International Institute for Environment and Development.

Almansi, Florencia, Ana Hardoy, Jorgelina Hardoy, Gustavo Pandiella, Leonardo Tambussi and Gaston Urquiza with Gordon McGranahan and David Satterthwaite (2011) *Los Limites de la Participacion; La Lucha por el Mejoramiento Ambiental en Moreno, Argentina*, Buenos Aires: IIED-America Latina Publications.

Amis, Philip (2002) *Thinking about Chronic Urban Poverty*, CPRC Working Paper, Manchester: Chronic Poverty Research Centre.

Angel, Shlomo (2000) *Housing Policy Matters: A Global Analysis*, New York: Oxford University Press.

Angel, Shlomo and Somsook Boonyabancha (1988) 'Land sharing as an alternative to eviction: the Bangkok experience', *Third World Planning Review* 10:2, 107–127.

Anwar, Nausheen H. (2012) 'State power, civic participation and the urban frontier: the politics of the commons in Karachi', *Antipode* 44:3, 601–620.

Appadurai, Arjun (2001) 'Deep democracy: urban governmentality and the horizon of politics', *Environment and Urbanization* 13:2, 23–43.

——(2004) 'The capacity to aspire: culture and the terms of recognition', in Vijayendra Rao and Michael Walton (eds), *Culture and Public Action*, Stanford, CA: World Bank/Stanford University, pp. 61–84.

Archer, Diane (2012) 'Finance as the key to unlocking community potenial: savings, funds and the ACCA programme', *Environment and Urbanization* 24:2, 423–440.

Arévalo T. Pedro (1997) 'Huaycán self-managing urban community: may hope be realized', *Environment and Urbanization* 9:1, 59–79.

Arputham, Jockin (2008) 'Developing new approaches for people-centred development', *Environment and Urbanization* 20:2, 319–338.

——(2012) 'How community-based enumerations started and developed in India', *Environment and Urbanization* 24:1, 27–30.

Arputham, Jockin and Sheela Patel (2010) 'Recent developments in plans for Dharavi and for the airport slums in Mumbai', *Environment and Urbanization* 22:2, 501–504.

Arrieta, Gerardo M. Gonzales (1999) 'Access to housing and direct housing subsidies: some Latin American experiences', *Cepal Review* 69, 141–163.

Audefroy, Joël (1994) 'Eviction trends worldwide and the role of local authorities in implementing the right to housing', *Environment and Urbanization* 6:1, 8–24.

Auyero, Javier (1999) '"From the client's point of view": how poor people perceive and evaluate political clientelism', *Theory and Society* 28, 297–334.

——(2000) *Poor People's Politics*, Durham, NC, and London: Duke University Press.

——(2010) 'Visible fists, clandestine kicks, and invisible elbows: three forms of regulating neoliberal poverty', *European Review of Latin American and Caribbean Studies* 89:October, 5–26.

Auyero, Javier, Pablo Lapegna and Fernanda Page Poma (2009) 'Patronage politics and contentious collective action: a recursive relationship', *Latin American Politics and Society* 51:3, 1–31.

Avritzer, Leonardo (2006) 'New public spaces in Brazil: local democracy and deliberative politics', *International Journal of Urban and Regional Research* 30:3, 623–637.

Baiocchi, G. (2003) 'Participation, activism and politics: the Porto Alegre experiment', in A. Fung and E.O. Wright (eds), *Deepening Democracy: Institutional Innovations in Empowered Participatory Governance*, London and New York: Verso, pp. 45–77.

Banks, Nicola (2010) 'Employment and mobility among low-income urban households in Dhaka', Ph.D. thesis, Manchester: University of Manchester Press.

——(2012) 'Employment and mobility among low-income urban households in Dhaka, Bangladesh', Institute for Development Policy and Management, Manchester: University of Manchester.

Banks, Nicola and David Hulme (2012) *The Role of NGOs and Civil Society in Development and Poverty Reduction*, BWPI Working Paper 171, Manchester: Brooks World Poverty Institute, University of Manchester.

Barbosa, R., Y. Cabannes and L. Morães (1997) 'Tenant today, posserio tomorrow', *Environment and Urbanization* 9:2, 17–46.

Barrientos, Armando and David Hulme (2008a) *Social Protection for the Poor and Poorest in Developing Countries: Reflections on a Quiet Revolution*, BWPI Working Paper No. 30, Manchester: Brooks World Poverty Institute, University of Manchester.

Barrientos, Armando and David Hulme (eds) (2008b) *Social Protection for the Poor and Poorest: Concepts, Policies and Politics*, London: Palgrave.

Barrientos, Armando, Miguel Niño-Zarazúa and Mathilde Maitrot (2010) *Social Assistance in Developing Countries Database (Version 5.0)*, Manchester: Chronic Poverty Research Centre, University of Manchester.

Barrientos, Armando and James Scott (2008) *Social Transfers and Growth: A Review*, BWPI Working Paper No. 52, Manchester: Brooks World Poverty Institute, University of Manchester.

Barrig, M. (1991) 'Women and development in Peru: old models, new actors', *Environment and Urbanization* 3:2, 66–70.

Baskin, Julian and Daniel Miji (2006) 'Partnership for infrastructure in the musseques of Luanda, Angola', in Lucy Stevens, Stuart Coupe and Diana Mitlin (eds), *Confronting the Crisis in Urban Poverty: Making Integrated Approaches Work*, Rugby: Intermediate Technology Publications, pp. 39–62.

Batliwala, S. (2002) 'Grassroots movements as transnational actors: implications for global civil society', *Voluntas: International Journal of Voluntary and Nonprofit Organizations* 13:4, 393–409.

Baumann, Ted (2003) *Harnessing People's Power: Policy Makers' Options for Scaling Up the Delivery of Housing Subsidies via the People's Housing Process*, Housing Finance Resource Programme Occasional Paper No. 10, Johannesburg: Urban Institute.

——(2007) *Shelter Finance Strategies for the Poor: South Africa*, background paper for meeting of the CSUD/Rockefeller Foundation Global Urban Summit, Bellagio: International Institute for Environment and Development.

Baumann, Ted, Joel Bolnick and Diana Mitlin (2004) 'The age of cities and organizations of the urban poor: the work of the South African Homeless People's Federation', in Diana Mitlin and David Satterthwaite (eds), *Empowering Squatter Citizens: Local Government, Civil Society and Urban Poverty Reduction*, London: Earthscan Publications, pp. 193–214.

Bawa, Zainab (2011) 'Where is the state? How is the state? Accessing water and the state in Mumbai and Johannesburg', *Journal of Asian and African Studies* 46:5, 491–503.

Bayat, Asef (2000) 'From "dangerous classes" to "quiet rebels": politics of the urban subaltern in the global south', *International Sociology* 15:3, 533–557.

Bebbington, Anthony (2005) 'Donor–NGO relations and representations of livelihood in nongovernmental aid chains', *World Development* 33:6, 937–950.

Bebbington, Anthony J., Sam Hickey and Diana Mitlin (eds) (2008a) *Can NGOs Make a Difference? The Challenge of Development Alternatives*, London and New York: Zed Books.

Bebbington, Anthony J., Sam Hickey, and Diana Mitlin (2008b) 'Introduction: Can NGOs make a difference? The challenge of development alternatives', in Anthony J Bebbington, Sam Hickey and Diana Mitlin (eds), *Can NGOs Make a Difference? The Challenge of Development Alternatives*, London and New York: Zed Books, pp. 3–35.

Bebbington, Anthony, Martin Scurrah and C. Bielich (2011) *Los Movimientos Sociales y la Politica de la Pobreza en el Peru*, Lima: IEP.

Behrman, J. and J. Hoddinott (2000) *An Evaluation of the Impact of PROGRESA on Preschool Child Height*, Washington, DC: International Food Policy Research Institute (IFPRI).

Behrman, J.R., J. Gallardo-García, S.W. Parker, P. Todd and V. Vélez-Grajales (2011) *Are Conditional Cash Transfers Effective In Urban Areas? Evidence From Mexico*, Penn Institute For Economic Research Working Paper 11–024, Penn Institute For Economic Research, University Of Pennsylvania.

Bénit-Gbaffou, Claire (2011) '"Up close and personal" – How does local democracy help the poor access the state? Stories of accountability and clientelism in Johannesburg', *Journal of Asian and African Studies* 46:5, 453–464.

——(2012) 'Party politics, civil society and local democracy – reflections from Johannesburg', *Geoforum* 43:2, 178–189.

Benjamin, Solomon (2000) 'Governance, economic settings and poverty in Bangalore', *Environment and Urbanization* 12:1, 35–56.

——(2004) 'Urban land transformation for pro-poor economies', *Geoforum* 35, 177–187.

Benjamin, S.A. and R. Bhuvaneswari (2001) *Democracy, Inclusive Governance and Poverty in Bangalore*, Urban Governance, Poverty And Partnerships Working Paper, Birmingham: University of Birmingham.

Bhan, Gautam (2009) '"This is no longer the city I once knew." Evictions, the urban poor and the right to the city in millennial Delhi', *Environment and Urbanization* 21:1, 127–142.

Biswas, Smita (2003) 'Housing is a productive asset – housing finance for self-employed women in India', *Small Enterprise Development* 14:1, 49–55.

Bolnick, Joel (1993) 'The People's Dialogue on land and shelter; community driven networking in South Africa's informal settlements', *Environment and Urbanization* 5:1, 91–110.

——(1996) 'uTshani Buyakhuluma (The grass speaks): People's Dialogue and the South African Homeless People's Federation (1994–6)', *Environment and Urbanization* 8:2, 153–170.

——(2008) 'Development as reform and counter-reform: paths travelled by Shack/Slum Dwellers international', in Anthony J. Bebbington, Sam Hickey and Diana Mitlin (eds), *Can NGOs Make a Difference? The Challenge of Development Alternatives*, London and New York: Zed Books, pp. 316–335.

Boonyabancha, Somsook (1999) 'The urban community environmental activities project and its environmental fund in Thailand', *Environment and Urbanization* 11:1, 101–106.

——(2004) 'A decade of change: from the Urban Community Development Office to the Community Organization Development Institute in Thailand', in Diana Mitlin and David Satterthwaite (eds), *Empowering Squatter Citizens: Local Government, Civil Society and Urban Poverty Reduction*, London: Earthscan, pp. 25–53.

——(2005) 'Baan Mankong: going to scale with "slum" and squatter upgrading in Thailand', *Environment and Urbanization* 17:1, 21–46.

——(2009) 'Land for housing the poor – by the poor: experiences from the Baan Mankong nationwide slum upgrading programme in Thailand', *Environment and Urbanization* 21:2, 309–330.

Boonyabancha, Somsook and Diana Mitlin (2012) 'Urban poverty reduction: learning by doing in Asia', *Environment and Urbanization* 24:2, 403–422.

Boonyabancha, Somsook, Norberto Carcellar and Tom Kerr (2012), 'How poor communities are paving their own pathways to freedom', *Environment and Urbanization* 24:2, 441–462.

Bovaird, T. (2007) 'Beyond engagement and participation: user and community co-production of public services', *Public Administration Review* 67:5, 846–860.

Bradlow, Benjamin, Joel Bolnick and Clifford Shearing (2011) 'Housing, institutions, money: the failures and promise of human settlements policy and practice in South Africa', *Environment and Urbanization* 23:1, 267–275.

Bredenoord, Jan and Paul van Lindert (2010) 'Pro-poor housing policies: rethinking the potential of assisted self-help housing', *Habitat International* 34, 278–287.

Britto, Tatiana (2005) *Recent Trends in the Development Agenda of Latin America: An Analysis of Conditional Cash Transfers*, paper presented at the Conference on Social Protection for Chronic Poverty hosted by the Institute for Development Policy and Management and the Chronic Poverty Research Centre, University of Manchester, United Kingdom, 23–24 February, University of Manchester.

Brown, Alison and Annali Kristiansen (2009) *Urban Policies and the City: Rights Responsibilities and Citizenship*, Management of Social Transformation (MOST 2 Policy Paper), UNESCO and UN-Habitat.

Brudney, J.L. and R.E. England (1983) 'Towards a definition of the co-production concept', *Public Administration Review* 43:1, 59–65.

Buckley, Robert M. and Jerry Kalarickal (2004) *Shelter Strategies for the Urban Poor: Idiosyncratic and Successful, but Hardly Mysterious, The World Bank*, Policy Research Working Paper Series No. 3427, Washington, DC: World Bank.

Buckley, Robert M. and Jerry Kalarickal (eds) (2006) *Thirty Years of World Bank Shelter Lending: What Have We Learned? Directions in Development: Infrastructure*, Washington, DC: The World Bank.

Budds, Jessica and Gordon McGranahan (2003) *Privatization and the Provision of Urban Water and Sanitation in Africa, Asia and Latin America*, Human Settlements Discussion Paper Series – Theme: Water No. 1, London: International Institute for Environment and Development.

Budds, Jessica, Paulo Teixeira and SEHAB (2005) 'Ensuring the right to the city: pro-poor housing, urban development and land tenure legalization in São Paulo, Brazil', *Environment and Urbanization* 17:1, 89–114.

Burgess, R. (1978) 'Petty commodity housing or dweller control? A critique of John Turner's views on housing policy', *World Development* 6:9/10, 1105–1133.

Burra, Sundar, Sheela Patel and Thomas Kerr (2003) 'Community designed, built and managed toilet blocks', *Environment and Urbanization* 15:2, 11–32.

Cabanas Díaz, Andrés, Emma Grant, Paula Irene del Cid Vargas and Verónica Sajbin Velásquez (2000) 'El Mezquital: a community's struggle for development', *Environment and Urbanization* 12:1, 87–106.

Cabannes, Yves (2004) 'Participatory budgeting: a significant contribution to participatory democracy', *Environment and Urbanization* 16:1, 27–46.

——(2013) *Participatory Budgeting; Reviewing the Last Decade*, forthcoming.

Cain, Allan (2007) 'Housing microfinance in post colonial Angola', *Environment and Urbanization* 19:2, 361–390.

Calderón, J. (2004) 'The formalisation of property in Peru 2001–2002: the case of Lima', *Habitat International* 28, 289–300.

Cambodia, Kingdom of (2002) *National Poverty Reduction Strategy*, Phnom Penh: Council for Social Development.

Campbell, Tim (2003) *The Quiet Revolution: Decentralization and the Rise of Political Participation in Latin American Cities*, Pittsburgh: University of Pittsburgh Press.

Casier, Bart (2010) 'The impact of microcredit programmes on survivalist women entrepreneurs in The Gambia and Senegal', in Sylvia Chant (ed.), *The International Handbook on Gender and Poverty*, Cheltenham: Edward Edgar, pp. 612–617.

Castells, Manuel (1983) *The City and the Grassroots: A Cross-cultural Theory of Urban Social Movements*, London and Victoria: Edward Arnold.

Cerdá, Magdalena, Jeffrey D. Morenoff, Ben B. Hansen, Kimberly J. Tessari Hicks, Luis F. Duque, Alexandra Restrepo and Ana V. Diez-Roux (2011) 'Reducing violence by transforming neighborhoods: a natural experiment in Medellín, Colombia', *American Journal of Epidemiology* 175:10, 1045–1053.

CGAP (2004) *Housing Microfinance*, Donor Brief No. 20, Washington, DC: CGAP.

Chakrabarti, Poulomi (2008) 'Inclusion or exclusion? Emerging effects of middle-class citizen participation on Delhi's urban poor', *IDS Bulletin* 38:6, 96–103.

Chant, Sylvia (2013) 'Cities through a "gender lens"; a golden "Urban Age" for women in the Global South?', *Environment and Urbanization* 25:1, 9–29.

Chapman, Jennifer, in collaboration with Valerie Miller, Adriano Campolina Soares and John Samuel (2009) 'Rights-based development: the challenge of change and power for development NGOs', in Sam Hickey and Diana Mitlin (eds), *Rights-based Approaches to Development: Exploring the Potential and Pitfalls*, Sterling, VA: Kumarian Press, pp. 165–185.

Chatterjee, Partha (2004) *The Politics of the Governed: Reflections on Popular Politics in Most of the World*, New York: Colombia University Press.

Chen, Martha, Renana Jhabvala, Ravi Kanbur and Carol Richards (eds) (2007) *Membership Based Organizations of the Poor*, Abingdon: Routledge.

Chitekwe-Biti, Beth (2009) 'Struggles for urban land by the Zimbabwe Homeless People's Federation', *Environment and Urbanization* 21:2, 347–366.

Cities Alliance (2002) 'Sewa Bank's housing Mirofinance Program in India', in *Shelter Finance for the Poor Series*, Washington, DC: Cities Alliance.

Cleaver, F. (2005) 'The inequality of social capital and the reproduction of chronic poverty', *World Development* 33:6, 893–906.

Cleaver, Frances (2009) 'Rethinking agency, rights and natural resource management', in Sam Hickey and Diana Mitlin (eds), *Rights-based Approaches to Development: Exploring the Potential and Pitfalls*, Sterling, VA: Kumarian Press, pp. 127–144.

Cohen, Michael (1983) *Learning by Doing: World Bank Lending for Urban Development 1972–82*, Washington, DC: World Bank.

——(2001) 'Urban assistance and the material world; learning by doing at the World Bank', *Environment and Urbanization* 13:1, 37–60.

COHRE (2006) *Forced Evictions: Violations of Human Rights*, Geneva: COHRE.

Collins, D., J. Morduch, S. Rutherford and O. Ruthven (2009) *Portfolios of the Poor: How the World's Poor Live on $2 a Day*, Princeton, NJ, and Oxford: Princeton University Press.

Connolly, Priscilla (2004) 'The Mexican National Popular Housing Fund', in Diana Mitlin and David Satterthwaite (eds), *Empowering Squatter Citizens: Local Government, Civil Society and Urban Poverty Reduction*, London: Earthscan, pp. 82–111.

Connors, Genevieve (2005) 'When utilities muddle through: pro-poor governance in Bangalore's public water sector', *Environment and Urbanization* 17:1, 201–218.

Cooke, Bill and Uma Kothari (eds) (2001) *Participation: The New Tyranny?* London: Zed Books.

Copestake, James (2002) 'Inequality and the polarising impact of micro-credit: evidence from Zambia's Copperbelt', *Journal of International Development* 14, 743–755.

Cross, J.C. (1998) 'Co-optation, competition, and resistance: state and street vendors in Mexico City', *Latin American Perspectives* 25:2, 41–61.

Crossa, Veronica (2009) 'Resisting the entrepreneurial city: street vendors' struggle in Mexico City's historic center', *International Journal of Urban and Regional Research* 33:1, 43–63.

Crossley, Nick (2002) *Making Sense of Social Movements*, Buckingham: Open University Press.

Cruz, Luis Fernando (1994) 'Fundación Carvajal; the Carvajal Foundation', *Environment and Urbanization* 6:2, 175–182.

Dagdeviren, Hulya (2008) 'Waiting for miracles: the commercialization of urban water services in Zambia', *Development and Change* 39:1, 101–121.

Dagnino, Evelina (2008) 'Challenges to participation, citizenship and democracy: perverse confluence and displacement of meanings', in Anthony J. Bebbington, Sam Hickey and Diana Mitlin (eds), *Can NGOs Make a Difference? The Challege of Development Alternatives*, London and New York: Zed Books, pp. 55–70.

Datta, Kavita (1999) 'A gendered perspective on formal and informal housing finance in Botswana', in G. Jones and K. Datta (eds), *Housing and Finance in Developing Countries*, London: Routledge, pp. 192–212.

Dávila, Julio D. (1990) 'Mexico's urban popular movements: a conversation with Pedro Moctezuma', *Environment and Urbanization* 2:1, 35–50.

——(2009) 'Being a mayor: the view from four Colombian cities', *Environment and Urbanization* 21:1, 37–57.

D'Cruz, C. and Mitlin, D. (2007) 'Shack/Slum Dwellers International: one experience of the contribution of membership organizations to pro-poor urban development', in M. Chen, R. Jhabvala, R. Kanbur and C. Richards (eds), *Membership Based Organizations of the Poor*, Abingdon: Routledge, pp. 221–239.

de la Brière, B. and L.B. Rawlings (2006) *Examining Conditional Cash Transfer Programs: A Role for Increased Social Inclusion?*, Social Protection Discussion Paper No. 0603, Washington, DC: World Bank.

De Soto, Hernando (2000) *The Mystery of Capital: Why Capitalism Triumphs in the West and Fails Everywhere Else*, New York: Basic Books.

Department of Housing, South Africa (2003) *A Social Housing Policy for South Africa: Towards an Enabling Environment for Social Housing (revised draft)*, Pretoria: Department of Housing (South Africa).

Department for International Development (DFID) (2005) *Social Transfers and Chronic Poverty: Emerging Evidence and the Challenge Ahead.* A DFID practice paper, London: DFID.

Desai, Vandana and Mick Howes (1995) 'Accountability and participation: a case study from Bombay', in Michael Edwards and David Hulme (eds), *Non-governmental Organizations: Performance and Accountability*, London: Save the Children and Earthscan Publications, pp. 83–94.

Devas, Nick (2004) *Urban governance, voice and poverty in the developing world*, London and Sterling VA: Earthscan Publications Ltd.

Devereux, Stephen (2002), 'Can social safety nets reduce chronic poverty?', *Development Policy Review* 20:5, 657–675.

Devereux, S., J. Marshall, J. MacAskill and L. Pelham (2005) *Making Cash Count: Lessons from Cash Transfer Schemes in East and Southern Africa for Supporting the Most Vulnerable Children and Households*, Brighton: Save the Children, HelpAge International, Institute of Development Studies University of Sussex.

Devine, Joseph (2007) 'Doing things different? The everyday politics of membership-based organizations', in Martha Chen, Renana Jhabvala, Ravi Kanbur and Carol Richards (eds), *Membership-based Organizations of the Poor*, New York: Routledge, pp. 297–312.

Díaz, A.C., E. Grant, P.I. del Cid Vargas and V.S. Velásquez (2001) 'The role of external agencies in the development of El Mezquital in Guatemala City', *Environment and Urbanization* 13:1, 91–100.

Dreyfus, H.L. and P. Rabinow (1982) *Michel Foucault: Beyond Structuralism and Hermeneutics*, Hemel Hempstead: Harvester Wheatsheaf.

Drinkwater, Michael (2009) ' "We are also human": identity and power in gender relations', in Sam Hickey and Diana Mitlin (eds), *Rights-based Approaches to Development: Exploring the Potential and Pitfalls*, Sterling, VA: Kumarian Press, pp. 145–162.

Duvendack, Maren, Richard Palmer-Jones, James G Copestake, Lee Hooper, Yoon Loke and Nitya Rao (2011) *What is the Evidence of the Impact of Microfinance on the Well-being of Poor People?* Systematic review, London: EPPI-Centre, Social Science Research Unit, Institute of Education, University of London.

Economist, The (2012) 'Boomtown slum; a day in the economic life of Africa's biggest shantytown', London, p. 73.

Edwards, Michael (2001) 'Introduction', in Michael Edwards and John Gaventa (eds), *Global Citizen Action*, Boulder, CO: Lynne Rienner, pp. 1–16.

Edwards, Michael and John Gaventa (eds) (2001) *Global Citizen Action*, Boulder, CO: Lynne Rienner.

Edwards, Michael and David Hulme (eds) (1992) *Making a Difference: NGOs and Development in a Changing World*, London: Earthscan Publications.

Environment and Urbanization (2012) Special issue on Mapping, Enumerating and Surveying Informal Settlements and Cities, 24:1.

Escobar, Arturo (1992) 'Planning', in Wolfgang Sachs (ed.), *The Development Dictionary: A Guide to Knowledge as Power*, London: Zed Books, pp. 132–145.

Escobar Latapí, Agustín and Mercedes Gonzáles De la Roche (2008) 'Girls, mothers and poverty reduction in Mexico: evaluating Progresa-Oportunidades', in S Razavi (ed.), *The Gendered Impacts of Liberalization: Towards Embedded Liberalism?*, New York: Routledge/ UNRISD, pp. 435–468.

Estache, Antonio and Eugene Kouassi (2002) *Sector Organization, Governance and the Inefficiency of African Water Utilities*, World Bank Policy Research Working Paper No. 2890, Washington, DC: World Bank.

Etemadi, Felisa U. (2000) 'Civil society participation in city governance in Cebu City', *Environment and Urbanization* 12:1, 57–72.

——(2001) *Case Studies of Three Urban Poor Communities and Five Marginalised Groups in Cebu City*, Urban Governance, Poverty And Partnerships Working Paper No. 25, Birmingham: University of Birmingham.

Evans, Peter (1996) 'Government action, social capital and development: reviewing the action on synergy', *World Development* 24:6, 1119–1132.

Eyben, Rosalind and Clare Ferguson (2004) 'How can donors become more accountable to poor people?', in Leslie Groves and Rachel Hinton (eds), *Inclusive Aid: Changing Power And Relationships In International Development*, London: Earthscan Publications, pp. 57–75.

Fahmi, Wael Salah (2005) 'The impact of privatization of solid waste management on the Zabaleen garbage collectors of Cairo', *Environment and Urbanization* 17:2, 155–170.

Farouk, Braimah R. and Mensah Owusu (2012) '"If in doubt, count": the role of community-driven enumerations in blocking eviction in Old Fadama, Accra', *Environment and Urbanization* 24:1, 47–58.

Ferguson, Bruce (1999) 'Micro-finance of housing: a key to housing the low or moderate-income majority', *Environment and Urbanization* 11:1, 185–200.

——(2003) 'Housing microfinance – a key to improving habitat and the sustainability of microfinance institutions', *Small Enterprise Development* 14:1, 21–31.

Fergutz, Oscar with Sonia Dias and Diana Mitlin (2011) 'Developing urban waste management in Brazil with waste picker organizations', *Environment and Urbanization* 23:2, 597–607.

Fernandes, Edesio (2007) 'Implementing the urban reform agenda in Brazil', *Environment and Urbanization* 19:1, 177–189.

Fernandes, Leela (2004) 'The politics of forgetting: class politics, state power and the restructuring of urban space in India', *Urban Studies* 41:12, 2415–2430.

Figueiredo, A., H. Torres and R. Bichir (2006) 'The Brazilian social conjuncture revisited [A Conjuntura Social Brasileira Revisitada]', *Novos Estudos* 75: July, 173–183.

Fiori, Jorge, Liz Riley and Ronaldo Ramirez (2000) *Urban Poverty Alleviation through Environmental Upgrading in Rio de Janeiro: Favela Bairro*, London: Development Planning Unit, University College.

Foweraker, Joe (1995) *Theorising Latin America, Critical Studies on Latin America*, London, and Boulder, CO: Pluto Press.

Fox, Jonathan (1994) 'The difficult transition from clientelism to citizenship: lessons from Mexico', *World Politics* 46:2, 151–184.

Freeland, N. (2007) 'Superfluous, pernicious, atrocious and abomnable? The case against conditional cash transfers', *IDS Bulletin* 38:3, 75–78.

Friedman, Milton (1962) *Capitalism and Freedom: A Leading Economist's View of the Proper Role of Competitive Capitalism*, Chicago, IL, and London: The University of Chicago Press.

Fung, A. and E.O. Wright (2003) *Deepening Democracy: Institutional Innovations in Empowered Participatory Governance*, London and New York: Verso.

Furedy, Christine (1992) 'Garbage: exploring non-coventional options in Asian cities', *Environment and Urbanization* 4:2, 42–61.

Gandy, Matthew (2006) 'Planning, anti-planning and the infrastructure crisis facing metropolitan Lagos', *Urban Studies* 43:2, 371–396.

Garrett, J. and A. Ahmed (2004) 'Incorporating crime in household surveys: a research note', *Environment and Urbanization* 16:2, 139–152.

Gaventa, John (2002) 'Introduction: Exploring citizenship, participation and accountability', *IDS Bulletin* 33:2, 1–11.

Gaye, Malick, and Fodé Diallo (1997) 'Community participation in the management of the urban environment in Rufisque (Senegal)', *Environment and Urbanization* 9:1, 9–29.

Gazzoli, Rubén (1996) 'The political and institutional context of popular organizations in urban Argentina', *Environment and Urbanization* 8:1, 159–166.

Gilbert, Alan (2002a) 'Power, ideology and the Washington Consensus: the development and spread of Chilean housing policy', *Housing Studies* 17:2, 305–324.

——(2002b) '"Scan globally; reinvent locally": reflecting on the origins of South Africa's capital housing subsidy policy', *Urban Studies* 39:10, 1911–1933.

——(2004) 'Helping the poor through housing subsidies: lessons from Chile, Colombia and South Africa', *Habitat International* 28, 13–40.

Gilbert, Alan G. and Julio D. Davila (2002) 'Bogota: progress within a hostile environment', in David J. Myers and Henry A. Dietz (eds), *Capital City Politics in Latin America: Democratization and Empowerment*, Boulder, CO, and London: Lynne Reinner, pp. 29–64.

Gledhill, John and Maria Gabriela Hita (2009) *New Actors, New Political Spaces, Same Divided City? Reflections on Poverty and the Politics of Urban Development in Salvador, Bahia*, BWPI Working Paper No. 102, Manchester: Brooks World Poverty Institute, University of Manchester.

Goldstone, Jack (2004) 'More Social Movements or Fewer? Beyond political opportunity structures to relational fields', *Theory and Society* 33:3/4, 333–365.

Gómez-Lobo, A. and D. Contreras (2004) 'Water subsidy policies: a comparison of the Chilean and Colombian schemes', *The World Bank Economic Review* 17:3, 391–407.

Goodland, Robina (1999) *Social Exclusion, Regeneration and Citizen Participation*, Issues Paper Number 1, December 1999, Urban Frontiers Program.

Goodwin, Jeff, James M. Jasper and Jaswinder Khattra (1999) 'Caught in a winding, snarling vine: the structural bias of political process theory', *Sociological Forum* 14:1, 27–54.

Hall, Anthony (2006) 'From Fome Zero to Bolsa Família: social policies and poverty alleviation under Lula', *Journal of Latin American Studies* 38:3, 689–709.

Hall, Nick, Rob Hart and Diana Mitlin (1996) *The Urban Opportunity: The Work of NGOs in Cities of the South*, Rugby: Intermediate Technology Publications.

Hanchett, Suzanne, Shireen Akhter and Mohidul Hoque Khan (2003) 'Water, sanitation and hygiene in Bangladeshi slums: an evaluation of Wateraid's urban programme', *Environment and Urbanization* 15:2, 43–56.

Hardoy, Jorge E. and David Satterthwaite (1981) *Shelter: Need and Response; Housing, Land and Settlement Policies in Seventeen Third World Nations*, Chichester: John Wiley.

——(1989) *Squatter Citizen: Life in the Urban Third World*, London: Earthscan Publications.

Hardoy, Jorge E., David Satterthwaite and Sandy Cairncross (eds) (1990) *The Poor Die Young*, London: Earthscan Publications.

Hardoy, Jorge E., Diana Mitlin and David Satterthwaite (2001) *Environmental Problems in an Urbanizing World*, London: Earthscan Publications.

Harriss, John (2006) 'Middle-class activism and the politics of the informal working class', *Critical Asian Studies* 38:4, 445–465.

Harvey, David (2012) *Rebel Cities: From the Right to the City to the Urban Revolution*, London and Brooklyn: Verso.

Hasan, Arif (2006) 'Orangi Pilot Project: the expansion of work beyond Orangi and the mapping of informal settlements and infrastructure', *Environment and Urbanization* 18:2, 451–480.

——(2007) 'The Urban Resource Centre, Karachi', *Environment and Urbanization* 19:1, 275–292.

——(2008) 'Financing the sanitation programme of the Orangi Pilot Project – Research and Training Institute in Pakistan', *Environment and Urbanization* 20:1, 109–119.

Hasan, Arif and Mansoor Raza (2011) 'The evolution of the microcredit programme of the OPP's Orangi Charitable Trust. Karachi', *Environment and Urbanization* 23:2, 517–538.

Heller, Patrick and Peter Evans (2010) 'Taking Tilly south: durable inequalities, democratic contestation, and citizenship in the Southern Metropolis', *Theory and Society* 39, 433–450.

Henry-Lee, A. (2005) 'The nature of poverty in the garrison constituencies in Jamaica', *Environment and Urbanization* 17:2, 83–99.

Heracleous, Maria, Mario González-Flores and Paul Winters (2010) *Conditional Cash Transfers and Schooling Choice: Evidence from Urban Mexico*, paper presented at the 10th Conference on Research on Economic Theory and Econometrics, Milos (Crete), 10–14 July 2011.

Hickey, Sam and Diana Mitlin (2009) 'Introduction', in Sam Hickey and Diana Mitlin (eds), *Rights-based Approaches to Development: Exploring the Potential and Pitfalls*, Sterling, VA: Kumarian, pp. 3–21.

Hickey, Samuel and Giles Mohan (2004a) *Participation: From Tyranny to Transformation?*, London and New York: Zed Books.

——(2004b) 'Towards participation as transformation: critical themes and challenges', in Samuel Hickey and Giles Mohan (eds), *Participation: From Tyranny to Transformation?* London and New York: Zed Books, pp. 3–24.

Hinton, Rachel and Leslie Groves (2004) 'The complexity of inclusive aid', in Rachel Hinton and Leslie Groves (eds), *Inclusive Aid: Changing Power and Relationships in International Development*, London: Earthscan Publications, pp. 3–20.

Hirschman, Albert O. (1970) *Exit, Voice and Loyalty: Responses to Decline in Firms, Organizations, and States*, Cambridge, MA: Harvard University Press.

Hordijk, Michaela (2005) 'Participatory governance in Peru: exercising citizenship', *Environment and Urbanization* 17:1, 219–236.

Hossain, Shahadat (2012) 'The production of space in the negotiation of water and electricity supply in a bosti of Dhaka', *Habitat International* 36, 68–77.

Houtzager, Peter (2005) 'Introduction: from polycentrism to the polity', in Peter Houtzager and Mick Moore (eds), *Changing Paths: International Development and the New Politics of Inclusion*, Ann Arbor: University of Michigan Press, pp. 1–31.

Houtzager, Peter P., Arnab Acharya and Adrian Gurza Lavalle (2007) *Associations and the Exercise of Citizenship in New Democracies: Evidence from São Paulo and Mexico City*, IDS Working Paper No. 285, Brighton: Institute for Development Studies.

Huchzermeyer, Marie (2003) 'A legacy of control? The capital subsidy for housing, and informal settlement intervention in South Africa', *International Journal of Urban and Regional Research* 27:3, 591–612.

Illich, Ivan, Kenneth Zola, John McKnight, Jonathan Caplan and Harley Shaiken (1977) *Disabling Professions*, London: Marion Boyars.

Inter-American Development Bank (IDB) (2006) *Sharpening the Bank's Capacity to Support the Housing Sector in Latin America and the Caribbean*, Background Paper for the Implementation of the Social Development Strategy, Washington, DC: Inter-American Development Bank.

Institute for Development Studies (IDS) (2003) *The Rise of Rights: Rights-based Approaches to International Development*, IDS Policy Briefing Issue 17, Brighton: IDS University of Sussex.

IFRC (2010) *World Disasters Report 2010: Focus on Urban Risk*, Geneva: IFRC (International Federation of Red Cross and Red Crescent Societies).

Johannsen, Julia, Luis Tejerina and Amanda Glassman (2009) *Conditional Cash Transfers in Latin America: Problems and Opportunities*, Inter-American Development Bank Social Protection and Health Division, Washington, DC: Inter-American Development Bank.

Johnson, Susan (2009) 'Microfinance is dead! Long live microfinance! Critical reflections on two decades of microfinance policy and practice', *Enterprise Development and Microfinance* 20:4, 291–303.

Jones, Linda and Alexandra Miehlbradt (2009) 'A 20/20 retrospective on enterprise development: in search of impact, scale and sustainability', *Enterprise Development and Microfinance* 20:4, 304–322.

Joshi, Anuradha (2008) 'Producing social accountability? The impact of service delivery reforms', *IDS Bulletin* 38:6, 10–17.

Joshi, Anuradha and Mick Moore (2004) 'Institutional co-production: unorthodox public service delivery in challenging environments', *Journal of Developing Studies* 40:4, 31–49.

Joshi, Sharadbala and M. Sohail Khan (2010) 'Aided self-help. The Million Houses Programme – revising the issues', *Habitat International* 34, 306–314.

Kabeer, Naila (2002) *The Power to Choose: Bangladeshi Garment Workers in London and Dhaka*, London and New York: Verso.

Kantor, Paula (2009) 'Women's exclusion and infavourable inclusion in informal employment in Lucknow, India: Barriers to voice and livelihood security', *World Development* 37:1, 194–207.

Kantor, Paula and Padmaja Nair (2003) 'Risks and responses among the urban poor in India', *Journal of International Development* 15:8, 957–967.

Karanja, Irene (2010) 'An enumeration and mapping of informal settlements in Kisumu, Kenya, implemented by their inhabitants', *Environment and Urbanization* 22:1, 217–240.

Keivani, Ramin and Michael Mattingly (2007) 'The interface of globalization and peripheral land in the cities of the South: implications for urban governance and local economic development', *International Journal of Urban and Regional Research* 31:2, 459–474.

Khan, Firoz and Edgar Pieterse (2006) 'The Homeless People's Alliance: purposive creation and ambiguated realities', in Richard Ballard, Adam Habib and Imraan Valodia (eds), *Voices of Protest: Social Movements in Post-Apartheid South Africa*, Durban: University of KwaZulu-Natal Press, pp. 155–178.

Khan, Mushtaq H. (2005) 'Markets, states and democracy: patron–client networks and the case for democracy in developing countries', *Democratization* 12:5, 704–725.

Kinyanjui, Michael and Janet Ngombalu (2006) 'Integrated development, squatters and tenants, in low-income settlements in Kenya', in Lucy Stevens, Stuart Coupe and Diana Mitlin (eds), *Confronting the Crisis in Urban Poverty: Making Integrated Approaches Work*, Rugby: Intermediate Technology Publications, pp. 117–142.

Kitschelt, Herbert and Steven Wilkinson (2007) 'Citizen–politician linkages: an introduction', in Herbert Kitschelt and Steven Wilkinson (eds), *Patrons, Clients and Policies*, Cambridge: Cambridge University Press, pp. 1–49.

Klink, Jeroen and Rosana Denaldi (2012) 'Metropolitan fragmentation and neolocalism in the periphery: revisiting the case of curitiba', *Urban Studies* 49:3, 543–561.

Klopp, Jacqueline M. (2008) 'Remembering the destruction of Muoroto: slum demolitions, land and demcratization in Kenya', *African Studies* 67:3, 295–314.

Köhler, Gabriele, Marta Cali and Mariana Stirbu (2009) *Social Protection in South Asia: A Review*, Kathmandu: United Nations Children's Fund (UNICEF), Regional Office for South Asia (ROSA).

Korten, David (1990) *Getting to the 21st Century: Voluntary Action and the Global Agenda*, West Hartford: Kumarian Press.

Lall Somik V., Ajay Suri and Uwe Deichmann (2006) 'Housing savings and residential mobility in informal settlements in Bhopal, India', *Urban Studies* 43:7, 1025–1039.

Lama-Rewal, Stéphanie Tawa (2011) 'Urban governance and health care provision in Delhi', *Environment and Urbanization* 23:2, 563–581.

Lantz, Maria and Jonatan Habib Engqvist (eds) (2008) *Dharavi: Documenting Informalities*, Stockholm: The Royal University College of Fine Arts, Art and Architecture.

Laurie, Nina, Robert Andolina and Sarah Radcliffe (2005) 'Ethno-development: social movements, creating experts and professionalising indigenous knowledge in Ecuador', *Antipode* 37:3, 470–496.

Lavalle, Adrian Gurza, Arnab Acharya and Peter P. Houtzager (2005) 'Beyond comparative anecdotalism: lessons on civil society and participation from São Paulo, Brazil', *World Development* 33:6, 951–964.

Leckie, Scott (1989) 'Housing as a human right', *Environment and Urbanization* 1:2, 90–108.

——(1992) *From Housing Needs to Housing Rights: An Analysis of the Right to Adequate Housing Under International Human Rights Law*, Human Settlements Programme, London: IIED.

Leftwich, Adrian (2005) 'Politics in command: development studies and the rediscovery of social science', *New Political Economy* 10:4, 573–606.

——(2008) *Developmental States, Effective States and Poverty reduction: The Primacy of Politics'*, UNRISD Project on Poverty Reduction and Policy Regimes, Geneva: United Nations Research Institute for Social Development (UNRISD).

Lemanski, Charlotte (2009) 'Augmented informality: South Africa's backyard dwellings as a by-product of formal housing policies', *Habitat International* 33, 472–484.

——(2011) 'Moving up the ladder or stuck on the bottom rung? Homeownership as a solution to poverty in urban South Africa', *International Journal of Urban and Regional Research* 35:1, 57–77.

Levy, Sanitago (2006) *Progress against Poverty: Sustaining Mexico's Progresa-Oportunidades Program*, Washington, DC: The Brookings Institution Press.

Levy, Sanitago and E. Rodríguez (2004) 'Education, Health and Nutrition Programme, PROGRESA – Human Development Programme Oportunidades' [El Programa de Educación, Salud y Alimentación, PROGRESA – Programa de Desarrollo Humano Oportunidades']', in Sanitago Levy (ed.), *Essays on the Economic and Social Development of Mexico [Ensayos Sobre Desarrollo Económico y Social de México]*, Mexico City: Fondo de Cultura Económica, pp. 181–379.

Lewis, David (2007) *Management of Non-governmental Development Organizations*, Abingdon: Routledge.

——(2008) 'Crossing the boundaries between "Third Sector" and state: life-work histories from the Philippines, Bangladesh and the UK', *Third World Quarterly* 29:1, 125–141.

Lindell, Ilda (ed.) (2010) *Africa's Informal Workers: Collective Agency, Alliances and Transnational Organizing in Urban Africa*, Sweden, London and New York: The Nordic Africa Institute and Zed Books.

Lindert, Kathy, Anja Linder, Jason Hobbs and Bénédicte de la Brière (2007) *The Nuts And Bolts Of Brazil's Bolsa Família Program: Implementing Conditional Cash Transfers in a Decentralized Context*, Social Protection Discussion Paper No. 0709, Washington, DC: World Bank.

Llanto, Gilberto M. (2007) 'Shelter finance strategies for the poor: Philippines', *Environment and Urbanization* 19:2, 409–425.

López Follegatti, Jose Luis (1999) 'Ilo: a city in transformation', *Environment and Urbanization* 11:2, 181–202.

Mageli, Eldrid (2004) 'Housing mobilization in Calcutta – empowerment for the masses or awareness for the few?', *Environment and Urbanization* 16:1, 129–138.

Maharaj, Brij (2010) 'The struggle for Warwick Market in Durban', *EchoGéo* 13: July/ August, 2–3

Malhotra, Mohini (2003) 'Financing her home, one wall at a time', *Environment and Urbanization* 15:2, 217–228.

Manda, Mtafu A. Zeleza (2007) 'Mchenga – urban poor housing fund in Malawi', *Environment and Urbanization* 19:2, 337–360.

Mandar, H. (2005) 'Right as struggle – Towards a more just and humane world', in P. Gready and J. Ensor (eds), *Reinventing Development? Translating Rights-based Approaches from Theory into Practice*, London and New York: Zed Books, pp. 233–253.

Mangin, William (1967) 'Latin American squatter settlements; a problem and a solution', *Latin American Research Review* 2:3 (summer), 65–98.

Marris, Peter (1979) 'The meaning of slums and patterns of change', *International Journal of Urban and Regional Research* 3:3, 419–441.

Mathey, K. (1997) 'Self-help approaches to the provision of housing: the long debate and the few lessons', in J. Gugler (ed.), *Cities in the Developing World. Issues, Theory and Policy*, Oxford: Oxford University Press, pp. 280–290.

Mayo, Marjorie (2005) *Global Citizens: Social Movements and the Challenge of Globalization*, London and New York: Zed Books.

Mayo, Stephen (1999) *Subsidies in Housing*, Sustainable Development Department Technical Papers Series, Washington, DC: World Bank.

McAdam, Doug, Sidney Tarrow and Charles Tilly (2001) *Dynamics of Contention*, Cambridge: University of Cambridge Press.

McFarlane, Colin (2008) 'Governing the contaminated city: infrastructure and sanitation in colonial and post-colonial Bombay', *International Journal of Urban and Regional Research* 32:2, 415–435.

McKinsey Global Institute (2011) 'Urban world: mapping the economic power of cities', McKinsey and Company.

Mehta, Lyla and Birgit La Cour Madsen (2003) *Is the WTO after your Water? The General Agreement on Trade and Services (GATS) and the Basic Right to Water*, Brighton: Institute for Development Studies.

Mehta, Meera (2008) *Assessing Microfinance for Water and Sanitation*, A study for the Bill & Melinda Gates Foundation, Ahmedabad.

Melo, Marcus with Flávio Rezende and Cátia Lubambo (2001) *Urban Governance, Accountability and Poverty: The Politics of Participatory Budgeting in Recife, Brazil*, Urban Governance, Partnerships and Poverty Research Working Papers No. 27, Birmingham: University of Birmingham.

Menegat, Rualdo (2002) 'Participatory democracy and sustainable development: integrated urban environmental management in Porto Alegre, Brazil', *Environment and Urbanization* 14:2, 181–206.

Milbert, Isabelle (1992) *Cooperation and Urban Development; Urban Policies of Bilateral and Multilateral Cooperation Agencies*, Geneva: IUED/SDC.

Milbert, Isabelle and Vanessa Peat (1999) *What Future for Urban Cooperation? Assessment of Post Habitat II Strategies*', Berne: Swiss Agency for Development and Cooperation.
Miraftab, Faranak (2003) 'The perils of participatory discourse: housing policy in post-Apartheid South Africa', *Journal of Planning Education and Research* 22, 226–239.
Mitlin, Diana (2008a) 'With and beyond the state: coproduction as a route to political influence, power and transformation for grassroot organizations', *Environment and Urbanization* 20:2, 339–360.
——(2008b) *Urban Poor Funds: Development by the People, for the People*, IIED Poverty Reduction in Urban Areas Series Working Paper No. 18, London: IIED.
——(2011a) 'Cities for citizens in the Global South: approaches of non-governmental organizations working in urban development', in B. Reinalda (ed.), *The Ashgate Research Companion to Non-State Actors*, Farnham: Ashgate, pp. 419–433.
——(2011b) 'Shelter finance in the age of neo-liberalism', *Urban Studies* 48:6, 1217–1234.
——(2013a) 'A Class Act: professional support to people's organizations in towns and cities of the global South', *Environment and Urbanization* 25:1, forthcoming.
——(2013b) 'Innovations in shelter finance', in Elliott D Sclar, Nicole Volavka-Close and Peter Brown (eds), *The Urban Transformation: Health, Shelter and Climate Change*, Abingdon and New York: Routledge, pp. 126–147.
Mitlin, Diana and Anthony Bebbington (2006) 'Social movements and chronic poverty across the urban-rural divide: concepts and experiences', *Chronic Poverty Research Centre Working Paper* No. 65, Manchester: University of Manchester.
Mitlin, Diana and David Satterthwaite (2007) 'Strategies for grassroots conrol of international aid', *Environment and Urbanization* 19:2, 483–500.
——(2013) *Urban Poverty in the Global South: Scale and Nature*, London and New York: Routledge.
Mitlin, Diana, David Satterthwaite and Sheridan Bartlett (2011) *Capital, Capacities and Collaboration: The Multiple Roles of Community Savings in Addressing Urban Poverty*, IIED Poverty Reduction in Urban Areas, Working Paper No. 34, London: International Institute for Environment and Development.
Mohan, Giles and Kristian Stokke (2000) 'Participatory development and empowerment: the dangers of localism', *Third World Quarterly* 21:2, 247–268.
Mohanty, Ranjita, Lisa Thompson and Vera Schattan Coelho (2011) *Mobilising the State? Social Mobilisation and State Interaction in India, Brazil and South Africa*, IDS Working Paper No. 359, Brighton: Institute for Development Studies.
Molyneux, Maxine and Sian Lazar (2003) *Doing the Rights Thing: Rights-based Development and Latin American NGOs*, London: ITDG Publications.
Moore, Mick (2005) 'Arguing the politics of inclusion', in Peter P. Houtzager and Mick Moore (eds), *Changing Paths: International Development and the New Politics of Inclusion*, Ann Arbor: University of Michigan Press, pp. 260–283.
Moore, M., J. Leavy and H. White (2005) 'How governance affects poverty', in P. Houtzager and M. Moore (eds), *Changing Paths: International Development and the New Politics of Inclusion*, Ann Arbor: University of Michigan Press, pp. 167–203.
Moser, Caroline (1993) *Gender Planning And Development: Theory, Practice And Training*, London and New York: Routledge.
——(1997) *Household Responses to Poverty and Vulnerability Volume 1: Confronting Crisis in Cisne Dos, Guayaquil, Ecuador*, Urban Management Programme Policy Paper No. 21, Washington, DC: World Bank.
——(2004) 'Editorial: Urban violence and insecurity: an introductory roadmap', *Environment and Urbanization* 16:2, 3–15.

——(2007) 'Asset accumulation policy and poverty reduction', in Caroline Moser (ed.), *Reducing Global Poverty: The Case for Asset Accumulation*, Washington, DC: The Brookings Institute, pp. 83–103.

——(2009) *Ordinary Families, Extraordinary Lives: Assets And Poverty Reduction In Guayaquil 1978–2004*, Washington, DC: Brookings Institution Press.

Moser, Caroline and Cathy McIlwaine (1997) *Household Responses to Poverty and Vulnerability (Volume 3): Confronting Crisis in Commonwealth, Metro Manila, the Philippines*, Urban Management Programme Policy Paper No. 23, Washington, DC: World Bank.

Moser, Caroline, Andy Norton, Tim Conway, Clare Ferguson and Polly Vizard (2001) *To Claim Our Rights: Livelihood Security, Human Rights And Sustainable Development*, London: Overseas Development Institute.

Mthembi-Mahanyele, S. (2001) Statement by the South African Minister of Housing, Sankie Mthembi-Mahanyele, at the Special Session of the United Nations General Assembly for an Overall Review and Appraisal of the Implementation of the Habitat Agenda, New York, 6–8 June 2001. Available at www.info.gov.za/speeches/2001/010802245p1001.htm (accessed 17 January 2006).

Mukhija, V. (2004) 'The contradictions in enabling private developers of affordable housing: a cautionary case from Ahmedabad, India', *Urban Studies* 41:11, 2231–2244.

Muller, Anna and Diana Mitlin (2004) 'Windhoek, Namibia – towards progressive urban land policies in Southern Africa', *International Development Planning Review* 26:2, 167–186.

——(2007) 'Securing inclusion: strategies for community empowerment and state redistribution', *Environment and Urbanization* 19:2, 425–441.

Muller, Mike (2008) 'Free basic water – a sustainable instrument for a sustainable future in South Africa', *Environment and Urbanization* 20:1, 67–89.

Myers, Garth (2003) *Verandahs of Power*, New York: Syracuse University Press.

——(2011) *African Cities: Alternative Visions of Urban Theory and Practice*, London and New York: Zed Books.

Nance, E. and L. Ortolano (2007) 'Community participation in urban sanitation: experiences in northeastern Brazil', *Journal of Planning Education and Research* 26, 284–300.

Nelson, J. (2005) 'Grounds for alliance?', in Peter P. Houtzager and Mike Moore (eds), *Changing Paths: International Development and the New Politics of Inclusion*, Ann Arbor: University of Michigan Press, pp. 119–138.

Niger, Republique du (2008) *Accelerated Development and Poverty Reduction Strategy 2008–2012*, Niamey: Prime Minister's Office, PRSP Permanent Secretariat.

Nigerian National Planning Commission (2004) *Meeting Everyone's Needs; National Economic Empowerment and Development Strategy*, Abuja: Nigerian National Planning Commission.

Nilsson, David (2006) 'A heritage of un-sustainability? Reviewing the origin of the large-scale water and sanitation system in Kampala, Uganda', *Environment and Urbanization* 18:2, 369–386.

Niño-Zarazúa, Miguel (2010) *Mexico's Progresa-Oportunidades and the Emergence of Social Assistance in Latin America*, BWPI Working Paper No. 142, Manchester: Brooks World Poverty Institute, University of Manchester.

Niño-Zarazúa, Miguel, Armando Barrientos, David Hulme and Sam Hickey (2010) *Social Protection in Sub-Saharan Africa: Will the Green Shoots Blossom?*, BWPI Working Paper No. 116, Manchester: Brooks World Poverty Institute, University of Manchester.

OECD (2003) *Improving Water Management: Recent OECD Experience*, Paris: Organization for Economic Cooperation and Development.

Oldfield, Sophie (2004) 'Urban networks, community organising and race: an analysis of racial integration in a desegregated South African neighbourhood', *Geoforum* 35:2, 189–201.

——(2008) 'Participatory mechanisms and community politics: building consensus and conflict', in Mirjam van Donk, Mark Swilling, Edgar Pieterse and Susan Parnell (eds), *Consolidating Developmental Local Government: Lessons from the South African Experience*, Cape Town: UCT Press, pp. 487–500.

Orangi Pilot Project – Research and Training Institute OPP-RTI (1995) 'NGO Profile: Orangi Pilot Project', *Environment and Urbanization* 7:2, 227–236.

——(1998) *Proposal for a Sewage Disposal System for Karachi*, Karachi: City Press.

——(2002) *Katchi Abadis of Karachi: Documentation of Sewerage, Water Supply Lines, Clinics, Schools and Thallas – Volume One: The First Hundred Katchi Abadis Surveyed*, Karachi: OPP-RTI.

Ostrom, Elinor (1996) 'Crossing the great divide: coproduction, synergy and development', *World Development* 24:6, 1073–1087.

Papeleras, Ruby, Ofelia Bagotlo and Somsook Boonyabancha (2012) 'A conversation about change-making by communities: some experiences from ACCA', *Environment and Urbanization* 24:2, 463–480.

Parks, Roger B., Paula C. Baker, Larry Kiser, Ronald Oakerson, Elinor Ostrom, Vincent Ostrom, Stephen L. Percy, Martha B. Vandivot, Gordon Whitaker and Rick Wilson (1981) 'Consumers as co-producers of public services: some economic and institutional considerations', *Policy Studies Journal* 9:7, 1001–1011.

Patel, Sheela (2004) 'Tools and methods for empowerment developed by slum dwellers federations in India', *Participatory Learning and Action 50*, London: International Institute for Environment and Development.

——2013 'Upgrade, rehouse or resettle? An assessment of the Indian Governments' Basic Services for the Urban Poor programme', *Environment and Urbanization* 25:1.

Patel, Sheela and Jockin Arputham (2007) 'An offer of partnership or a promise of conflict in Dharavi, Mumbai?', *Environment and Urbanization* 19:2, 501–608.

Patel, Sheela and Jockin Arputham (2008) 'Plans for Dharavi: negotiating a reconciliation between a state-driven market redevelopment and residents' aspirations', *Environment and Urbanization* 2:1, 243–253.

Patel, Sheela and Celine d'Cruz (1993) 'The *Mahila Milan* crisis credit scheme: from a seed to a tree', *Environment and Urbanization* 5:1, 9–17.

Patel, Sheela and Diana Mitlin (2002) 'Sharing experiences and changing lives', *Community Development* 37:2, 125–137.

——(2004) 'Grassroots-driven development: the alliance of SPARC, the National Slum Dwellers Federation and *Mahila Milan*', in Diana Mitlin and David Satterthwaite (eds), *Empowering Squatter Citizens*, London: Earthscan Publications, pp. 216–244.

——(2009) 'Reinterpreting the rights based approach: a grassroots perspective on rights and development', in Sam Hickey and Diana Mitlin (eds), *Rights-based Approaches to Development: Exploring the Potential and Pitfalls*, Sterling, VA: Kumarian Press, pp. 107–125.

——(2010) 'Gender issues and slum/shack dweller federations', in Sylvia Chant (ed.), *The International Handbook on Gender and Poverty*, Cheltenham: Edward Edgar, pp. 379–384.

Patel, Sheela and Kalpana Sharma (1998) 'One David and three Goliaths: avoiding anti-poor solutions to Mumbai's transport problems', *Environment and Urbanization* 10:2, 149–160.

Patel, Sheela, Celine d'Cruz and Sundar Burra (2002) 'Beyond evictions in a global city: people-managed resettlement in Mumbai', *Environment and Urbanization* 14:1, 159–172.

Payne, Geoffrey, Alain Durand-Lasserve and Carole Rakoki (2009) 'The limits of land titling and home ownership', *Environment and Urbanization* 21:2, 443–462.

Pearson, Ruth (2004) 'Organizing home-based workers in the global economy: an action-research approach', *Development in Practice* 14:1 and 2, 136–148.

Peattie, Lisa (1990) 'Participation: a case study of how invaders organize, negotiate and interact with government in Lima, Peru', *Environment and Urbanization* 2:1, 19–30.

Perlman, Janice (2010) *Favela: Four Decades of Living on the Edge in Rio de Janeiro*, New York: Oxford University Press.

Perreault, Thomas (2006) 'From the Guerra Del Agua to the Guerra Del Gas: resource governance, neoliberalism and popular protest in Bolivia', *Antipode* 38, 150–173.

Pervaiz, Arif and Perween Rahman with Arif Hasan (2008) *Lessons from Karachi: The Role of Demonstration, Documentation, Mapping and Relationship Building in Advocacy for Improved Urban Sanitation and Water Services*, Human Settlements Discussion Paper Series – Theme: Water No.6, London: International Institute for Environment and Development.

Phonphakdee, Somsak, Sol Visal and Gabriela Sauter (2009) 'The urban poor development fund in Cambodia: supporting local and citywide development', *Environment and Urbanization* 21:2, 569–586.

Pieterse, Edgar (2006) 'Building with ruins and dreams: some thoughts on realising integrated urban development in South Africa through crisis', *Urban Studies* 43:2, 285–304.

Piper, Laurence and Cherrel Africa (2012) 'Unpacking race, party and class from below: surveying citizenship in the Msunduzi municipality', *Geoforum* 43:2, 219–229.

Piper, Laurence, and Bettina von Lieres (2011) *Expert Advocacy for the Marginalised: How and Why Democratic Mediation Matters to Deepening Democracy in the Global South*, Development Research Centre: Citizenship, Participation and Accountability Working Paper No. 364, Brighton: Institute for Development Studies.

Porio, Emma with Christine S. Crisol, Nota F. Magno, David Cid and Evelyn N. Paul (2004) 'The Community Mortgage Programme: an innovative social housing programme in the Philippines and its outcomes', in Diana Mitlin and David Satterthwaite (eds), *Empowering Squatter Citizens*, London: Earthscan Publications, pp. 54–81.

Pornchokchai, S (1992), *Bangkok Slums: Review and Recommendations*', Bangkok: School of Urban Community Research and Action, Agency for Real Estate Affairs.

Portes, Alejandro (1979), 'Housing policy, urban poverty and the state: the favelas of Rio de Janeiro', *Latin American Research Review* 14: summer, 3–24.

Posner, Paul (2012) 'Targeted assistance and social capital: housing policy in Chile's neoliberal democracy', *International Journal of Urban and Regional Research* 36:1, 49–70.

Rabinow, Paul (ed.) (1991) *The Foucault Reader: An Introduction to Foucault's Thought*, London: Penguin. Original edition, Pantheon Books.

Racelis, Mary (2003) *Begging, Requesting, Demanding, Negotiating: Moving Towards Urban Poor Partnerships in Governance*, paper presented at the World Bank Urban Research Symposium, Washington, DC, 15–17 December.

——(2008) 'Anxieties and affirmations: NGO–donor partnerships for social transformation', in Anthony J. Bebbington, Sam Hickey and Diana Mitlin (eds), *Can NGOs Make a Difference? The Challenge of Development Alternatives*, London and New York: Zed Books, 196–219.

Rahman, Perween (2004) *Katchi Abadis of Karachi; A Survey of 334 Katchi Abadis*, Karachi: Orangi Pilot Project – Research and Training Institute.

——(2008) *Water Supply in Karachi; Situation/Issues, Priority Issues and Solutions*, Karachi: Orangi Pilot Project – Research and Training Institute.

Rahman, Perween and Anwar Rashid (1992) *Low Cost Sanitation Programme Maintenance/Rectification*, Karachi: Orangi Pilot Project – Research Training Institute.

Rakodi, Carole (2006a) 'Social agency and state authority in land delivery processes in African cities', *International Development Planning Review* 28:2, 263–285.

——(2006b) 'State society relations in land delivery in five African Cities', *International Development Planning Review* 28:2, 127–136.

Rakodi, Carole with Tony Lloyd-Jones (eds) (2002) *Urban Livelihoods: A People-centred Approach to Reducing Poverty*, London: Earthscan Publications.

Ravallion, Martin, Shaohua Chen and Prem Sangraula (2007) *New Evidence on the Urbanization of Global Poverty*, WPS4199, Washington, DC: World Bank.

Riofrio, Gustavo (1996) 'Lima: mega-city and mega-problem', in Alan Gilbert (ed.), *The Mega City in Latin America,* Tokyo: United Nations University Press, pp. 155–172.

Roberts, John Michael and Nick Crossley (2004) 'Introduction', in Nick Crossley and John Michael Roberts (eds), *After Habermas: New Perspectives On The Public Sphere*, Oxford: Blackwell, pp. 1–28.

Robins, Steven (2008) *From Revolution to Rights in South Africa*, Pietermaritzburg: UKZN Press.

Robins, Steven, Andrea Cornwall and Bettina Von Lieres (2008) 'Rethinking "citizenship" in the Postcolony', *Third World Quarterly* 29:6, 1069–1086.

Rodgers, Dennis (2009) 'Slum wars of the 21st century: gangs, mano dura and the new urban geography of conflict in Central America', *Development and Change* 40:5, 949–976.

——(2010) 'Contingent democratisation ? The rise and fall of participatory budgeting in Buenos Aires', *Journal of Latin American Studies* 42, 1–27.

Rodríguez, Alfredo and Ana Sugranyes (2007) 'The new housing problem in Latin America', in Fernando M. Carrion and Lisa M. Hanley (eds), *Urban Regeneration and Revitalization in the Americas: Towards a Stable State*, Washington, DC: Woodrow Wilson Center for International Scholars, pp. 51–66.

Rolnik, Raquel (2011) 'Democracy on the edge: limits and possibilities in the implementation of an urban reform agenda in Brazil', *International Journal of Urban and Regional Research* 35:2, 239–255.

Rowlands, Jo (1997) *Questioning Empowerment: Working with Women in Honduras*, Oxford: Oxfam.

Roy, Ananya (2004) 'The gentlemen's city: urban informality in the Calcutta of new communism', in Ananya Roy and Nezar Alsayyad (eds), *Urban Informality: Transnational Perspectives from the Middle East, Latin America and South Asia*, Oxford: Lexington Books, pp. 147–170.

Roy, A. N., Jockin Arputham and Ahmad Javed (2004) 'Community police stations in Mumbai's slums', *Environment and Urbanization* 16:2, 135–138.

Russell, Steven and Elizabeth Vidler (2000) 'The rise and fall of government–community partnerships for urban development: grassroots testimony from Colombo', *Environment and Urbanization* 12:1, 73–86.

Sabry, Sarah (2008) *Poverty Lines in Greater Cairo: Underestimating and Misrepresenting Urban Poverty*, Poverty Reduction in Urban Areas Series Working Paper No. 21, London: International Institute for Environment and Development.

Saha, Satya Ranjan and Habibur Rahman (2006) 'Understanding urban livelihoods in a secondary town in Bangladesh', in Lucy Stevens, Stuart Coupe and Diana Mitlin (eds), *Confronting the Crisis in Urban Poverty: Making Integrated Approaches Work*, Rugby: Intermediate Technology Publications, pp. 209–224.

SAIRR (2008) 'Press release From bare fields to the backyards of properties: the shifting pattern of informal dwelling erection', dated 24 November 2008, Johannesburg: SAIRR (South African Institute of Race Relations).

Satterthwaite, David (1997) *The Scale and Nature of International Donor Assistance to Housing, Basic Services and Other Human Settlements Related Projects*, Helsinki: WIDER.

——(2001) 'Reducing urban poverty: constraints on the effectiveness of aid agencies and development banks and some suggestions for change', *Environment and Urbanization* 13:1, 137–157.

——(2002) 'Lessons from the experience of some urban poverty-reduction programmes', in Carole Rakodi with Tony Lloyd-Jones (eds), *Urban Livelihoods: A People-centred Approach to Reducing Poverty*, London: Earthscan Publications, pp. 257–271.

Satterthwaite, David and Cecilia Tacoli (2002) 'Seeking an understanding of poverty that recognizes rural–urban differences and rural–urban linkages', in Carole Rakodi with Tony Lloyd-Jones (eds), *Urban Livelihoods: A People-centred Approach to Reducing Poverty*, London: Earthscan Publications, pp. 52–70.

Scheper-Hughes, Nancy (1992), *Death Without Weeping: the Violence of Everyday Life in Brazil*, Berkeley: University of California Press.

SDI (Shack/Slum Dwellers International) (2006) *The Rituals and Practices of Slum/Shack Dwellers International*' Cape Town: Shack/Slum Dwellers International.

——(2007) *Voices from the Slums*, Cape Town: Shack/Slum Dwellers International.

Shah, Anwar (ed.) (2006) *Local Governance in Developing Countries*, Public Sector Governance and Accountability Series, Washington, DC: World Bank.

Shatkin, Gavin (2007) *Collective Action and Urban Poverty Alleviation: Community Organizations and the Struggle for Shelter in Manila*, Aldershot: Ashgate Publishing.

Shea, Michael (2008) *Multilateral and Bilateral Funding of Housing and Slum Upgrading Development in Developing Countries*, Washington, DC: International Housing Coalition.

Shenya, S. A. (2007) 'Reconceptualizing housing finance in informal settlements: the case of Dar es Salaam, Tanzania', *Environment and Urbanization* 19:2, 441–456.

Simelane, Hloniphile Yvonne (2012) *The Interplay of Urban Land Tenurial Systems and its Effects on the Poor: A Case Study of Manzini in Swaziland*, Ph.D. thesis, Brighton: Institute of Development Studies.

Simone, AbdouMaliq (2004) *For the City Yet to Come: Changing African Life in Four Cities*, Durham, NC, and London: Duke University Press.

——(2006) 'Pirate towns: reworking social and symbolic infrastructures in Johannesburg and Douala', *Urban Studies* 43: 2, 357–370.

——(2011) 'The ineligible majority: urbanizing the postcolony in Africa and Southeast Asia', *Geoforum* 42, 266–270.

Simone, AbdouMaliq and Vijayendra Rao (2012) 'Securing the majority: living through uncertainty in Jakarta', *International Journal of Urban and Regional Research* 36:2, 315–335.

Sinha, Pravin (2004) 'Representing labour in India', *Development in Practice* 14:1, 127–135.

Sinwell, Luke (2012) 'Transformative left-wing parties' and grassroots organizations: unpacking the politics of "top-down" and "bottom-up" development', *Geoforum* 43:2, 190–198.

Sisulu, Lindewe (2006) 'Partnership between government and shack/slum dwellers', *Environment and Urbanization* 18:2, 401–405.

Smith, Adam ([1776] 1982) *The Wealth of Nations*, London: Penguin.

Solo, Tova Maria (2008) 'Financial exclusion in Latin America or the social costs of not banking the urban poor', *Environment and Urbanization* 20:1, 47–66.

Souza, Celina (2001) 'Participatory budgeting in Brazilian cities: limits and possibilities in building democratic institutions', *Environment and Urbanization* 13:1, 159–184.

SPARC (1985) *We, the Invisible: A Census of Pavement Dwellers*, Mumbai: SPARC.

SPARC and KRVIA (2010) *Reinterpreting, Re-imagining, Redeveloping Dharavi*, Mumbai: SPARC and KRVIA (Kamla Raheja Vidyanidhi Institute for Architecture and Environmental Studies).

Stein, Alfredo (2001) 'Participation and sustainability in social projects: the experience of the Local Development Programme (PRODEL) in Nicaragua', *Environment and Urbanization* 13:1, 11–35.

Stein, Alfredo with Luis Castillo (2005) 'Innovative financing for low-income housing improvement: lessons from programmes in Central America', *Environment and Urbanization* 17:1, 47–66.

Stein, Alfredo and Irene Vance (2007) 'The role of housing finance in addressing the needs of the urban poor: lessons from Central America', *Environment and Urbanization* 20:1, 13–30.

Stren, Richard (2008) 'International assistance for cities in developing countries; do we still need it?', *Environment and Urbanization* 20:2, 377–392.

——(2012) *Donor Assistance and Urban Service Delivery in Africa*, Working Paper No. 2012/49, Helsinki: UNU-WIDER.

Suh, Doowon (2011) 'Institutionalizing social movements: the dual strategy of the Korean women's movement', *The Sociological Quarterly* 52, 442–471.

Tannerfeldt, Göran and Per Ljung (2006) *More Urban, Less Poor: An Introduction to Urban Development and Management*, London: Earthscan Publications.

Tarrow, Sidney (1998) *Power in Movement: Social Movements and Contentious Politics*, New York: Cambridge University Press.

Teichman, J. (2008) 'Redistributive conflict and social policy in Latin America', *World Development* 36:3, 446–460.

Thorbek, Susanne (1991) 'Gender in two slum cultures', *Environment and Urbanization* 43:2, 71–91.

——(1994) *Gender and Slum Culture in Asia*, London: Zed Books.

Thorn, Jessica and Sophie Oldfield (2011) 'A politics of land occupation: state practice and everyday mobilization in Zille Rain Heights, Cape Town', *Journal of Asian and African Studies* 46:5, 518–530.

Thorp, R., F. Stewart and A. Heyer (2005) 'When and how far is group formation a route out of chronic poverty', *World Development* 33:6, 907–920.

Tibaijuka, Anna Kajumulo (2005) *Report of the Fact-finding Mission to Zimbabwe to Assess the Scope and Impact of Operation Murambatsvina by the UN Special Envoy on Human Settlements Issues in Zimbabwe*, Nairobi: UN-Habitat.

Tilly, Charles (2004) *Social Movements 1768–2004*, Boulder, CO: Paradigm Publishers.

Tironi, Manuel (2009) 'The lost community? Public housing and social capital in Santiago, Chile, 1985–2001', *International Journal of Urban and Regional Research* 33:4, 974–997.

Todes, Alison, Aly Karam, Neil Klug and Nqobile Malaza (2010) 'Beyond master planning? New approaches to spatial planning in Ekurhuleni, South Africa', *Habitat International* 34, 414–420.

Tomas, A. (2005) 'Reforms that benefit poor people – practical solutions and dilemmas of rights-based approaches to legal and justice reform', in P. Gready and J. Ensor (eds), *Reinventing Development? Translating Rights-based Approaches From Theory Into Practice*, London and New York: Zed Books, pp. 171–184.

Tomlinson, Richard (2003) 'HIV/AIDS and urban disintegration in Johannesburg', in Philip Harrison, Marie Huchzermeyer and M Mayekiso (eds), *Confronting Fragmentation: Housing and Urban Development in a Democratizing Society*, Cape Town: UCT Press, pp. 78–87.

Turner, John F.C. (1966) *Uncontrolled Urban Settlements: Problems and Policies*, Pittsburg: Report for the United Nations seminar on Urbanization.

——(1968) 'Housing priorities, settlement patterns and urban development in modernizing countries', *Journal of the American Institute of Planners* 34, 354–363.

——(1976) *Housing By People – Towards Autonomy in Building Environments, Ideas in Progress*, London: Marion Boyars.

——(1996) 'Seeing tools and principles within "best" practices', *Environment and Urbanization* 8:2, 198–199.

Turner, Sarah and Laura Schoenberger (2012) 'Street vendor livelihoods and everyday politics in Hanoi, Vietnam: the seeds of a diverse economy?', *Urban Studies* 49:5, 1027–1044.

UN-Habitat (2005) *The Global Report on Human Settlements 2005: Shelter Finance*, London: Earthscan for UN Habitat.

——(2007) *Forced Evictions – Towards Solutions? Second Report of the Advisory Group on Forced Evictions to the Executive Director of UN-Habitat*, Nairobi: UN-Habitat.

——(2012) *State of the World's Cities Report 2012/2013: Prosperity of Cities*, Nairobi: United Nations Human Settlements Programme.

UN-ISDR (2012) *Making Cities Resilient Report 2012; A Global Snapshot of how Local Governments Reduce Disaster Risk*, Geneva: The United Nations Office for Disaster Risk Reduction.

Urban Management Programme for Latin America and the Caribbean (2001) 'Towards participatory urban management in Latin American and Caribbean cities: a profile of the Urban Management Programme for Latin America and the Caribbean', an Institutional Profile, *Environment and Urbanization* 13:2, 175–178.

Urban Poor Fund International UPFI (2012) *Urban Poor Fund International: Financing Facility of Shack/Slum Dwellers International (SDI), 2011 Annual Report*, Cape Town: UPFI.

Urban Resource Centre URC (1994) 'The Urban Research Centre: NGO profile', *Environment and Urbanization* 6:1, 158–163.

Uvin, P. (2004) *Human Rights and Development*, Bloomfield, CT: Kumarian Press.

Valença, Márcio Moraes (2007) 'Poor politics – poor housing. Policy under the Collor government in Brazil (1990–2)', *Environment and Urbanization* 19:2, 391–408.

Valença, Márcio Moraes and Mariana Fialho Bonates (2010) 'The trajectory of social housing policy in Brazil: from the National Housing Bank to the Ministry of the Cities', *Habitat International* 34,165–173.

Vandermoortele, Jan (2011) 'The MDG story: intention denied', *Development and Change* 42:1, 1–21.

van der Linden, Jan (1997) 'On popular participation in a culture of patronage; patrons and grassroots organization in a sites and services project in Hyderabad, Pakistan', *Environment and Urbanization* 9:1, 81–90.

Vaquier, Damien (2010) *The Impact of Slum Resettlement in Urban Integration in Mumbai: The Case of the Chandivali Project*, CSH Occasional Paper No. 26, New Delhi: Centre de Science Humaines.

Venkatesh, Sudhir (2009) *Gang Leader for a Day*, London: Penguin Books.

Von Weizsäcker, Ernest Ulrich, Oran R. Young and Matthias Finger with Marianne Beisheim (2005) *Limits to Priviatization: How to Avoid Too Much of a Good Thing*, A report to the Club of Rome, London and Sterling, VA: Earthscan Publications.

Waage, Jeff, Rukmini Banerji, Oona Campbell, Ephraim Chirwa, Guy Collender, Veerle Dieltiens, Andrew Dorward, Peter Godfrey-Faussett, Piya Hanvoravongchai, Geeta Kingdon, Angela Little, Anne Mills, Kim Mulholland, Alwyn Mwinga, Amy North, Walaiporn Patcharanarumol, Colin Poulton, Viroj Tangcharoensathien and Elaine Unterhalter (2010) 'The Millennium Development Goals: a cross-sectoral analysis and principles for goal setting after 2015', *Lancet and London International Development Centre Commission.*

Wain, Ross (2011) *Cash Transfers Background Paper*, London: International Institute for Environment and Development.

Ward, Peter (ed.) (1982) *Self-help Housing; A Critique*, London: Mansell Publishers.

Watson, Vanessa (2007) 'Urban planning and twenty-first century cities: can it mean the challenge?', in Allison M. Garland, Mejgan Massoumi and Blair A. Ruble (eds), *Global Urban Poverty: Setting the Agenda*, Washington, DC: Woodrow Wilson International Center for Scholars, pp. 205–237.

Wegelin, Emiel (1994) 'Everything you always wanted to know about the urban management programme (but were afraid to ask). Comment on "the World Bank's 'new' urban management programme: paradigm shift or policy continuity?" by Gareth A. Jones and Peter M. Ward', *Habitat International* 18:4, 127–137.

Weinstein, Liza (2008) 'Mumbai's development mafias: globalization, organized crime and land development', *International Journal of Urban and Regional Research* 32:1, 22–39.

Weru, Jane (2004) 'Community federations and city upgrading: the work of Pamoja Trust and Muungano in Kenya', *Environment and Urbanization* 16:1, 47–62.

Whitaker, Gordon P. (1980) 'Coproduction: citizen participation in service delivery', *Public Administration Review* 40:3, 240–246.

Whitehead, Laurence and George Gray-Molina (2005) 'Political capabilities over the long run', in Peter P. Houtzager and Mick Moore (eds), *Changing Paths: International Development and the Politics of Inclusion*, Ann Arbor: University of Michigan, pp. 32–57.

WHO and UNICEF (2011) *Progress on Sanitation and Drinking Water: 2010 Update*, Geneva: WHO/UNICEF Joint Monitoring Programme for Water Supply and Sanitation.

Wilson, Gordon (2006) 'Beyond the technocrat? The professional expert in development practice', *Development and Change* 37:3, 501–523.

Wood, Geof (2003) 'Staying secure, staying poor: the "Faustian bargain"', *World Development* 31:3, 455–471.

World Bank (2008) *Reshaping Economic Geography; World Development Report 2009*, Washington, DC: World Bank.

World Bank and International Monetary Fund (2013) *Rural Urban Dynamics and the Millennium Development Goals*, Washington, DC: World Bank.

World Bank Kenya (2006) *Kenya Inside Informality: Poverty, Jobs, Housing and Services in Nairobi's Slums*, Report No. 36347-KE, Washington, DC: World Bank.

Wust, Sebastien, Jean-Claude Bolay and Thai Thi Ngoc Du (2002) 'Metropolitization and the ecological crisis: precarious settlements in Ho Chi Minh City', *Environment and Urbanization* 14:2, 211–224.

Yahya, Saad, Elijah Agevi, Lucky Lowe, Alex Mugova, Oscar Musandu-Nyamayaro and Theo Schilderman (2001) *Double Standards, Single Purpose: Reforming Housing Regulations to Reduce Poverty*, London: ITDG Publishing

Zack, Tanya and Sarah Charlton (2003) *Better Off But ... Beneficiaries' Perceptions of the Government's Housing Subsidy Scheme*, Housing Finance Resource Programme No. 12, Johannesburg: Urban Institute.

Index